THE ALGORITHM
OF CREATION

Universalism's Algorithm of the Infinite and
Space-Time, the Oneness of the Universe and the
Unitive Vision, and a Theory of Everything

JOHN HUNT PUBLISHING

First published by O-Books, 2023
O-Books is an imprint of John Hunt Publishing Ltd., 3 East St., Alresford,
Hampshire SO24 9EE, UK
office@jhpbooks.com
www.johnhuntpublishing.com
www.o-books.com

For distributor details and how to order please visit the 'Ordering' section on our website.

Text copyright: Nicholas Hagger 2022

ISBN: 978 1 78535 137 2
978 1 78535 138 9 (ebook)
Library of Congress Control Number: 2022938251

A CIP catalogue record for this book is available from the British Library.

Design: Stuart Davies

UK: Printed and bound by CPI Group (UK) Ltd, Croydon, CR0 4YY
Printed in North America by CPI GPS partners

We operate a distinctive and ethical publishing philosophy in all areas
of our business, from our global network of authors to production and
worldwide distribution.

THE ALGORITHM
OF CREATION

**Universalism's Algorithm of the Infinite and
Space-Time, the Oneness of the Universe and the
Unitive Vision, and a Theory of Everything**

Nicholas Hagger

BOOKS

Winchester, UK
Washington, USA

Also by Nicholas Hagger

The Fire and the Stones
Selected Poems
The Universe and the Light
A White Radiance
A Mystic Way
Awakening to the Light
A Spade Fresh with Mud
Overlord, books 1–2
The Warlords
Overlord
A Smell of Leaves and Summer
Overlord, books 3–6
Overlord, books 7–9
Overlord, books 10–12
The Tragedy of Prince Tudor
The One and the Many
Wheeling Bats and a Harvest Moon
The Warm Glow of the Monastery Courtyard
The Syndicate
The Secret History of the West
The Light of Civilization
Classical Odes
Overlord, one-volume edition
Collected Poems 1958–2005
Collected Verse Plays
Collected Stories
The Secret Founding of America
The Last Tourist in Iran
The Rise and Fall of Civilizations
The New Philosophy of Universalism
The Libyan Revolution
Armageddon
The World Government

The Secret American Dream
A New Philosophy of Literature
A View of Epping Forest
My Double Life 1: This Dark Wood
My Double Life 2: A Rainbow over the Hills
Selected Stories: Follies and Vices of the Modern Elizabethan Age
Selected Poems: Quest for the One
The Dream of Europa
The First Dazzling Chill of Winter
Life Cycle and Other New Poems 2006–2016
The Secret American Destiny
Peace for our Time
World State
World Constitution
King Charles the Wise
Visions of England
Fools' Paradise
Selected Letters
The Coronation of King Charles
Collected Prefaces
A Baroque Vision
The Essentials of Universalism
Fools' Gold
The Promised Land
The Building of the Great Pyramid
The Fall of the West

"'Algorithm', maths, a process or set of rules used for calculation or problem-solving, especially with a computer."

Concise Oxford Dictionary

"'Creation', the act of creating; the creating of the universe; (usually Creation) everything so created, the universe; a product of human intelligence, especially of imaginative thought or artistic ability."

Concise Oxford Dictionary

*

"I asked [John Barrow, author of *Theories of Everything*, over breakfast on 11 April 1992]: 'Where do love and order come into your mathematics?' John Barrow was not prepared to go outside his area of cosmology, the materialist level.... Everything beyond that is speculative. He isn't interested in a theory of everything; only in a theory of cosmology."

Nicholas Hagger, *My Double Life 2:*
A Rainbow over the Hills, p.389

"Question: You have a Grand Unified Theory. Will there be a Theory of Everything?
Answer: A Grand Unified Theory is a partial theory. In physics, a Grand Unified Theory unifies three forces, and a Theory of Everything unifies four forces. The position is that the weak and electromagnetic forces have definitely been unified, and an attempt to integrate these with the strong force works in theory but there is no evidence that it works in practice. There is no proof that gravity can be 'quantised' like the other three forces, and that there are gravitons. In physics, a Grand Unified Theory is disputed, and there is no sign of a Theory of Everything. Hawking is seeking to unify the four physical materialistic forces, but a true Theory of Everything takes the higher metaphysical levels into account. 'Everything' includes love and conversation. I have done my Grand

Unified Theory: *The Fire and the Stones* is subtitled 'A Grand Unified Theory of World History and Religion'. That was the hard part, to integrate mysticism, metaphysics, religion and world history in precise historical stages, all of which have precise dates. I am now seeking to formulate a Theory of Everything by integrating physics, biology, philosophy and other disciplines. The Theory of Everything I see is very simple: the Light is behind everything and shapes it, the Light many people have called God. It is simple enough to be expressed in a simple mathematical formula such as $E = mc^2$. Are the mathematics already there in the Universe or are they human constructs we invent? I believe they're there, and that they just have to be uncovered, like scraping away sand to reveal a ruined Roman temple."

Nicholas Hagger, interviewed on 8 February 1993 on the publication of *The Universe and the Light*, in *The One and the Many*, 1999, pp.157–158

The front cover shows a picture of the cosmic microwave background (CMB) radiation and ripples, an echo from the Big Bang first emitted 380,000 years afterwards, taken from the European Space Agency's Planck satellite spacecraft's telescope in 2013 and not released until 2018. This is the clearest picture of the CMB since the COBE (Cosmic Background Explorer) 1992 picture whose ripples influenced Nicholas Hagger's Form from Movement Theory and the 2003 WMAP (Wilkinson Microwave Anisotropy Probe) with resolution 35 times better, and shows small variations (or anisotropies) in the uniform temperature of 2.7K, with cooler temperature shown as blue and warmer as orange-red, see p.59.

Acknowledgments

To the memory of Junzaburo Nishiwaki, who in October 1965 wrote down for me an algebraic formula that became part of the algorithm of Creation; and of John Barrow, with whom I discussed a Theory of Everything (see pp.vi-vii). I was fortunate to have discussions with Einstein's collaborator David Bohm (on order in the universe and hidden variables) and with Sir Roger Penrose (on the singularity before the Big Bang). I am grateful to my grandson Ben Hagger for acting as a consultant mathematician regarding my algorithm and for preparing matrices for my Theory of Everything. And once again I am grateful to my PA Ingrid Kirk for her invaluable help as I set out my Theory of Everything. Without her I could not have completed this work.

CONTENTS

CONTENTS

showing section headings and structure

Nothingness

Preface

The Origin, Development and End of the Universe: An Algebraic Formula, an Algorithm of Creation and a Theory of Everything

Rational and intuitive ways of seeing

The universe can be viewed in different ways. The rational, social ego analyses and sees differences. When it examines the universe it fragments and reduces it to its parts so it is in specialised pieces and separate disciplines, like fragments of an urn from the Roman Empire discovered in an archaeologist's trench. The deeper part of the self, the core, sees intuitively. It collects fragments and pieces them together with what Coleridge called "the esemplastic power of the imagination" (*eis en plattein* in Greek meaning 'shape into one'). The approach of the self's core is Expansionist, not reductionist, and it can piece together and restore the universe from its fragments into a whole that can be placed on show in a curator's museum.

We are in contact with our deeper self at the beginning of the Mystic Way. The long journey can start with bewildering intensity and unfamiliar images, even visions, and an awakening to our deeper, more spiritual self is accompanied by a centre-shift from our rational, social ego to our deeper core and a first surprising glimpse of the inner Light which can be found in the 17th-century Metaphysical poets. Purgation follows and, after a Dark Night of the Soul that sets the bewildered mystic onto a clear path free from distractions, in due course full illumination. There is then a long period of further purgation, a Dark Night of the Spirit and more ordeals. Finally, the self emerges into unitive living, when it *instinctively sees* the universe as a unity.

My Unitive Way and the self's deeper way of seeing is behind my Universalism, my philosophy of the unity of the universe and humankind.

My Mystic Way and an algebraic formula

I underwent an awakening in 1965 (see pp.5–6 and 325–327). Shortly before my centre-shift I had a conversation in Japan with Junzaburo Nishiwaki, Japan's T.S. Eliot, during which I asked him, "What is the wisdom of the East?" (I had gone to live in Japan to find the wisdom of the East.) He wrote down an algebraic formula: "+A + −A = 0" (see pp.4–5). He explained, "The Absolute is where there is no difference." From my deeper self I immediately saw that all pairs of opposites are reconciled in an underlying harmony, but, bewildered by the images and visions that were happening to me and trying to understand, I recorded my awakening in the long poem I was writing at the time, 'The Silence'. The pentameter "(+A) + (−A) = Nothing" found its way into 'The Silence' (see the manuscript of 'The Silence' on pp.306 and 317–332), along with the line "the Absolute is where there is no difference".

The algebraic formula "+A + −A = 0", which Nishiwaki derived from Confucius and includes *yang* (+A) and *yin* (−A) (see p.4), has stayed with me for 56 years. I went on to have my full illumination six years later, on 10 September 1971 (see pp.341–343), after a Dark Night of the Soul, and journeyed through a Dark Night of the Spirit to emerge nearly 30 years after my awakening into unitive living, in 1993 (see p.6). My long journey can be found in my *Collected Poems 1958–2005*. (I am haunted by the name of the house we moved into in July 1945, which became the family home, 'Journey's End'.)

By the time I was ready to publish books with some understanding of what happened to me, in the early 1990s, I was already *instinctively seeing* the unity of the universe. My philosophical writings – *The Universe and the Light* (1993), *The One and the Many* (1999) and *The New Philosophy of Universalism* (2009) – all have this unitive view I reached at my journey's end.

A number of the 56 Prefaces in my *Collected Prefaces* (2022) refer to the algebraic formula +A + −A = 0 that is behind my Unitive Way, encapsulates the Oneness of the universe and its reconciling of all opposites, contradictions and differences within an underlying harmony, and spans my 30-year-long writing career: the Prefaces to *Selected Poems: A Metaphysical's Way of Fire* (1991); *The New Philosophy of Universalism* (2009); *My Double Life 1: This Dark Wood* (2015); *Life Cycle and Other New Poems* (2016); *The Secret American Destiny* (2016);

Peace for our Time (2018); *King Charles the Wise* (2018); *Fools' Paradise* (2020); *The Coronation of King Charles* (2021); *Fools' Gold* (2022); and *The Essentials of Universalism* (2023). This formula has been central to the dialectical method of my thinking for the last 56 years.

The formula symbolises my Unitive Way of seeing the universe: behind all the apparent differences between species is an underlying unity, the One, as the Presocratic Greeks knew, the forebears of my Universalism. The formula can be seen in *The New Philosophy of Universalism* (pp.5–6, 264, 265, 268, 271, 278, 281, 283, 303, 305, 325, 362 and 363) as being behind my dialectical method which unites philosophical traditions.

My algebraic formula as the algorithm of Creation

I did not see $+A + -A = 0$ as an algorithm until March 2021, when, while I was putting the finishing touches to *The Promised Land*, I received an email from a reader mentioning an 'algorithm'. It set me thinking, and I wrote in my diary for 27 March 2021 that an "algorithm may flesh out my $+A + -A = 0$ and take it into a Theory of Everything". I could see a book coming, and at first thought it might be called *The Algorithm of Harmony*, or *The Algorithm of Oneness* or *The Algorithm of the Oneness of the Universe*.

Then on 28 March I received *The Algorithm of Creation* in my sleep. On 28 March I recorded in my diary: "Saw the oval cloud last night (vision)." (I had had 19 experiences of seeing an oval cloud of Light covered in swansdown, see Appendix 2 ('19 experiences of the oval cloud') of *The Promised Land*, and in several 'oval cloud' poems thought it might be my Muse, or perhaps even my soul.) I continued in my diary: "Woke with the title *The Algorithm of Creation*." I had received most of my titles in my sleep, which I saw as coming from the beyond, perhaps my Muse.

I sent off *The Promised Land* and on 6 April wrote to John Hunt, then my publisher, with a proposal for *The Algorithm of Creation*. I worked on another book (*The Fall of the West*), and eventually drafted *The Algorithm of Creation* from 3 to 19 September 2021, and amended it in Cornwall, working intensely before an upstairs window overlooking the sea and sky, and stars, from 22 to 26 October 2021. I worked on a print-out down in Cornwall from 14 to 16 December 2021. I worked on *The Fall of the West* for much of January and was not able to get back to *The Algorithm of Creation* until 24 January 2022.

As I described the origin, development and end of the universe, I found that my algebraic formula, +A + −A = 0, also worked as an algorithm as it gave a set of rules or instructions to the universe with four variables for 0 =, +A, −A and = 0 in each of 20 stages.

My algorithm of Creation as a Theory of Everything

I had long thought about a Theory of Everything. As I wrote *The Algorithm of Creation*, it dawned on me that a Theory of Everything should include many of the ingredients (elements or conditions) in my 20-stage account of the origin, development and end of the universe. My algorithm of Creation, the variations within +A + −A = 0 and the four variables with ingredients (or conditions) in 20 stages, included the origin of the universe and of life; evolution; order; and the 40 bio-friendly conditions surrounding the earth that all had to work in conjunction with each other to make life possible. These were the ingredients that I saw as needing to be in a Theory of Everything.

I was aware that in mathematics a theorem is a 'conjecture' until it is proved, and that in the sciences only a well-tested hypothesis can be a 'theory'. I saw that many of the ingredients (the algorithm's elements or conditions) had been tested, enough for my theory to be regarding as a Theory of Everything.

I had been thinking about a Theory of Everything since reading John Barrow's *Theories of Everything* in 1991, and especially since the weeks before April 1992, when I gave an evening lecture to 500 mystics and scientists in Winchester (which can be read as the first part of *The Universe and the Light*). The next morning, 11 April 1992, I had breakfast with John Barrow. I asked him, "Where do love and order come into your mathematics?" He was a Professor of Astronomy at the University of Sussex, and he said he was not prepared to go outside his area of cosmology, the materialistic level (see p.vi). I had been convinced since that day that a Theory of Everything must include love, order and harmony.

I now saw that I could set out a Theory of Everything based on my algorithm of Creation. I would have more than 50 ingredients, the algorithm's elements or conditions, and give them mathematical symbols.

I worked on the mathematical symbols in early February 2022, and added 26 on 7–8 February to reach 72 symbols. On 8 February my mathematician grandson Ben unexpectedly arrived from Exeter University, saying "I felt Grandpa needed me." In the course of our discussion on the mathematical perspective of my project I confirmed that the algorithm of Creation was $0 = +A + -A = 0$. This now seems to be a Providential encounter as he appeared while he was ruminating on the mathematics of my algorithm. On 11 February we drove to Cornwall, and it so happened that Ben was returning to Exeter that day. We met for lunch at Leigh Delamere service station in Wiltshire, and he confirmed that the mathematics for a universe before ours, our universe and a universe after ours could be $0 = +A + -A = 0 \equiv 0 = +A + -A = 0 \equiv 0 = +A + -A = 0$ (the $\equiv$ meaning that the universes are equivalent to our universe but not necessarily equal). Again, this 'accidental' meeting now seems to be equally Providential. He sent me through the model for a matrix (see p.293) while I was in Cornwall.

By 13 February I had added more mathematical symbols and was up to 88 symbols. I was working on the end of the universe on 16 February. My PA Ingrid was typing up 21 tapes I had given her and she had now reached the mathematical symbols. On 2 March I amended the mathematical symbols and added more, and there were now exactly 100, but unbeknown to me the $+A$ and $-A$ boxes were not equal. Ben unexpectedly appeared on 3 March and studied my symbols in downward columns. He took away several pages to turn my downward columns into matrix form.

He returned with the three variations of the four matrices of our universe on pp.294, 296–297 and 299–300 on 4 March. He had identified there were 51 and 48 symbols in the $+A$ and $-A$ boxes, and I found repetition and omissions and evened them out so a perfect symmetry was revealed (14 symbols in the $0=$ and $=0$ boxes, and 50 in each of the $+A$ and $-A$ boxes). By 4 March I had an algorithm of 4 symmetrical variables and 100 key conditions that drew on the 20 stages of the origin, development and end of the universe in section 2. I edited my

PA's daily transcriptions of my tapes as they came through, and made the last amendments to sections 2 and 3, working intensely, on 17 and 18 March 2022.

All along it was my intention to invite mathematical physicists and astrophysicists to test any of the 100 ingredients (the algorithm's elements or conditions) not tested so my proposed Theory of Everything, which would "unite all the laws of Nature into a single statement that reveals the inevitability of everything that was, is and is to come in the physical world",[1] could be confirmed.

My algorithm of Creation had shown that the essential structure of the universe as a whole can be "algorithmically compressed".[2] As Barrow put it:

> We can replace sequences of facts and observational data by abbreviated statements which contain the same information content. These abbreviations we often call 'laws of Nature'. If the world were not algorithmically compressible, then there would exist no simple laws of Nature.[3]

Barrow's *Theories of Everything* attempts to show "what a Theory of Everything might hope to teach about the unity of the Universe and the way it may contain elements that transcend our present compartmentalized view of Nature's ingredients".[4] Barrow's approach is rational and scientific, and so is mine. But my approach is also intuitional as it includes perception from within unitive living along the Unitive Way at the end of the Mystic Way when the deeper self *instinctively sees* the universe as a unity.

I therefore offer my partially-tested algorithm of Creation's four variables and 100 conditions as a Theory of Everything, and wait for mathematicians and mathematical physicists such as Roger Penrose, and astrophysicists and the findings of the Webb telescope (see pp.xxi, 24 and 50), to test each ingredient or condition that has not yet been tested so my hypothesis becomes a fully tested Theory of Everything. As many of the ingredients or conditions have already been tested, and as the four forces including gravity were all unified at the start of the first second after the Big Bang, my Theory of Everything is not a misnomer. The Webb space telescope, which has a mirror seven times larger than Hubble's, was sent in 2021 to the Lagrange point, where the sun's and earth's gravity cancel each other out, and it will remain stationary as it follows the earth's orbit round the sun, and with MIRI (mid-infrared wavelength range) is seeing through dust to peer much further back in time than Hubble and is detecting the earliest stars and galaxies in the universe, and is assembling evidential data and proofs that will test variables within stage (6) of my account of Creation, and in my Theory of Everything. My Theory of Everything is within the definition and function of what a Theory of Everything should show that is outlined in Barrow's *Theories of Everything*, as we have just seen.

Universalism's algorithm and Theory of Everything

My algorithm of Creation ($0 = +A + -A = 0$) has come out of my Universalism. Universalism developed from my dialectical handling of opposites following my adoption of $+A + -A = 0$ in Japan in 1965, and it surfaced in my poem 'Night Visions in Charlestown' (1983) and can be found in my Introduction to *The Fire and the Stones* (1991).

Universalism is the new philosophy which I outlined in *The Universe and the Light* (1993), *The One and the Many* (1999) and *The New Philosophy of Universalism* (2009): a philosophy that focuses on the unity of the universe and humankind, and the interconnectedness of all disciplines through a new way of perceiving following a centre-shift from the rational, social ego to the deeper centre of consciousness known in meditation, the core that sees all opposites and differences as being reconciled within an underlying unity. The seven disciplines on which Universalism are focused, all seen as wholes and one, are mysticism, literature, philosophy and the sciences, history, comparative religion, international politics and statecraft, and world culture. Each of these seven disciplines seems separate, like the seven bands of a rainbow that seem different to the rational, social ego, but they are one rainbow when seen by the deep consciousness that perceives their unity behind differences.

Universalism has focused on the unity of the universe as I set out in *The New Philosophy of Universalism*, and it is appropriate that *The New Philosophy of Universalism* should be reflected within *The Algorithm of*

Creation, which grew out of that work. The algorithm of Creation is Universalism's algorithm, and is therefore Universalism's Theory of Everything: a Theory of Everything that is only possible because of the perceptions and thinking within Universalism and *The New Philosophy of Universalism*.

My intuitional Unitive Way and a Theory of Everything

I think back to the beginning of my journey to Universalism in Japan where, as a young Professor of English Literature, I was taken to experience a meditation in a Zen *dojo* (meditation centre) in Ichikawa City, Tokyo and ten days later spent a night in Koganji Temple, near Sugamo in Tokyo, and meditated before dawn. I found I was undergoing an awakening that I found bewildering, a succession of images and visions in the course of which I underwent a centre-shift from my rational, social ego to a deeper centre within myself and had my first experience of the Light that features so much in the poems of the 17th-century Metaphysical poets. I did not know at the time that I had started on a Mystic Way that would last nearly 30 years, during which I would experience a Dark Night of the Soul, full illumination and, after a Dark Night of the Spirit and further ordeals, a Unitive Way along which I would *instinctively* perceive, from my deep centre, the unity of the universe and humankind. I did not know that my Universalism, which is behind my 58 books, would grow out of my unitive perception of the Oneness of the universe – an intuitional approach to the One and my view of Creation that would reinforce my philosopher's rational approach and bring me to my Theory of Everything.

And I think back to Japan again, but this time to the room in the Bank of Japan where I gave its Executive Director, Mr Fukuchi, a weekly one-to-one English class. He had spent his life studying the hexagrams in *The I Ching* (or *Book of Changes*, c.1143 or 1142BC), and on 17 November 1966 he brought in 50 bamboo sticks in a cup to find a hexagram that would tell my fortune and Providential destiny. He put one stick on the table (to signify God) and shook the rest over his head and then divided the sticks into two groups, one of 25 and one of 24 (to represent Heaven and Earth). He divided the group of 25 by

8 for the bottom trigram and repeated the process for the top trigram. Then he divided the top trigram by six for the variation. There was a remainder of 5 and 0 and the sticks formed hexagram 46, "*Shang*". The variation worked out as number 32, "*Hang*".[5]

The I Ching says of hexagram 46: "*Shang* indicates that under its conditions there will be great progress and success. Seeking (by the qualities implied in it) to meet with the great man, its subject need have no anxiety." It adds that there will be "advancing upwards with the welcome (of those above him). There will be great good fortune." It speaks of "sincerity" and "no error", of "this subject ascending upwards (as into) an empty city", as "employed by the King to present his offerings on Mount Khi" (serving spiritual beings), as "firmly correct, and therefore enjoying good fortune. He ascends the stairs (with all due ceremony)", glossed as "he grandly succeeds in his aim". And he is "advancing upwards blindly", glossed as being "in the highest place".[6] The *Khang-hsi* editors say, "All is done in accordance with the ordinances of Heaven".[7]

The I Ching says of hexagram 32: "*Hang* indicates successful progress and no error (in what it denotes). But the advantage will come from being firm and correct; and movement in any directions will be advantageous."[8]

The "*Shang*" trigram means "wood and earth"[9] and the "*Hang*" trigram means "thunder and wind".[10] Together the two trigrams mean "wind and earth" (top-bottom). Hexagram 46 and 32 are associated with "great progress and success", with great insistence on ascent, "advancing upwards blindly" as into "an empty city" (two themes in 'The Silence' I was then writing, see pp.317–332). "Advance upwards", "great good fortune", "he grandly succeeds in his aim" – the Executive Director, Mr Fukuchi, interpreting hexagram 46 for me, said, "This must be one of the best hexagrams in *The I Ching*." Hexagram 46 was what the Will of Heaven had in mind for me.

Hexagrams 46 and 32 revealed that I would be governed by wind and earth, by the inspiration of Heaven and by the appeal of woods on earth. I have always been inspired in my writings from beyond, and I have always been drawn to woods and have lived much of my life by Epping Forest, outside London. As a Gemini torn between two conflicting twins within, I have lived with +A (inspiration) and −A (grounded by forest trees). All my life has been a dialectical working-out of inspiration and grounding, which I have at last reconciled into a harmonious unity (= 0).

Looking back on my life at 82, I recognise the "advancing upwards blindly" as I wrote one book after another, endlessly climbing a mountain. Although doing the bamboo sticks may seem a random activity of chance, the results brought me back to the pairs of opposites and underlying harmony of my algorithm. "Wind and earth" reminds me of the title of Volume 21 of my *Collected Poems*, *The Wind and the Earth* (1981). I do not think I was thinking of hexagram 46 when this title came to me. Certainly much of my work has been inspired from the beyond, and many of my titles have been received in sleep and are about the interplay of "Heaven and earth", *yin* (darkness) and *yang* (light), +A + −A = 0.

The algorithm of Creation has resonated in my own life as an instruction that will help me to calculate the balance of wind/inspiration/Heaven and centring on the earth in my life as an answer to the problem of how I should live: +A (inspiration from Heaven) + −A (centring on earth blindly) has given me an equilibrium, a oneness, a unity (= 0) in my living, for which I am grateful. And my intuitional Unitive Way, which has opened to inspiration from Heaven and filled me with an instinctive perception of the unity of the universe, has brought me to a Theory of Everything.

The fundamental theme of Universalism

I have always had a warm feeling about my algebraic formula as it reflects my own self-unification after my early self-division between materialist accident and nothing surviving the brain and therefore death (−A) and metaphysical order and the soul crossing over to an after-life (+A). I reconciled this division into +A (the metaphysical view) + −A (the physical world) = 0 (Oneness). Being a Gemini, I was aware that I had been dominated by two different inner twins (+A + −A), whom I had reconciled into harmony (= 0).

An algebraic formula, the algorithm of Creation and now a Theory of Everything, +A + −A = 0 (the One, the unity of the metaphysical and physical universe and humankind), now reflects the fundamental theme of my Universalism – and of the universe: +A (the oneness of consciousness and all humankind at a metaphysical level) + –A (the entire physical universe) = 0 (the One, the unity of the metaphysical and physical universe and humankind). The One, a metaphysical non-duality, gave rise to both matter and consciousness (a diversity from a common origin), and the infinite manifesting Reality philosophers have sought for millennia and have seen as the ordering Light.

6–7, 10–11 November 2021; 17–18, 21–22 March 2022

Algorithms and the Universe

Background to the universal algorithm of Creation

An algorithm as a set of instructions to answer a problem

An algorithm is a set of step-by-step mathematical instructions or rules that will help calculate an answer to a problem. At its simplest it is a shopping list for groceries: a set of instructions to a shopper as to what to buy to answer the problem of low stocks in the larder, fridge and freezer. A more sophisticated algorithm would be a set of instructions to a grocery chain as to what stocks to provide, based on a prediction of what products consumers will buy again or for the first time and in what store, and the consumers' personalised shopping lists, to answer the problem of what provisions to stock on their branches' shelves.

Algorithms are also used to give a set of instructions in a recipe for cooking a particular food, to predict property crimes such as housebreaking, and to show which Instagram posts to display at the top of a feed, based on browsing habits. They can be used in the everyday world to give a set of instructions to help a teacher sort papers into alphabetical order, enable facial recognition, effect a Google search, make traffic-lights work, and create bus timetables. In all these there is a sequence of steps that give a set of instructions that help calculate an answer to a problem.

There have been algorithms since the beginning of recorded history. A Sumerian clay tablet from c.2500BC in Shuruppak near Baghdad gives a set of instructions for division, and Babylonian clay tablets from c.1800–1600BC give sets of instructions for formulas. Algorithms were used in Babylonian astronomy. Algorithms on papyrus give instructions for Egyptian arithmetic from c.1500BC that were still used in Hellenistic Greece. Euclid's *Elements* gives instructions for calculating the greatest common divisor (GCD) or highest common factor (HCF) of two integers (numbers), the largest number that divides them both without a remainder: the Euclidian algorithm (c.300BC). Nicomachus of Gerasa's *Introduction to Arithmetic* describes the sieve of Eratosthenes, numbers 1 to 120 in 12 rows of 10 'boxes' to find prime numbers, which accurately calculated the earth's size c.240BC.

Al-Khwarizmi, whose name is behind 'algorithm' and first book behind 'algebra'

Algebra came to the fore in the 9th century AD. The word 'algebra' comes from the Arabic *al-jabr* ('completion' or 'restoration'), which is in the title of the first work of a Iranian mathematician, Muhammad ibn Musa al-Khwarizmi (c.780–850): *Kitab al-jabr wa al-muqabalah, The Compendious Book on Calculation by Completion and Balancing,* c.820). It set out the rules for cancelling like terms on opposite sides of an equation, as in $+A + -A = 0$. It compiled rules for arithmetical solutions of linear and quadratic equations, for elementary geometry and for distributing inherited money in proportions. It was based on Babylonian mathematics in the 2nd millennium BC that descended through Hellenistic, Hebrew and Indian-Hindu traditions.

Al-Khwarizmi's second work *Concerning the Hindu Art of Reckoning*, was translated into Latin as *Algoritmi de Numero Indorum*. The Latin

Statue of al-Khwarizmi in Amir Kabir University, Tehran, Iran.

title includes the Latinised word for his name al-Khwarizmi: *Algoritmi*, which is our 'algorithm'. Al-Khwarizmi used sets of rules and techniques to solve algebraic equations, and his rules and techniques were called his 'algorithm'. So al-Khwarizmi's two titles provided our words 'algebra' and 'algorithm', which is indicative of the influence his works have had.[1]

Al-Khwarizmi, the founder of algebra and algorithms, lived in Baghdad under the caliphate of Harun al-Rashid and his successor al-Ma'moun, during the Golden Age of the Abbasid Muslims when the Islamic Empire stretched from the Mediterranean to India. He worked as head of the Library in the

Two views of the House of Wisdom, Dar al-Hikmah
(above and below), Baghdad in the 9th century.

House of Wisdom (Dar al-Hikmah), also known as the Great Library
of Baghdad. The House of Wisdom acquired and translated scientific
and philosophical treatises, and by focusing on Greek, Jewish and
Indian treatises al-Khwarizmi drew on the eastern and western ends of

the Islamic Empire. The House of Wisdom was destroyed by Mongols during the siege of Baghdad in 1258, and the Library's books were thrown into the Tigris. The ink from the books turned the river black. On a personal note, my life's path seems to have been Providential. It was as though I 'was sent' to the places where those I would need to focus on in the future once lived. I left Oxford in 1961 to live in Baghdad – I lectured at the University of Baghdad – and every day I walked along Rashid Street on my way back from my university. I was told about the House of Wisdom and tried to find its location near the Tigris. Similarly, every day I passed a medieval building associated with Suhrawardi (1154–1191), another Iranian drawn to Baghdad, who founded the Iranian Illuminationist school of philosophy, and again knew he would be significant in my future. I thought about him many times after my own illumination on 10 September 1971.

Nishiwaki's algebraic thinking, Taoism and +A + −A = 0

Algebraic thinking was widespread in the second half of the 19th century, and it influenced Junzaburo Nishiwaki (1894–1982), a Japanese thinker and poet regarded as Japan's T.S. Eliot as he brought out 'January in Kyoto' a year after Eliot's 'The Waste Land', which came out in 1922.[2] Nishiwaki introduced me to part of the algorithm behind this work, the algorithm after the Big Bang: +A + −A = 0.

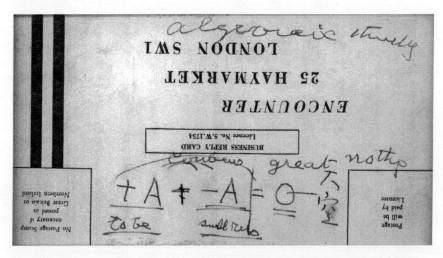

Card on which Junzaburo Nishiwaki wrote +A + −A = 0 on 5 October 1965.

He was Professor Emeritus at Keio University, Tokyo (one of my universities) when I was a Professor of English Literature in Japan, and I was taken to see him on 13 December 1963 and met him again on 5 October 1965 in a *saké* bar with sawdust on the floor near Keio University. He was then 71, with silky white hair and spectacles, and he received seven nominations for the Nobel Prize in Literature between 1963 and 1966. In the course of our conversation I asked him, "What is the wisdom of the East?" He told me he would express it in "algebraic thinking", "witty combinations".

He said "algebraic thinking" derived from al-Khwarizmi's equations, and before that from Confucius, who had included Taoist (Daoist) *yin* and *yang* in his thinking. He reached out and took the business-reply card from a copy of *Encounter* I had with me and had placed on the table, and wrote on it (see picture), "algebraic thinking" and (in scrawled handwriting that can be misread as "Coribeus") "Confucius". And then "+A + −A = 0". He wrote beneath +A "to be" (one's own life) and beneath −A "small zero" (one's own death), and above 0 "great nothing", the Absolute. He said to me, "The Absolute is where there is no difference."I was stunned. I immediately saw that +A + −A = 0 was a variation of *yin* + *yang* in the *Tao*, the Nothingness from which Creation came. *Yin* was in fact associated with metaphysical Reality (+A) in the beginning, Being, the infinite, eternity and dark; whereas *yang* was associated with physical reality (−A), existence, the material and social, the finite, time and natural light. I also saw that +A and −A stood for pairs of opposites – (now putting −A first in each pair) day and night, war and peace, life and death, time and eternity, the finite and infinite – all of which were reconciled in the One. For the "0" was the Oneness of the universe, Oneness. I wrote the algebraic formula and the Absolute into my poem 'The Silence', which I completed between 1965 and 1966 (see pp.317–332): "(+A) + (−A) = Nothing,/The Absolute is where there is no difference." (See p.306 for manuscript page.)

The Absolute is the Void of the *Tao*, the Nothingness from which the universe and life came and to which it will one day return. It is ultimate Reality, and within it there is no difference between "to be" and "small zero" (one's own death). (Hamlet – and, indeed,

Shakespeare – would have appreciated advice from Nishiwaki on the wisdom of the East during his "To be or not to be" speech.) In Taoism the Absolute Nothingness is in fact also a Plenitude and gave rise to Creation.

Of Taoism, I would write in *The New Philosophy of Universalism* (2009):[3]

The early Chinese were influenced by shamanistic Central Asia, which was active in nearby Mongolia. The early Chinese Supreme Ruler from c.1766BC united man and Heaven (*T'ien*). In due course the Chinese civilisation expressed the ordered unity of the universe in the *Tao*. The founder of Taoism Lao-Tze or Lao-Tzu, is traditionally thought to have been born c.570BC. The *Tao* (Way) is a Void out of which all creation came. It is the One, the unity under the plurality and multiplicity of the universe. "We look at it and do not see it; its name is The Invisible. We listen to it and do not hear it; its name is The Inaudible. We touch it and do not find it; its name is The Subtle (formless)....Infinite and boundless, it cannot be given any name." (*Tao Te Ching, Book of the Way and its Power*, poem 14.) It anticipates the Greek "infinite and boundless". And: "*Tao* produced the One. The One produced the two. The two produced the three. And the three produced the ten thousand things." (Poem 42.)

We are enjoined to live in harmony with the *Tao*, the One, and with eternal energy, *ch'i*. The *Tao* "flows everywhere like an ocean" (poem 34). It is Non-Being or *wu*: "All things in the world come from being. And being comes from non-being." (Poem 40.)[4]

Nishiwaki reiterated that the algebraic formula was known to the Arab mathematician al-Khwarizmi, and that it contained a set of rules about all the opposites in the universe: +A + –A can be day and night, life and death, time and eternity, the finite and the infinite. The Absolute (= 0) reconciles all these opposites and is where there is no difference as opposites cease to be opposite and are part of the Ultimate Reality, which is the *Tao*.

Instinctive perception of unity

Rational analysis sees differences and makes distinctions, and reduces a whole to its parts, like breaking an urn into pieces. What Coleridge called "the esemplastic power of the imagination" ('esemplastic' being Greek for 'shaping into one') restores the fragments into a whole, sees the One behind the many, and like an archaeologist in a trench pieces together the fragments into a whole urn that can be stuck together and displayed in a museum. The long Mystic Way, which in my case took from 1964 until 1993 (nearly 30 years), involved a centre-shift from the rational, social ego ($-A$, *yang*) to the deeper self ($+A$, *yin*) which can see unity ($= 0$), and in the case of those who have traversed the Mystic Way, leaves the self instinctively perceiving the unity of the universe and humankind. To see *this* unity ($+A + -A = 0$) all the time is why disciples meditate in Eastern temples to achieve Eastern wisdom.

The *yin-yang* symbol, *taijitu* or 'diagram of the Great Ultimate'.

On a personal note, I knew at Oxford that I had to get myself to Japan, but I did not understand why. My life's path again seems to have been Providential as I progressed from seeing approximately where al-Khwarizmi worked in the 9th century in Baghdad to seeing that his algebraic thinking reflected Taoism and provided a formula for describing, and "balancing", all the opposites in the universe within the Oneness that pervades Creation. The *yin* and *yang* of Taoism can be found in the *I Ching* (c.1143 or 1142BC), and the Executive Director of the Bank of Japan, a devotee of the *I Ching*, one day brought in a handful of sticks to my weekly class with him, which he shook and told my fortune. He explained it was "wind and earth", the best fortune in the *I Ching*. (See pp.xxii–xxiv for details.) This was a further illustration of how Japan set me on the right course for the rest of my life. *Yin* and *yang* can also be found in the 2,500-year-old *Tai Chi* (slow, meditative body movements), and the *yin* and *yang* symbol, *taijitu* or 'diagram of the Great Ultimate' (see picture), is attributed to the philosopher Zhou Dunyi (1017–1073).

The algebraic formula used as an algorithm to solve the problem of the birth of the universe

This realisation I had on 5 October 1965 that all opposites are reconciled within the One can be found in all my works. And for more than 55 years I regarded it as a helpful algebraic formula rather than as an algorithmic tool – with one exception. In 1993 I brought out *The Universe and the Light* with my Form from Movement Theory of Creation at the back. And I manipulated the algebraic formula into a set of rules. I reversed the formula: 0 = +A + –A. And I saw +A and –A as the first two particles thrown up before the Big Bang, which came out of the endless night of the One, the Plenitude "0".

My Form from Movement Theory (1992, see pp.14–23) is based on the ripples in the cosmic microwave background radiation first detected between 1989 and 1992, and on the Presocratic Anaximander of Miletus. I wrote it up in the dining-hall at Jesus College, Cambridge while attending a conference on Reductionism. Roger Penrose was present. His work on the singularity that gave rise to the Big Bang resulted in the Nobel Prize in Physics in 2020. I got to know him and sat with him and had many chats, and we got lost the first evening in Jesus College looking for a phone, and found ourselves in a don's room. Penrose stood in the dining-hall at Jesus on 4 September 1992 and watched, beaming, as the Norwegian Henning Broten supplied me with mathematics as I wrote on the crowded table, sitting next to me near the end of dinner. (See *My Double Life 2: A Rainbow over the Hills*, pp.413–414 for details.)

John Barrow, the author of *Theories of Everything*, was also there. I had had breakfast with him at an earlier conference on 11 April 1992 (see **p.vi**), the morning after I gave the keynote Friday-evening speech that is in the first section of *The Universe and the Light*.

In my Form from Movement Theory I was unconsciously using +A + –A = 0 as an algorithm, as a set of instructions to solve the problem of the birth of the universe. Yet for 28 more years I continued to see it as an algebraic formula, as a way of presenting the Oneness of the universe as surpassing all opposites and contradictions. As I developed my Universalist outlook in my literature and philosophy, in my history and religion and international politics, I drew on this formula. I had framed the business-reply card on which Nishiwaki wrote the formula, and it looked down on me from my wall and is mentioned in many of my books. I saw it as a convenient way of describing the process of reconciliation and unification in my works.

Realisation that the algebraic formula can be used as an algorithm
It was only in 2021, 56 years after my conversation with Nishiwaki, when I was asked if I had an algorithm of Creation that would explain the universe, that it dawned on me that +A + −A = 0 was more than an algebraic formula that surfaced in Golden-Age Baghdad where al-Khwarizmi worked in the Library on treatises, and was in fact an algorithm (the word that al-Khwarizmi gave to the world with his Latinised name and title, *Algoritmi de Numero Indorum*). It dawned on me that the formula contained a set of mathematical instructions or rules that would calculate an answer to the problem: how did the universe come into existence, what sustains it and how will it end?

Drawing on The New Philosophy of Universalism
I had already worked on these problems in other books, especially in *The New Philosophy of Universalism*, and as I explain how my algorithm of Creation works out and applies to the universe I will draw on passages from that book and other books of mine that focus on the universe. We can now consider the progress of Creation in terms of the instructions that will help calculate an answer to the problem of Ultimate Reality, the emergence, development and future prospects of the universe, and in particular to the question of whether the universe is an accident or has a purpose.

2

The Origin, Development and End of the Universe in 20 Stages

Data: ingredients, elements or conditions for the algorithm of Creation's 4 variables

Nothingness

The algorithm of Creation after the Big Bang, $+A + -A = 0$, gave instructions to the latent universe, which seemed to follow mathematical laws, to bring into existence its 2 trillion galaxies, each with billions of stars and countless planets born from a cosmic web of interconnected filaments of interlocking dark matter.

In setting out how this happens at every stage I follow my work *The New Philosophy of Universalism*, which at every stage gives all the options before settling evidentially on the scheme I follow here. As I said on p.9 it makes sense to interpret the algorithm of Creation in relation to each numbered stage of the development of the universe by quoting the relevant passages from *The New Philosophy of Universalism*. In fact, this book on the algorithm of Creation should ideally be read in conjunction with *The New Philosophy of Universalism*.

Contrary to Stephen Hawking's assertion[1] that space-time began with the Big Bang and so we cannot think of time *before* the Big Bang, there was a Nothingness at the very beginning of Creation which was also potentially a Void of Non-Being and Being. Like the *Tao*, in which both *yin* and *yang*, $+A + -A$, were latently present, this Nothingness was a Plenitude, a Fullness, containing latent Non-Being and Being. The algorithm can be stated in reverse to include the origin of Creation before the Big Bang and before the beginning of space-time: $0 = +A + -A$, Nothingness = potential Non-Being + potential Being. This Nothingness was an infinite darkness.

We can now proceed to the 20 stages of the origin, development and end of the universe:

(1) The structure of what is

In *The New Philosophy of Universalism* I set out the concept and structure of what is in a way that would bring together the two strands of metaphysical philosophy: the intuitional approach to the infinite and the rational approach to the finite and science. In my metaphysical thinking manifestation, the process of Creation, was an emergence of the universe from the infinite to the finite, and Existence emerged from Nothingness, or the One, through Non-Being and then Being: 0 (Nothingness, the One) = +A (Non-Being) + −A (Being). To put it another way, out of this Nothingness, which contains the potentiality of Non-Being, came Non-Being, which contains the potentiality of Being; and out of Being, which contains the potentiality of Existence, came Existence with the Big Bang.[2] In other words, out of Nothingness, or the One, came Non-Being, Being and then Existence: 0 (the One's Non-Being) = +A (Being) + −A (Existence).

The *New Philosophy of Universalism*'s section 'The Concept and Structure of What Is' says:[3]

> The sum total of everything, both infinite and finite, can be referred to as the All or the One.[4] In the Prologue I referred to algebraic thinking as a dialectical method that reconciles all contradictions, and I set out the simple formula +A + −A = 0. If 0 is the One, then the infinite + the finite = 0, the All or the One. The All or One includes all things concrete and abstract, natural and supernatural, known and unknown, probable and improbable, orderly and chaotic, temporal and eternal, material and immaterial, post-Big-Bang and pre-Big-Bang.
>
> The concept of the All or One holds all possibilities within itself. It includes every possible concept including the unmanifest and the manifested. Within the One are all the potentialities of Non-Being and Being, and so the One has been described as Supreme Being but is more accurately to be described as Nothingness.
>
> Within the All or One, the infinite has two aspects: the Void that preceded the Big Bang and which is outside the universe, or Non-Being; and the void within the finite universe after the Big Bang, or Being. Non-Being is the Nothingness within infinity that the surfer [see pp.54, 56] breasts and is outside the conical universe. It includes

all the potentialities of Being but in an unmanifested form. Being, on the other hand, is the manifestation of Non-Being which contains all potentialities of everything manifested and all possibilities of Existence. It is the quantum vacuum, which appears to be empty but in fact seethes within the activity of virtual particles like an invisible sea of energy. If we were to accept Plato's Ideas as being virtual, they would be found in Being along with the blueprints, templates and germs of existing things in their potential form. In our algebraic, dialectical thinking we can say that just as $+A + -A = 0$, so Non-Being + Being = the infinite All or One.

Within the All or One, the finite has two aspects: Being, which as we have just seen contains the invisible potentialities of Existence and is an intersection between the infinite and finite, between the timeless and time; and Existence, the manifestation of Being, the multiplicity which is contained within the unity of Being, all the phenomenal forms of Nature, all atoms and cells, all matter and organisms, all existing things we can see or touch in the universe of space-time. In our algebraic, dialectical thinking, we can say that just as $+A + -A = 0$, so Being + Existence = the finite.

Within a philosophical category, the process of manifestation from the metaphysical to the scientific proceeds from the infinite to the finite, from the hidden Reality to the universe, from the boundless the surfer breasts to the phenomena of Nature and its ecosystems. Manifestation originates in Nothingness and becomes Non-Being and then, like a nothingness becoming a virtual particle, becomes an idea in Being, whence, like a virtual particle becoming a real particle, it becomes a form in Existence, the world of evolution and organisms. Existence pours out of Being like the multitudinous plenty pouring from a *cornucopia* (or horn of plenty). The totality – the infinite *and* the finite – is the All or One which permeates everything, both infinite and finite, as a Nothingness or Emptiness that is also a Fullness, a *pleroma*, a latent Non-Being and Being – an idea that is found in Eastern thinking and in Gnosticism – and that includes their manifestation into Existence.

For sake of clarity I can sum up the process of manifestation as four tiers:

- Nothingness/Fullness, or the All or One – the potentialities of infinite Non-Being, infinite/finite Being and finite Existence;
- Non-Being – the infinite Void outside the universe that preceded the Big Bang, the potentialities of Being and manifestation of Nothingness;
- Being – the infinite/finite void which succeeded the Big Bang within the finite universe, the potentialities of Existence and manifestation of Non-Being; and
- Existence – the finite manifestation of Being.

At the philosophical level, within a philosophical category, the rational approach to Reality therefore recognises that the infinite the surfer breasts which lies outside our universe is Nothingness and Non-Being, and that within the finite universe it manifests into Being. At the philosophical level, the rational approach to the system of the universe focuses on finite Existence and space-time which began after the Big Bang. At the philosophical level, we are looking at a manifestational process from Nothingness or the One, through the Void of Non-Being into the quantum vacuum or unity of Being and then to the diversity and multiplicity of Existence.

To sum up, the algorithm of Creation, +A + −A = 0 moves through steps from the infinite to the finite, with variations of +A and −A as follows:

+A (Nothingness) + −A (Fullness) = 0 (a Plenitude of Nothingness/ Fullness, potential Non-Being, the One);
0 (the One) = +A (Nothingness) + −A (Non-Being);
0 (the Infinite All or the One) = +A (Non-Being) + −A (Being);
0 (the Infinite and Finite) = +A (Being) + −A (Existence);
0 (the All or the One) = +A (the infinite) + −A (the finite).

(2) The origin and creation of the universe: Form from Movement Theory (1992)

Before the Big Bang 13.82 billion years ago[5] there was a moving infinite, an ocean-like vacuum in which there were repulsive gravitational effects that correspond to the ripples or wrinkles in the cosmic microwave background (CMB) radiation discovered by NASA's COBE (Cosmic Background Explorer) between 1989 and 1992 (see p.23).

The ripples or wrinkles in the CMB were an 'echo' of the Big Bang that were first emitted 380,000 years after the Big Bang, and although space scientists focus on their being an aftershock and evidence of the Big Bang, a relic of the universe's earliest phase, its first light (in the microwave part of the energy spectrum) and quantum fluctuations or gravitational waves that became the universe's earliest structure of stars, galaxies and clusters of galaxies, they correspond to and may offer evidence for an infinite movement within the nearly-still Nothingness (or 'ocean of energy', a Plenitude, a Fullness, see p.11) in the vacuum or Void that existed *before* the Big Bang, a virtual structural plan before the vacuum's energy was transformed into matter by the Big Bang and the ripples or wrinkles became the universe's real structure of stars, galaxies and clusters of galaxies *after* the Big Bang. I reflected this infinite movement before the Big Bang in my Form from Movement Theory (1992) on the origin and creation of the universe, in which the ripples or wrinkles of the infinite movement correspond to the ripples or wrinkles of the CMB, and it would be dramatically confirmed to me by my deep source in my sleep on 2 December 2021 (see pp.347–350).

To put it in terms of +A + −A = 0, +A (the vacuum or Void or 'ocean of energy' in Nothingness, a Plenitude, a Fullness) + −A (virtual ripples or wrinkles, infinite movement within Nothingness, a virtual or potential structural plan for the coming universe) = 0 (the universe whose aftershock after the Big Bang materialised the virtual structural plan into real ripples or wrinkles, the structure of stars, galaxies and clusters of galaxies).

The gravitational effects within the initial vacuum state created a high-density region which threw up a singularity, one of a pair of particles that began the Big Bang.[6] The Big Bang resulted in a rapid expansion of space (inflation) that led to the 'ocean of energy' being transformed into matter that shaped our universe. The energy in the ocean of Nothingness was dumped into expanding space as matter, in which the ripples became denser regions of gas that collapsed into the first galaxies and stars, and would fill the expanding universe with 2 trillion galaxies.[7]

The method by which this structure became Creation I set out in *The New Philosophy of Universalism* as 'The Origin and Creation of the Universe', in which I follow the Presocratic Ionian Greeks Anaximander of Miletus and Heracleitus of Ephesus, and Aristotle, in seeing an eternal movement in which two pre-particles were thrown up, +p and –p. So 0 (the infinite and finite One) = +A (+p) + –A (–p). My full Form from Movement Theory is as follows:[8]

> The view that subtle and physical light comprise the essential stuff of the universe has arisen from and is supported by the Universalist proposal regarding the origin of the universe: a Form from Movement Theory which looks back to Heracleitus's eternally moving Fire c.500BC, the Fire of Reality which has always existed: "This world-order did none of gods or men make, but it always was and is and shall be: an ever-living fire, kindling in measures and going out in measures."[9]
>
> The Form from Movement Theory differs from most theories and ideas of Creation, which begin with stillness and simplicity that later become movement and complexity. Xenophanes' motionless unity is an early Presocratic example of this stillness. The infinite which threw up the universe with the Big Bang is a moving timeless "boundless" that has always existed, and I begin with an eternal and infinite movement and complexity as an ontological basic condition. This infinite movement is like Anaximander's eternally moving "boundless" *apeiron*[10] and Heracleitus' eternally moving "ever-living Fire" which may also have been *aither*, Newton's ether.[11] Aristotle was half-way to this position in seeing an eternal and immutable "unmoved mover" (my Nothingness) that is infinite and excited a continuous motion in the infinite (my Non-Being), a motion that Bohm called a "holomovement".[12] The eternal movement was how it always has been, and rational objections to a moving beginning by Xenophanes and Parmenides ignore the basic ontological condition.
>
> The view of Anaximander, Heracleitus and Aristotle that the infinite is eternally moving is echoed at the beginning of *Genesis*: "In

the beginning God created the heaven, and the earth. And the earth was without form, and void; and darkness was upon the face of the deep. And the Spirit of God moved upon the face of the waters. And God said, Let there be light: and there was light" (1, 1–3). The formless Void and dark deep were moving like waters, and God acted as an unmoved mover. This text "echoes" Anaximander and Heracleitus as *Genesis*, like the rest of the Pentateuch, has Yahwist, Elohist and Priestly strains. Scholars are united in attributing *Genesis* 1 to the Priestly strain, and whereas the Yahwist strain can be dated as early as 950BC and the Elohist strain to 900–700BC, the Priestly strain is usually dated to the 5th century BC, *after* Anaximander and Heracleitus.[13]

Manifestation is a reduction or limitation of this infinite movement rather than an additional quality, and is passed through four distinct stages: from Nothingness to Non-Being, to Being and Existence.

Nothingness had always been a completely self-entangled energy of virtual, infinite, irregular movement in all directions (although it preceded space-time and therefore direction), a real nothing or Oneness, a latent "ever-living Fire" or metaphysical Light. I postulate that this Nothingness was infinitely self-aware and a non-locality. I can formally denote this infinite movement as M,[14] which is absolutely the most subtle substance.

All limitations of M are potentialities within M, and in a region of M which I can denote as S a limitation of movement occurs so that there is a regular movement, possibly in the form of a multidimensional (in the sense that it had more than three dimensions) spiral (Non-Being). This regular movement arises as all irregular movement dies away. S, or Non-Being, can be said to be a subset of M, where both M and S are infinite. Everything in M not in S is denoted M–S (M minus S). There is therefore now in place of Oneness a duality, that of M–S versus S (or pre-manifestation versus first manifestation). Whereas M–S entails all sorts of infinitely complex movement, S is a fairly simple and symmetrical form of movement. As in set theory both M and S are infinite, yet M is larger than S and contains the whole of S. In the same way there is a more definite yet still infinite self-entanglement (or non-locality) in all of S. There has to be a creative tension or pressure between M–S and S, between irregularity and regularity, and as a result of this interaction new forms of order arise between M–S and S, one such possible order being symmetric points or pre-particles. I postulate that a pair of pre-particles arises in S as a result of the pressure of M–S on S. These pre-particles, which can be termed +p and −p, are symmetrically entangled (i.e. in an unbroken relationship) and are both opposite and complementary. I can state this entanglement succinctly by drawing on algebraic, dialectical thinking (+A + −A = 0). In sum, the two pre-particles equal zero. Thus, S: +p + −p = 0.

I now postulate a reduction of symmetry because of the disturbing influence of the infinitely complex and irregular movement of M–S upon S. This manifests an annihilation of one of the pre-particles, −p. In other words, the regular movement constituting a pre-particle is absorbed back into the vast movement of M and ceases to have any presence in S. The symmetry is broken, for the other pre-particle, +p, is not absorbed into M but derives energy from the movement of M–S upon S. This pre-particle is an empty point or vacuum or singularity. This point potentially contains Non-Being. I can say that the point is "a defined Non-Being" in contrast to the undefined Non-Being which is S, and that both aspects of Non-Being were always potentials within the unmanifest M. This point, +p, begins to spread like a multi-dimensional (or more-than-three-dimensional) wave in all directions in the field of S, and expands. The point, +p, is gradually becoming Being, the first subtle structure in the vacuum field of S: +p → B. I can thus say that from Non-Being happens Being.

Due to the interaction between B and M–S in the field of S, B evolves more and more structure. M–S, Nothingness, is the source of infinitely complex impulses, a self-aware Non-Being, primitively intelligent, apparently random – but perhaps fundamentally purposive – generator. S, in contrast, provides a background of silence and a field in which the reverberations of the structure can manifest into form, and be displayed, within M, without disturbance. In the language of physics I can say that the emergence from Nothingness (M) of Non-Being (the vacuum field of S) and then Being (B within S) is activity within the quantum vacuum, and that B, which is implicit (in the sense that it is not plainly expressed but implied as it is hidden), evolves more and more explicit (in the sense that it is expressly stated or manifested) orders within its implicitness. I can denote this series of more and more explicit implicit orders which B contains by $I_1, I_2, I_3, \ldots$ I_n; where I_n corresponds to the most explicit order that is still implicit to us.

The process of Being (formerly +p) becoming Existence (E) is one in which potentialities become actualities, pre-particles become particles, pre-matter becomes matter, pre-organisms become organisms, pre-consciousness becomes consciousness and possibility becomes factuality, completing the process of manifestation. I can denote Existence or the explicit order by I_{n+1}. I can say that $E = I_{n+1}$, which is an event in M. I can postulate the possibility that there are one or more loops between these implicit orders, such as between I_{n+1} (the explicit order) and I_{n-2} (an earlier implicit order). These loops provide the rudiments of primitive consciousness entangled in and manifesting from within matter. There is a possibility that there is "horizontal" division into two manifesting orders, one eventually leading up to explicit matter and the other leading up to aware consciousness and creative intelligence, with one or more "vertical" bridges between these two orders.

As a result of +p deriving energy from M–S, the infinite and implicit +p or Being, which has spread in all directions in the field of S and expanded, has become pre-matter (and perhaps as we have just seen also pre-consciousness) in place of a first pre-proton (+p), and form is ready to arise from the infinite movement. Virtual particles emerge from the quantum vacuum (B in the field of S) in pairs, and one particle in each pair has the potentiality to become a real particle in explicit Existence if it draws energy from the pressure of M–S on S. Through quantum processes, the emergence of virtual particles happens in many regions of S, all over the expanded field of +p or B, and, deriving energy from the pressure of M–S on S, many ephemeral virtual particles become enduring real particles.

As a result of the proliferation of simultaneously emerging real particles, the process of the Big Bang or hot beginning – the actual Creation of the universe as we know it – takes place.

Space-time emerged as part of the physical universe as Einstein has shown in his general theory of relativity which holds that gravitation is an effect of the curvature of the space-time continuum. The anti-Newtonian Leibniz, in his second letter to Clarke, saw space and time in terms of events. He held that space is an order of co-existence of events, and that time is an order of succession of events. However, Einstein's work supersedes this event-based view which preceded, and in itself cannot account for, the curvature of space causing gravitation.

To the Universalist philosopher, then, the process of the origin and Creation of the universe begins with a first principle or first cause, an eternal and infinite moving Nothingness or latent "ever-living Fire" or metaphysical Light. From this formed a spiral (Non-Being), from which emerged or manifested a pre-particle that became a point or empty vacuum or singularity (Being). Out of this, Existence – Form – manifests as energy becoming matter through the Big Bang, inflation and the expanding universe Hubble discovered. Matter and consciousness have thus both manifested from ever-living Fire or subtle Light.

This Form from Movement Theory can be set down mathematically: $M \rightarrow M-S + S$. In S, $+p \rightarrow B$, which evolves $I_1, I_2, I_3, \dots I_n$. $B \rightarrow E$, which is I_{n+1}. (M=movement; S=spiral; p=pre-particle; B=Being; I=implicit order; E=Existence.)

Beginning with an infinite, self-aware, self-entangled movement like Anaximander's boundless raises the question as to whether there is an arbitrariness about the process I have described, which could in theory be one of infinitely many possible processes unless the self-aware movement is purposive – in which case this origin and Creation was the only possible process that could have happened, and life too had to happen. The universal order principle may indicate that the process may not have been arbitrary and random.

This account of Creation and causality – which draws on the tradition of scientific cosmology and ontology – postulates that the invisible substance behind the universe manifests into radiation and physical light, and into the *yin-yang*-like symmetry of neutrinos and photons and the ordered system of Nature.

Anaximander of Miletus flourished c.570BC and saw the world as emerging from the eternal, infinite "boundless" (*to apeiron*), an eternal and eternally-moving boundless Reality from which the universe began as a finite germ (*gonimon*). He said that the boundless has no distinguishable qualities. His germ is like Penrose's singularity, or a point which contains infinity. Anaximander was close to the view of physicists in our time.

Heracleitus of Ephesus, who flourished c.500BC, held that Reality was Fire, "An ever-living Fire", the metaphysical Fire that "ever was, and is, and shall be". And so the world was in a permanent flux: *panta rhei*. He saw unity behind all opposites, and all opposites reconciled in underlying unity, so that "the way up and the way down are one and the same". +A + –A = 0 is behind his unity-in-opposites in a world of contradictions. To Heracleitus, the opposites were essential features of the unity.

These early Presocratic philosophers saw Ultimate Reality, or primeval Nothingness, as an ever-moving Fire or Light, and it is therefore understandable that to them Ultimate Reality should be moving. Living in Ionia, they were influenced by the Persian religion of Ahura Mazda, god of Light, who created the two spirits of good and evil, Spenta Mainyu and Angra Mainyu, also known as Ahriman. Again the algorithm of Creation is at work: 0 (Ahura Mazda) = +A (Spenta Mainyu) + –A (Angra Mainyu), a unification of Iranian dualism.

I have already said (on pp.8, 16) that I wrote the first version of the Form from Movement Theory steeped in the Presocratic philosophers after writing *The Fire and the Stones*, at a conference on Reductionism at Jesus College, Cambridge on 4 September 1992, for which I was invited to write a paper. I had sat with Roger Penrose during many of the lectures and had talked with him about the singularity before the Big Bang, and he had asked to read a draft of *The Universe and the Light*, which I had been reading through in my room at the end of each day. A young Norwegian mathematician, Henning Broten, confronted me and said he had read my paper and had rung David Bohm (who collaborated with Einstein) about it and wanted to do the mathematics for my conception of Creation, and this happened over dinner in the dining-hall at Jesus College on 4 September. As I described on p.8, word got round as to what we were doing, and Penrose came and stood behind me, near my left shoulder, and observed what I was writing with a big approving smile on his face as I wrote each step-by-step line down after conferring with Henning Broten. Freeman Dyson, Mary Midgley and John Barrow were also present. (See *My Double Life 2: A Rainbow over the Hills*, pp.409–416 for further details.) When I got home I wrote up my notes into the Form from Movement Theory.

While I was dictating it on 15 September two Jehovah's Witnesses came to the door, and one said, "Can I tell you how the universe was created?" I said, "I'm just dictating my theory on it now, you'll be able to read it before the end of the year." They went away evidently startled. An early version of the theory appeared as an Appendix in *The Universe and the Light*, together with a comment on the back cover by Penrose.

I am confident in my account of Creation, and of the mathematics, and can now see it in terms of the structure in (1) and can now see the manifestation of the structure in terms of the algorithm of Creation, +A + −A = 0. The movement of M (infinite movement, Nothingness) is limited by a symmetrical movement S (a spiral, a limitation of movement, symmetrical Non-Being). The irregular movement of M (Nothingness) – S (Non-Being) upon S (Non-Being) creates a duality: + (M–S) + −S = 0, or +A (+ M–S) + −A (−S) = 0. Because of the pressure of M–S on S (Non-Being), a pair of symmetrically-entangled pre-particles arises in S: +p and −p (the first two particles which are opposites but complementary and symmetrically entangled) = 0 (Nothingness/Non-Being). Another way of putting it is: +A (+p) + −A (−p) = 0 (Nothingness/Non-Being).

The irregular movement of M–S upon S annihilates −p and the symmetry is broken, leaving +p independent and deriving energy from the movement of M–S upon S. Now S + p + −S (within M, Non-Being within infinite movement, Nothingness) = 0, or +A (S + p) + −A (−S) = 0 (within M, infinite movement/Nothingness, and Non-Being). This +p is an empty point or vacuum, or singularity, which is why Penrose was so interested in standing and watching as we worked in the crowded hall in 1992. Now +p expands and becomes Being.

As a result of the infinite +p deriving energy from M–S the virtual particle +p is ready to arise from infinite movement as the first pre-proton. Virtual particles arise in pairs in +p or B (Being) and derive energy from the pressure of M–S on S, and one particle becomes a real particle (Being). So +p in B + M–S on S (Non-Being) = 0 (the One). Or +A (+p in B) + –A (– M–S on S) = 0 (the One).

Existence (E) emerges through quantum processes within Being after simultaneously emerging real particles create a hot beginning, the Big Bang. So +p in B (Being) +E (Existence) = 0 (the One), or +A (+p in B) + –A (–E) = 0 (the One).

(3) The Big Bang

The Big Bang took place 13.82 billion years ago.[15] The gravitational effects within the initial vacuum state created a high-density region which threw up a singularity, one of a pair of particles that began the Big Bang.[16]

We have just seen that in the beginning, before the Big Bang which created matter, there was an infinite Nothingness, an 'ocean of energy' that was nearly still but seems to have been gently rippling, ripples that correspond to the ripples in the cosmic microwave background (CMB) radiation that was first emitted 380,000 years after the Big Bang. The ripples in CMB which became galaxies were in the echo of the Big Bang because they preceded it (see p.15). Hence my movement (M) in Nothingness and my Form from Movement Theory, in which gravitational effects within the initial vacuum state created a high-density region which threw up a singularity, one of a pair of particles that began the Big Bang (see previous paragraph and pp.15–16).

In the course of and as a result of the Big Bang, energy in the ocean of Nothingness was dumped into expanding space as matter, in which ripples, now denser regions of gas, collapsed into the first galaxies and stars. The CMB is key evidence for the Big Bang as it is its echo.

The CMB was first discovered accidentally by Penzias and Wilson in 1964. Its ripples were first discovered by Smoot and Mather's investigations into the COBE (Cosmic Background Explorer) map of CMB produced by NASA between 1989 and 1992. The image was refined in the 2003 WMAP (Wilkinson Microwave Anisotropy Probe), and further refined in the Planck space telescope images of 2013, released in 2018 (see front cover).

Powerful space telescopes can look back into time and reveal images that left the distant universe more than 13 billion years ago (now more than 13 billion light-years away). Clearer images have followed from the Hubble space telescope and will follow from the James Webb space telescope, which will see through murky dust clouds to the forming of stars near the beginning of time. All the space telescopes have confirmed the predictions of the scientists regarding the Big Bang.

In setting out the next stage of the algorithm of Creation, I am not setting out the alternative models to the Big Bang (Steady-State, the multiverse, String-Theory and the Big-Wave models), all of which are covered in *The Philosophy of Universalism*, but am sticking with my conclusion in *The New Philosophy of Universalism*.

First I must be clear on the meaning of the infinite, from which the Big Bang came. In *The New Philosophy of Universalism* I distinguish between Absolute Infinity (the totality of everything, which is beyond mathematics), to which the infinity within the singularity that became the Big Bang belongs; infinity in the physical universe (such as a singularity), which is transfinite; and mathematical infinity (transfinite numbers). (See p.25.)

As we shall soon see, the progress of the Big-Bang theory since Einstein incorporates some key discoveries along the way, notably Hubble's 1929 discovery that the universe is expanding; Penzias and Wilson's discovery of cosmic microwave background radiation (CMB) on 20 May 1964; the 1992 COBE (Cosmic Background Explorer) image and findings; and the 2003 WMAP (Wilkinson Microwave Anisotropy Probe) image of the same radiation but with a 35-times-better resolution.

The New Philosophy of Universalism has a section 'The Big Bang and Infinite Singularity', which describes how the Big Bang theory took hold and is associated with Penrose's singularity:[17]

The notion of the Big Bang began as a cosmological theory and has received some confirmation from astrophysicists' evidence. The theory of the Big Bang goes back to Einstein. One class of potential solutions to Einstein's equations allowed for the possibility that the universe was expanding or contracting. Einstein himself dismissed this possibility as there was a general belief that the universe was not in motion but static.

The knot-garden at Otley Hall, Suffolk, commissioned by Nicholas Hagger and created by Sylvia Landsberg to reflect one of the six knots on the front cover of Elizabeth I's 1544 embroidered prayer-book, and showing the origin of the universe: the infinite in the centre of the knot manifesting through four infinity (∞) symbols (if the centre is taken as half a symbol for each of the four loops) of Nothingness, Non-Being, Being and Existence, suggesting the conjunction between the infinite and finite space-time in the process of Creation, into history (25 herb beds representing 25 civilisations).

It is amazing that Einstein laid the foundations of the Big Bang theory despite believing in a static rather than an expanding universe and without knowing about the strong and weak forces, which had not been discovered. In 1905 he transformed physics with his special theory of relativity, his theory on the equivalence of mass and energy, his theory of Brownian motion and his photon theory of light.

In his special theory of relativity Einstein held that the speed of light is the same for all observers, and that all physical laws are the same for two frames of reference moving with uniform speed relative to each other. This theory was verified in 1919. Meanwhile his general theory of relativity of 1915 showed that space, time and gravitation are based on frames of reference – for example the earth's surface or an interplanetary spaceship – whose motions are accelerated in relation to each other.

Einstein confirmed Newton's laws for bodies travelling at low speeds but gave different results for bodies travelling at high speeds, such as stars, galaxies and elementary particles. In the general theory Einstein found that when an observer or frame of reference accelerates in relation to another when gravitational forces are present, gravity is seen not as a force, but as a curvature in space-time. (The geometrical concepts for this theory had all been developed in the 19th century and Einstein was able to adopt them ready-made.) Thus the earth orbits the sun in a curved path because the sun distorts space-time in the sun's vicinity. The general law explains why Mercury's orbit differs from how Newton saw it in his gravitational theory. Newton's theory did not take account of the bending of stellar light and the gravitational redshift (in which an atom near a star has a longer wavelength and looks redder than an atom on earth). Einstein's theory that large objects can distort space and time around them was confirmed to within 95.95 per cent certainty in 2006 when Jodrell Bank researchers showed that a beam of radio waves from one pulsar (the remains of an exploded star) was affected by a distortion of time and space caused by another pulsar.

Einstein's theory of gravity introduced a new conception of space-time. According to Einstein gravity is an effect of the curving of space-time.[18] If matter, which is formed by the mass and energy it contains, is of too great a density (more than 6 hydrogen atoms per cubic metre of space, emptier than the most "empty" vacuum), the universe will be curved up into a finite volume while emptier spaces can extend forever. A universe not exceeding a density of 6 hydrogen atoms per cubic metre of space can expand forever.[19]

The Big-Bang hypothesis was first put forward in 1927 by Georges Lemaître.[20] It was opposed on the grounds that a universe created by a Big Bang would be less than the age of the earth suggested by geological evidence, and the theory fell out of favour.

In 1929 the Russian cosmologist Alexander Friedmann worked out that the universe was not static. Seven years earlier he had formulated a model of the universe deriving from Einstein's general theory of relativity, in which the average mass density is constant and all parameters are known except the expansion factor. In late 1922 and 1924 he postulated a "Big-Bang" model for the evolution of the universe. Now, realising that the universe was identical in whichever direction we look, he came up with three models. Either the universe expands and then collapses and comes to an end in a Big Crunch, in which case space is finite. Or the universe expands forever, in which case space is finite until the expansion reaches a point beyond which it will last forever and so becomes infinite. Or, if the universe's density is just less than the critical 6 hydrogen atoms per cubic metre of space, the critical rate of expansion, space is flat, balanced between expansion and collapse, in which case space is finite but will be infinite if the equilibrium lasts forever.[21]

The first evidence of the Big Bang came in 1929 when the American astrophysicist Edwin Hubble found that the universe was expanding, and that the recession speed of a galaxy is proportional to its distance from the earth. Until then astronomical distance had been measured by triangles, the apex being a distant star and the base a known length with known angles, and by "brightness candles" ("Cepheids"), which allowed enormous distances to be estimated. Hubble discovered that redshifts of nebulae (which the US astronomer Vesto Slipher had been measuring) were receding from our galaxy, and that brightness candles were correlated to the recession velocities. (The relationship became known as Hubble's law.) The Hubble constant states that there is a constant of proportionality between distance and velocity. Hubble's data showed that the universe was expanding at an accelerating rate to within a 10-per-cent uncertainty margin. The picture was still far from clear. Calculations based on Hubble's work bewilderingly suggested that the age of the oldest stars (up to 20 billion years old assuming an even accelerating rate, a wrong premiss) was older than the universe (13.7 [since updated to 13.82] billion years old).[22]

In the 1940s the Russian-born, US-based nuclear physicist and cosmologist George Gamow proposed a series of reactions from neutrons which produced the light elements after a hot beginning. He and two colleagues predicted the background temperature of the universe. John Archibald Wheeler, the American physicist, coined the phrase "Big Bang" as it corresponded to Einstein's general theory of relativity. (He also coined the phrase "black hole".) Fred Hoyle, the British cosmologist, is also credited with the first use of the term in the 1950s after becoming an advocate of the Steady-State theory formulated by the Austrian-British mathematician Hermann Bondi and the Austrian-British Thomas Gold in 1948. The Steady-State theory holds that the universe is expanding and that new matter is being created at a rate that keeps its mean density constant and balances the expansion. To Hoyle, the Big Bang was a term of derision for he was deeply critical of the Big-Bang theory.[23]

In 1964 the US radioastronomers Arno Penzias and Robert Wilson discovered a constant background noise level that originated in the cosmos while they were trying to improve microwave antennae at their laboratory. The cosmic microwave background was interpreted as the residual radiation from the hot beginning. It had a temperature of 2.725K (about −270°C).[24] The COBE satellite measurements confirmed the uniformity of this radiation to a millionth of a degree. In other words, the radiation is the same everywhere throughout the universe. This suggests that the Big Bang had no centre in the universe, for otherwise the sky around its centre would glow with primordial radiation and the sky a long way away would be cold.[25]

By 1970 the Big Bang had won over most of the supporters of the Steady-State model proposed by Bondi, Gold and Hoyle. This was a variation of Friedmann's third model, the flat universe. It had infinite space and no beginning or end.

Einstein's theory of gravity requires that the universe exploded from a state of infinite density. At the beginning of the Big Bang, according to Friedmann's original 1922/1924 model, the density of the universe and the curvature of space-time were infinite, and the distance between galaxies was zero. Mathematics comes to a halt in the face of the infinite, and the general theory of relativity predicts that at its beginning the universe was contained in a point, in which the theory of relativity breaks down. This point is called a singularity (an atom-size point in space-time at which the space-time curvature becomes infinite). Interestingly, Dante writes of the infinitesimal "point" in *Paradiso*, canto 28.

The idea of a singularity stemmed from the US theoretical physicists Robert Oppenheimer and Hartland Snyder, who had shown in 1939 that a spherical collapse of a massive star could lead to a central space-time singularity (or black hole) at which the classical theory of relativity is stretched beyond its limits.[26] It was taken up in the British mathematician and physicist Roger Penrose's 1965 theorem, which stated that any body undergoing gravitational collapse must eventually form a singularity.[27] For a singularity or black hole the size of an atom to form, 10^{36} atoms must be squeezed into the space occupied by one atom.[28]

Reading this in 1965, the British theoretical physicist Stephen Hawking realised that if time were reversed Penrose's theorem would still hold, and prove that an expanding universe must have begun with a singularity.[29] In 1970 Penrose and Hawking proved in a theorem that the general theory of relativity implied that the universe must have a beginning and possibly an end and that at the beginning there must have been a Big-Bang singularity of infinite density and space-time curvature. The initial space-time singularity (the Big Bang) represents the Creation of space-time and matter in "a white hole" (the hypothetical time-reverse of a black hole) rather than its destruction in a black hole. Thus the Big Bang began in one atom-size point and happened everywhere within that point. Everything was connected in that one point, and was One....[30]

The infinity within the singularity is not a part of finite space-time but is part of the totality of the infinite Absolute. This may be an uncomfortable fact for some cosmologists and physicists who want the Big Bang to have begun within the finite, beginning finite space-time, but to the philosopher that is a fact, a truth that cannot be ignored as being inconvenient. Mathematician engineers doing calculations involving air speed find that a speed faster than the speed of sound (750 mph), which causes a sonic boom, appears in equations as an infinity, and particle physicists' equations are frequently part-finite and part-infinite. They subtract the infinite part in a process called "renormalisation". The mathematical infinities may be due to their being constructed in the human mind, perhaps in a clumsy way of seeing things.

To put it bluntly, the distinction between metaphysical infinity and mathematical infinity allows mathematicians to study geometry and arithmetical series without being enjoined to reserve the term "infinite" for the Absolute.[31] But mathematical infinity leads to metaphysical infinity because the infinity in the singularity is essentially a metaphysical infinity as it pre-exists space-time and the physical world, and is therefore beyond physics ("metaphysical").

The mathematics of cosmologists, then, suggest that the Big Bang began as a singularity in one point of infinite density in which all matter and energy were concentrated. The trouble is, a singularity is an unphysical state[32] and there does not seem to be a physics of singularity. Hawking therefore changed his mind and came to believe that there was no singularity at the beginning of the universe once quantum effects are taken into account. He argued that just as there is nothing north of the North Pole, so there is no singularity before the Big Bang.[33] He maintained that the gravitational field was so strong that quantum gravitational effects became important, and cited the US physicist Richard Feynman's proposal to formulate quantum theory in terms of a sum over histories. His mathematical construction, in conjunction with the American James Hartle, of a quantum state, for which there was no observation or evidence, sees time as finite in the past, not stretching back into timelessness, yet having no beginning. It sees the universe as a natural event in terms of a framework that is supranatural (or supernatural).[34]

In classical theory of gravity, the universe either existed for an infinite time or it began as a singularity at a finite time in the past. In the quantum theory of gravity, space-time can be finite in extent yet have no singularities that form a boundary or edge, on which the laws of science break down, and no edge of space-time. Hawking argued that the universe would be self-contained and have no beginning or end, but just be.[35] Penrose did not accept Hawking's view and wrote: "Though ingenious, his suggestion involves severe theoretical difficulties, and I do not myself believe that it can be made to work."[36] Universalists, anyway sceptical of aspects of "random" quantum theory, do not accept Hawking's view, which ignores the beginning of the universe.

Einstein, Penrose and Hawking have regarded singularities as arbitrary as the laws of physics break down when a singularity appears. Space and time are destroyed at the places where infinite densities appear, and the laws of gravity cease to hold at physical infinities. A physical infinity is more serious than mathematical infinity as encountered by particle physicists and engineers. Cosmologists are reluctant to admit physical infinities into the universe and prefer to see them as mathematical problems, but the fact remains that the universe began in an infinity when there was no space-time or gravity, merely an infinite Void.[37]

Then, in the 1990s there was a development. The cosmic microwave background radiation seemed too smooth and uniform, and suggested a smooth universe 380,000 years after the Big Bang which would have resulted in a universe with no stars or planets, or people. In 1992 the COBE (Cosmic Background Explorer) satellite launched by NASA found ripples in the cosmic microwave background radiation, showing that 380,000 years after the Big-Bang singularity there were already wispy clouds of matter stretching in excess of 500 million light-years in length. These could be ripples or clouds that collapsed in upon themselves, broke up and formed clusters of galaxies. The size of the clouds seemed to confirm that 96 per cent of the universe is missing and may be in the form of invisible matter or energy, which is different from bright stars and galaxies,[38] and that there must be a hundred times more invisible matter in the universe than the matter we can see.[39]

In short, there was not enough visible matter to support the experientially-measured expansion rate. In fact, the hypothetical concepts of dark matter and dark energy accounted for the missing mass of the universe: 23 per cent dark matter and 73 per cent dark energy. (These figures were estimated by the WMAP – Wilkinson Microwave Anisotropy Probe – mapping team of 2003 which revised the previously accepted 30 per cent dark matter and 66 per cent dark energy.) It is thought that dark matter was needed to help galaxies to form. In September 2008 it was announced that the European space probe Pamela had discovered a surge of high-energy particles (positrons, a form of antimatter) from the heart of the Milky Way which closely matches the radiation signature predicted for dark matter.[40]

All this follows a variation of $+A + -A = 0$: $+A$ (the infinite singularity in Being's quantum vacuum) $+ -A$ (the finite Existence's real particles) $= 0$ (the One expanding universe after the Big Bang).

To put it another way, $+A$ (the metaphysical oneness, a non-duality that gave rise to both matter and consciousness, and humankind) $+ -A$ (the entire physical universe) $= 0$ (the One, unity of the metaphysical universe and humankind). (See p.xxiii.)

(4) The quantum vacuum and infinite order

The post-1896 notion of a quantum vacuum to explain radiation, which goes back to at least 13.42 billion years ago, throws light on the Void from which the first virtual particles $+p$ and $-p$ arose 13.82 billion years ago. (Please note, the age of the universe was estimated as being 13.7 billion years when *The New Philosophy of Universalism* was published in 2009, see pp.28, 50, 55, 58 and 62, but has since been updated to 13.82 billion years.) As we shall see on p.34 a quantum vacuum contains virtual photons that can emerge and become real. The energy within the quantum vacuum may have supplied energy to the emerging universe, and Bohm proposed that the quantum vacuum (Being) contains infinite order behind the uncertainty principle.

In *The New Philosophy of Universalism* I have a section 'The Quantum Vacuum and the Expanding Force of Light', which links quanta to virtual photons:[41]

Just as cosmology and astrophysics came out of a point, so in physics the subatomic and galactic worlds are one and interconnected in the point. Now I must introduce a complication to our account of the Big Bang: quantum physics and the workings of the quantum vacuum.[42]

In a sense, quantum theory began with the discovery of radiation. In 1896 the French scientist Henri Becquerel found that uranium compounds emitted radiation, and after a search for other radioactive substances Marie Curie extracted polonium and radium. The wave theory of light could not explain the interaction between electromagnetic radiation and matter. In 1900 Max Planck proposed that emission of radiation was intermittent rather than continuous, in multiples of an atom or quantum of energy. Using his constant, Planck proposed that the energy of a quantum is inversely proportional to the wavelength of radiation but directly proportional to the frequency. Quantum mechanics began with Planck's theory of black-body radiation in 1900.

Quantum theory was based on seeing radiation as multiples of quanta or particles as well as rays. Einstein extended Planck's ideas in 1905. He saw light and electromagnetic radiation emitted as quanta, and showed that light behaves as if all its energy is concentrated quantum packets of energy called photons. Light could be both a wave and particles: this electromagnetic radiation could behave like waves in some circumstances and like particles in others. Quantum theory developed with Niels Bohr's proposal of the quantum theory of spectra in 1913.

In 1923 Louis de Broglie (pronounced "de Broy") suggested that matter might also have wave-particle duality and have wave properties. Within a few years his speculation was confirmed by experiments: moving particles of matter also have wave-like properties that can be described in quantum theory.

In the 1920s physicists found that in some cases an electron acts like a particle, occupying only one position in space at a time, whereas in other cases it acts like a wave, occupying several places in space at the same time. The wave-particle duality of matter means that there is uncertainty in Nature (as to whether matter is going to act as particles or waves) and Nature therefore can only be described by probabilities according to quantum mechanics. But the uncertainty impacted on gravity. As we have seen, Einstein's theory of gravity requires that the universe exploded from a state of infinite density, but high densities of matter present problems for traditional classical physics.

True quantum mechanics[43] appeared in 1926 with the matrix theory of Max Born and Werner Heisenberg, the wave mechanics of Louis de Broglie and Erwin Schrödinger and the transformation theory of P.A.M. Dirac and Pascal Jordan, all of which were different aspects of a single body of quantum law. Heisenberg, a supporter of Bohr, held that Schrödinger's model of an atom surrounded by waves was wrong as what an atom looks like, and its speed and position at any one time, are unknowable and can only be understood through mathematics.

In 1931 Dirac's equation describing the motion and spin of electrons incorporated both quantum theory and the special theory of relativity. Dirac laid the foundation for quantum electrodynamics (QED),[44] a quantum theory of the interactions of charged particles with the electromagnetic field. In the 1940s Richard Feynman developed QED together with Julian Schwinger and Shinichiro Tomonaga with his idea that charged particles (electrons and positrons) interact by emitting and absorbing photons, the particles of light that transmit electromagnetic forces. (Photons can produce electrons and positrons in pairs so long as the photons' energy is greater than the total mass-energy of the two particles.)

This theory led to knowledge of virtual photons, and the way they may behave in a quantum vacuum where the emptiness of the vacuum in fact contains virtual photons that can emerge and become real.

In *The New Philosophy of Universalism* there is also a section 'The Quantum Vacuum and Infinite Order', which links the quantum vacuum's virtual particles to order and the emergence of real particles:[45]

Bohm saw the quantum vacuum as a reservoir of order. To Bohm, quantum theory has certain basic assumptions although it has been formulated differently by Heisenberg, Schrödinger, Dirac, von Neumann and Bohr. These assumptions are that the laws of quantum theory are to be expressed in terms of a wave function, that the physical results are to be calculated with the aid of "observables" which are sharply defined in certain circumstances but fluctuate at random (lawlessly) in others. As we have just seen, Bohm countered Heisenberg's indeterminacy principle with his theory of hypothetical hidden variables, which he believed explained essential features of quantum mechanics.

He proposed a hidden order beneath the apparent chaos and lack of continuity of the particles of matter of quantum mechanics. He saw a hidden dimension behind the surface, an "implicate order" which is the source of all visible, "explicate" matter in our space-time universe.[46] This implicate order has "infinite depth".[47] The term is another way of describing infinity.

Bohm saw our three-dimensional world as an explicate order of objects, space and time. He believed we live in a multi-dimensional world[48] – an aspect of the multiverse – and that the implicate order gives rise to our physical, psychological and spiritual experience and is in a fifth dimension of which we are largely unaware. (Time is the fourth dimension.) A devotee of Krishnamurti, he held that there is also a superimplicate order, a more subtle sixth dimension which is the source of our spiritual experiences.

Bohm does not appear in physics textbooks because he introduces hypothetical, unevidenced new dimensions, and textbooks rightly restrict themselves to space-time. However, his view is in keeping with the philosopher's view of the infinite. In his implicate order, the order is unfolded like a radio wave containing the sound or like a television wave containing the picture, and it needs an explicate order to unfold it (like a radio or TV set).[49] If we leave to one side Bohm's view of the "implicate" as a folding process, like folding over a piece of paper, and also his hypothetical multi-dimensional view, for which there is no evidence, then his order can be restated in terms of an infinite order from which an energy surges into chemical composition (via DNA) and plant growth, as Newton proposed.

To Bohm every individual unit contains within itself the totality of existence. Individual units of matter (nuclei) contain the totality within themselves, and the same is true of individual human beings. To Bohm Reality is an unending process of movement. It is not static, but a wholeness or totality from which the thinking observer and his or her ego cannot be separated.[50]

To Bohm the rules of quantum theory applied to gravity result in a gravitational field comprising wave-particles, each having zero-point energy (the energy of the totality), and so the gravitational field cannot be completely defined.[51] So-called empty space contains energy, and he held that matter is a small quantized excitation on top of a sea of energy, like a tiny ripple in a vast sea. He held that we calculate the difference between the energy of empty space and the energy of space with matter in it, and we miss the energy in empty space.[52] This may be the dark energy which was hypothetically postulated in 1998, six years after Bohm died in 1992. Space is full rather than empty, as Parmenides and Zeno held and as Democritus opposed.[53]

In quantum theory atoms affect each other simultaneously in non-local ways when on opposite sides of the universe and display order across space. To Bohm, the apparent randomness, probabilities and uncertainties of quantum theory have order behind them, as the tides order the sea. Bohm saw matter and consciousness as "not distinct" in the sense that we are in the same sea as physical objects.[54] The universe is a quantum vacuum from which we appear like particles and to which we return like particles.

To Bohm the Big Bang was a little ripple in the middle of an ocean. Its ripples spread outwards, Bohm maintained, constituting our expanding universe.[55] This vast sea of cosmic energy has to be on a scale larger than the critical length of 10^{-33} cms.[56] Most physicists agree that there is some kind of granularity on a scale of 10^{-33} cms, which is 10^{20} smaller than an atomic nucleus,[57] and Bohm's sea would therefore be composed of the tiniest conceivable, ether-like, granular structures. The Leibnizian philosopher Geoffrey Read has proposed that myriads of tiny particles arise from an underlying reality below quantum mechanical waves and particles in space, and has attempted to describe this level of reality in terms of a "hidden variable" theory.[58] In conversation with me in 1991, Bohm said: "The Big Bang was not a big event, it was a local excitation within the infinite totality, like a ripple or wave in a sea." He said that one cannot have a vision of the Whole as it is infinite and our minds are finite. But, as I pointed out, mystics hold that there is a part of our consciousness that is universal and capable of cosmic or "unity" consciousness, which is able to see Reality beyond space-time. (Whitehead put it slightly differently: "Our minds are finite, and yet even in the circumstances of finitude we are surrounded by possibilities that are infinite, and the purpose of life is to grasp as much as we can of that infinitude.") Of his theory of wholeness and the implicate order, Bohm said to me with disarming modesty, "These are just ideas." He meant, "There is no evidence." But Bohm is of interest as he was an expert in quantum mechanics continuing Einstein's tradition, having had discussions with Einstein, and he was proposing that there is an infinite order behind the uncertainty principle of quantum mechanics and within the quantum vacuum.

The energy within the quantum vacuum has applications for the emerging universe, an angle that has been taken up by a Belgian theoretical physicist in Brussels. Edgard Gunzig[59] of the Belgian school of physicists has done work on how a universe emerges from the quantum vacuum by quantum processes through virtual particles. Real nothing is emptiness. By principle, a vacuum is non-empty. It has an underground life but is the most possible emptiness that can be imagined. It is as close to an absence of particles and radiation as can be imagined, and is the lowest possible energy state. Virtual particles do not exist, they are potentialities of existence. A vacuum with zero energy trembles due to the gravitation of the curvature of space, which is a reservoir of energy that can be given to virtual particles. Virtual particles are expelled from the vacuum in pairs, live for a split second and then return to the quantum vacuum.

If one virtual particle of a pair attracts and picks up energy from the geometric background of the curvature of space and enters space-time, then it becomes a real particle which could be a first proton and could turn into photons which cannot then be undone. This happened in different parts of the quantum vacuum and there were numerous real particles with positive energy created by negative energy from the curvature of space, emerging from zero energy. Through quantum processes a quantum vacuum emerged and as virtual particles became real, so did space-time. Out of this came a hot beginning and inflation as the universe expanded from zero energy. On this view there did not need to be a Big Bang, just a hot beginning for a pre-quantum vacuum could have pre-existed the Big Bang. Inflation would do the rest, expanding seeds in the microworld to the galaxies of our universe.

Bohm, who I knew and talked with, had discussions with Einstein and wrote of "implicate order" as can be seen above, and my 'order principle' may be in the quantum vacuum where Bohm found his "implicate order".

In Eastern thinking, Nothingness is also a Fullness, a Plenitude, the Void of Taoism, and this Taoist view of the Void anticipated quantum theory. See Fritjof Capra's *The Tao of Physics* for an exploration of the parallels between modern physics and Eastern mysticism.

All this follows a variation of $+A + -A = 0$: $+A$ (the order principle/Being) $+ -A$ (quantum vacuum/Existence) $= 0$ (the One universe).

(5) The immediate aftermath of the Big Bang: the shaping of the universe through expanding light and contracting gravity

There is now a consensus among scientists as to what happened immediately after the Big Bang, during the cooling and later expansion and the shaping of the universe during the matter era, which began 400,000–500,000 million years after the Big Bang.

In *The New Philosophy of Universalism* I have a section 'The Shaping of the Universe', which describes how the universe took shape:[60]

There is a consensus on the known facts and timings of the early universe.[61] Many of the timings are minute fractions of the first second. In one second the infinitely small, hot and dense singularity expanded like a balloon inflating to the size of our universe, and outside the balloon there was no space, matter or light. The inflation was not centred within the universe, as we have seen. Rather, the universe *is* the inflation, which is why the cosmic microwave background radiation is uniform throughout the universe. Within that same second the expanded singularity cooled from trillions of degrees K to a few billion. (K is an abbreviation for Kelvin, a scale of temperature named after the British physicist Lord Kelvin with absolute zero as zero.)

Particle physicists have focused on the first three minutes after the Big Bang. The four fundamental forces (the electromagnetic, strong and weak forces and gravity) were unified at this stage. The gravitational force separated at 10^{-43} seconds after the Big Bang (Planck time, the shortest time interval that can ever be measured, an infinitesimal amount of a split second), when the temperature was 10^{32}K (the beginning of the Grand Unified Theory era). Mass-energy became the observed universe at 10^{-43} seconds. Space-time is regarded as having begun when the universe began.

Within 10^{-36} seconds (a trillionth of a trillionth of a trillionth of a second) of the Big Bang the microscopic compact, high-density point that resembled an atom expanded extremely rapidly and inflated large enough to encompass the solar system's size, the horizon we see, while keeping gravitational and kinetic energy in balance. It doubled in size every 10^{-34} seconds, so there were a hundred doublings within the space of 10^{-32} seconds. It has continued to expand ever since.

The expansion of the universe was like the surface of a balloon being blown up, and if the galaxies were spots on the balloon's surface, they all receded from each other as the balloon inflated.[62] Another analogy is raisins in a loaf. As the dough rose the raisins were carried away from each other.[63] This rapid expansion was attributed to "inflation" in a book by Alan Guth, *The Inflationary Universe* (1981). Spacecraft are being prepared to test this theory of inflation empirically by looking for primordial microscopic vibrations. During this period of early rapid expansion within the first minute, the density and temperature of the universe decreased and radiation and matter evolved. For the first 10^{-35} seconds everything was hotter than 10^{28} degrees.

The temperature had dropped quickly, and around 10^{-35} seconds after the Big Bang, at a temperature of 10^{27}K, it seems there was very rapid, exponential expansion and inflation by a factor of 10^{54}, faster than the speed of light. This was the quark era. Leptons and quarks were formed from radiation energy as energy passed into matter in accordance with Einstein's $E=mc^2$ (the beginning of the quark era).

Further cooling to 10^{15}K around 10^{-12} seconds (a trillionth of a second) separated strong and electroweak forces and quarks combined to form protons, neutrons and their antiparticles. Radiation in photons lacked the energy to break up the new particles. Matter and antimatter collided and annihilated each other, leaving a residue of matter which makes up the universe. At 10^{-5} seconds there were temperature fluctuations that seeded galaxies. Further expansion and cooling around 10^{-4} seconds after the Big Bang resulted in the electroweak force separating into electromagnetic and weak forces (the beginning of the lepton era). Lighter particles (such as electrons, muons, taus and neutrinos) were prevalent.

Ten seconds after the Big Bang the temperature was 10^{10}K (10 billion degrees). There was rapid cooling. From 10 seconds after the Big Bang, radiation dominated the universe (the beginning of the radiation era). After a brief balance between radiation and matter when there was an equal number of particles, antiparticles and photons, antimatter was eliminated and the universe consisted of photons and neutrinos.

Three minutes after the Big Bang the temperature was 10^9K (1 billion degrees).

From 1 to 20 minutes after the Big Bang nucleosynthesis took place: colliding protons and neutrons fused and formed light nuclei (deuterium, lithium, helium and hydrogen, the building blocks from which all elements would eventually form). Nucleosynthesis ceased after a few minutes, after which the ratio of hydrogen to helium was 3:1, as it is today. The Big Bang, or expanding singularity, was over within 30 minutes.

Some 380,000 years after the Big Bang the universe cooled to 3,000K, a temperature at which helium nuclei could absorb electrons. Electrons remained attached to nuclei and atoms formed (the beginning of the matter era), giving out light which is the cosmic microwave background radiation we can monitor today. It had uniformity of direction, suggesting that all matter in the universe was uniformly distributed. Before atoms formed, matter existed in a plasma (or gas of positive ions and free electrons with roughly equal positive and negative charges). The plasma of free nuclei and electrons absorbed photons, which gave radiation and matter the same temperature. Now atoms could travel without colliding with photons, clumps of atoms formed, and matter decoupled from energy. Atoms collected together through gravitation into clouds of gas.

Just over 13 billion years ago the energy in cosmic microwave background radiation cooled from 3,000K and became less than the energy in matter in the matter-dominated universe. The background heat now pervading the universe is 2.725K (about –270C, absolute zero being –273.16C).

380,000 years after the Big Bang photons (15 per cent) outnumbered atoms (12 per cent) and neutrinos (10 per cent), with dark matter forming 63 per cent of the matter in the universe – according to NASA's WMAP Mission Results. There were now three kinds of matter: radiation, including photons and neutrinos, massless or nearly massless particles that move at the speed of light and exert a large positive pressure; baryonic matter or "ordinary matter" composed of protons, neutrons and electrons that exert virtually no pressure; and dark matter, non-baryonic matter that interacts weakly with ordinary matter and which, though never directly observed in a laboratory, is suspected to exist. (Dark energy, a fourth kind of matter or perhaps a property of the vacuum, which has a large negative pressure, only took over after 8 billion years.)

Newton proposed an expanding force of light – photons outnumbered atoms 380,000 years after the Big Bang (see above) – that shaped the expanding universe and was counteracted by a contracting force of gravity, and so the universe was in balance. Einstein favoured a "cosmological constant", a static universe that countered gravity (see p.59) and though Hubble found the expansion of the universe was accelerating, a co-existence of expansion and gravitational contraction seems to be at work.

In *The New Philosophy of Universalism* there is a section 'Newton's Ordering Light', which links Einstein's "cosmological constant" to Newton's balancing force:[64]

Einstein was surprisingly influenced by Newton, whose view of the universe he is thought to have obliterated with his own work. Newton saw light as associated with ether. Ether was thought to fill space. Having made his Materialist discoveries in gravity, light and calculus in 1664–1666, from 1669 when he was twenty-six Newton spent nearly thirty years researching into alchemy to find an expanding force of light that would counteract and balance gravity, an idea that would interest Einstein. Gravity drew bodies together. Newton's expanding force would drive objects apart and keep the universe looking static. In 1687, in his *Principia*, Newton set out an absolute space and time and the laws of motion and gravity.

At the same time he was working on light. Newton deduced that "every Ray" of light had immutable properties and "may be considered as having four Sides or Quarters". In Query 30 of *Opticks* he wrote of "Particles of Light", which seems to anticipate wave-particle duality: "Are not gross Bodies and Light convertible into one another, and may not Bodies receive much of their Activity from the Particles of Light which enter their Composition?"[65] Newton held that this expanding force operates in the radiation of light, in chemical composition and biological growth, and also governs the mind – consciousness – and behaviour of human beings.

In Query 31 Newton asked: "Have not the small Particles of Bodies certain Powers, Virtues or Forces by which they act at a distance, not only upon the Rays of Light for reflecting, refracting, and inflecting them, but also upon one another for producing a great Part of the Phaenomena of Nature?"[66] In an alchemical manuscript now in the Burndy Library he states: "Ether is but a vehicle to some more active spirit & the bodies may be concreted of both (i.e. ether and spirit) together, they may imbibe ether well (*sic*) as air in generation & in the ether the spirit is entangled. This spirit is the body of light because both have a prodigious active principle."[67] In other words light, whose body is spirit, combines with ether, and there is a single unified system which includes both internal (spiritual) and external (physiological) systems.

Ether had been regarded as a tenuous substance that filled the vacuum and was a means of carrying light since the Presocratic Greeks' *aither*. By the 19th century physicists had proposed that the entire universe was permeated by "luminiferous ether" (i.e. light-bearing ether) which acted as a medium for carrying light. Lord Kelvin, the Victorian scientist who gave his initial to the temperature measurement K, suggested its properties:

"Now what is the luminiferous ether? It is matter prodigiously less dense than air – millions and millions and millions of times less dense than air. We can form some sort of idea of its limitations. We believe it is a real thing, with great rigidity in comparison with its density: it may be made to vibrate 400 million million times per second; and yet be of such density as not to produce the slightest resistance to any body going through it."

In other words ether was very strong yet insubstantial. Theosophy took it up, referring to "the etheric level" on which telepathic thoughts could be transmitted.

However, in 1880 the idea came to grief when Albert Michelson, the US's first Nobel Laureate in physics, set up an experiment to prove its existence involving two perpendicular beams of light, one of which was to be slowed by ether. The two beams of light arrived at the same time, and the experiment failed to detect ether. In 1887, seven years later, Michelson announced his results, which ironically "proved" that ether did not exist, a result Einstein had predicted. The science writer Banesh Hoffmann wittily wrote:

"First we had the luminiferous ether,

Then we had the electromagnetic ether,

And now we haven't e(i)ther."

But what if ether, the Greek *aither*, is a network of dark energy which is composed of tinier particles than neutrinos, perhaps of photinos? Could it be that Newton was right in associating light with ether, and that Michelson's experiment was inadequate and disproved nothing? We shall return to the possibility of an undetected tiny-particle form of light in due course.

Einstein took up Newton's idea of a balancing force in 1917 when he proposed his "cosmological constant": an expanding force that repels and balances contracting gravity. He saw this as a repulsive force of unknown origin which exactly balances the attraction of gravitation in all matters and keeps the universe static (as, like Newton, he thought the universe to be). Einstein believed the constant gave all particles stable interconnectedness for all time. We have seen that he abandoned this constant in 1929 when Hubble established that the universe is expanding and not static, and that he later called his abandonment "the biggest blunder of my life". We have seen that in our time Einstein's constant has returned as hypothetical, unevidenced dark energy, the force which repulses gravity within our accelerating universe. Dark energy may have been the expanding force Newton was seeking. No one knows if this dark energy is conveyed by light (perhaps mixed with "ether" and "spirit", Newton thought), which, Newton believed, governs human and plant growth.

Perhaps Newton was right and the expanding force that counterbalances gravity is light. Perhaps it contains a universal principle of order which stimulates chemical composition through DNA and growth through photosynthesis, and affects consciousness and human behaviour. Perhaps there is a wider spectrum of light than we realise. The electromagnetic spectrum is currently known to have gamma rays at the shortwave end, radio waves at the longwave end and physical light in the middle. Perhaps beyond gamma and virtually undetectable is metaphysical Light (Newton's "ether" and "spirit").

The mystical traditions claim that metaphysical Light is infinite Being from before the Big Bang, from within the timeless nothing that preceded the quantum vacuum, long before the cosmic microwave background radiation. According to many mystical traditions it can be received in consciousness below the rational, social ego. Does metaphysical Light manifest into the universe from timeless infinity, travel with physical light of space-time and, arriving among hosts of gamma rays, ultraviolet rays and radio waves, pervade Nature with order that is intermixed with physical light? If so, there can be a synthesis of physics, mysticism and metaphysics.

Einstein rejected the randomness of the universe and searched unsuccessfully for a "hidden variable" that would bring order to the uncertainty principle of quantum mechanics:[68]

Einstein's work on his cosmological constant must be seen alongside his unfinished search for "hidden variables". After a quarter of a century of improvised quantum theory, in 1925 Werner Heisenberg proposed a new quantum mechanics. In the new quantum theory, randomness or indeterminacy is fundamental, and identical electrons in identical experiments behave with apparent randomness. Many physicists claim that all nuclei were originally in an identical state, and that nuclei decay or do not decay on a random basis.

Einstein did not accept the randomness – "God does not play dice with the universe"[69] – and he proposed a hypothetical, unevidenced, unknown property, referred to as a "hidden variable" which causes some nuclei to decay but not others. He held that this hidden principle underlies the indeterminacy of the uncertainty principle.

Einstein never found a hidden principle of variability. The Danish theoretical physicist Niels Bohr and his Copenhagen school rejected hidden variables as qualities that could not be measured. After rejecting quantum theory at the fifth international conference on electrons and photons at Solvay, Brussels, in October 1927 (which marked the triumph of quantum mechanics), in 1930 Einstein had a bruising meeting with Bohr which ended in his grudgingly accepting quantum mechanics.[70] Einstein worked on quantum gravity and proposed hypothetical, unevidenced "gravitons", gravity particles which may also act as gravity waves. His general theory of relativity had predicted black holes from whose intense gravitational fields light cannot escape, and also massless gravitons travelling like photons at the speed of light.[71] If gravitons are proved to exist, the gravitational force could be described in terms of an exchange mechanism. It would operate like the electromagnetic force of repulsion between two electrons due to the exchange of a virtual photon between them, like the weak force in which the decay of a neutron is caused by the exchange of a virtual W-particle and like the strong force between quarks, due to the exchange of a gluon. There would then be a unified theory of all force, a huge step to a Theory of Everything.

Einstein was sure the universe was ordered rather than a random accident, and he hated the random uncertainty principle of quantum mechanics which had overtaken his own work. Bohr and Einstein talked throughout the 1930s but eventually had no more to say to each other.[72] The split between them was reflected in a split in physics that has lasted to this day.

Einstein spent the last thirty years of his life working unsuccessfully, trying to re-integrate gravity with the other three forces, attempting a unified field theory in which the strong and weak forces, electromagnetism and gravity would be united in an underlying field.[73]

But because Einstein never found a hidden principle of variability does not mean that it cannot be found. The Universalist philosopher detects within the hitherto undiscovered hidden variability workings of a universal principle of order that is behind the apparent randomness of all nuclei and the uncertainty principle. May this hidden principle also include the workings of Newton's expanding force? And may it be associated with metaphysical Light?

David Bohm, an American theoretical physicist and expert on quantum mechanics and Einstein's *protégé*, supported hidden variability. He had discussions with Einstein and improved the Einstein-Podolsky-Rosen thought experiment. Bohm wrote a book on quantum theory in 1951 and in January 1952 took Einstein's side against other physicists in an article in *Physical Review*, two papers published together entitled 'A Suggested Interpretation of the Quantum Theory in Terms of "Hidden" Variables'.

Bohm said that in quantum theory a system physically functions as waves that can only be measured in terms of probable results. "Hidden" variables would determine the precise behaviour of a system. Bohm admitted the internal consistency of quantum theory, but regarded its present form as incomplete as hidden variability is missing.[74] Roger Penrose has said that there *can* be hidden variability if it is non-local (meaning that the hidden parameters must be able to affect parts of the system in arbitrarily distant regions instantaneously),[75] a view he confirmed to me in conversation in 1991.

The universal principle of order can be found in sunlight's stimulating of the growth of flowers through photosynthesis (chemical composition through DNA), and may also be found in Light's stimulating of consciousness (and its chemical composition through DNA).

Both Newton's view and Einstein's view are supported by +A + −A = 0: +A (expanding force of light) + −A (contracting force of gravity) = 0 (united in the One, which is constant and in balance).

All this follows a variation of +A + −A = 0: +A (the expanding of the inflationary universe) + −A (the contracting force of gravity, which separated at 10^{-43} seconds after the Big Bang) = 0 (the One, the cooling universe).

(6) The formation of galaxies of stars and the observable universe

The formation of galaxies and stars began around 400 million years after the Big Bang, during cooling and gravitational attraction. The gravitational energy of some 5,000 known black holes, including Sagittarius A* (A-Star), may have contributed to the shape of all galaxies, including our galaxy the Milky Way. In the 1970s the only known planets were those in our solar system. There are now known to be 4,000 planets in the observable universe. Not long ago it was thought there were 10^{23} (1 with 23 noughts, 100 billion trillion, or 100 sextillion) stars. Now there are thought to be 200 billion trillion (or 200 sextillion) stars: the one universe with traditionally 10^{23} stars now has 200 trillion stars. There may be 20 billion earth-like worlds. (NASA's Perseverance Rover has already started looking for signs of life on Mars, drilling rock cores at its ancient lake in the hope of finding fossils.)

The New Philosophy of Universalism says of the formation of galaxies:[76]

A billion years after the formation of atoms, atoms continued to be brought together by gravity. Gravitational attraction between hydrogen atoms collected them into clouds of gas that condensed as stars grouped in clusters and galaxies. Gravity exerted pressure on their cores, creating mass and high temperature that produced helium as colliding hydrogen atoms fused. Over a period of 30 million years stars contracted and achieved stability as their cores were heated when protons and also carbon were converted to helium to become hydrogen for 10^{10} (10 billion) years. As the hydrogen fuel reserves were used up in thermonuclear fusion, gravity caused the cores to contract and as gravitational potential energy became kinetic energy, the cores became hotter and stars became red giants that burnt helium. The cores of smaller stars burnt to carbon and shrank to white dwarfs (stars with closely-packed matter and a high temperature but low luminosity) and then could be black dwarfs (hypothetical, unevidenced stars thought to have cooled and to have no luminosity and to be about to "go out" near the end of their stellar evolution). Our sun will "go out" as a black dwarf in 5 billion years' time.

In larger stars, thermonuclear fusion continued and the cores' temperature increased, combining electrons with protons to form neutrons. The stars then collapsed, blowing off their outer layers in giant supernovae explosions (like the Crab nebula, which was seen by the Chinese in 1054 and is still seen today). The radio waves from the rotation of neutron stars are detected as pulses; hence they are known as "pulsars". Sometimes two stars moved in orbit round a common mass, and these are known as "binary stars".

When massive stars contracted, their density was somehow so high and gravity so great that they became black holes (hypothetical, unevidenced cosmic bodies with such intense gravitational fields that no matter, radiation or light can escape them, thought to be often formed by the death of a massive stars). A black hole with a radius of just 1 cm could contain the mass of the earth. Black holes are thought to have been at the centre of some galaxies and to have powered luminous objects such as quasars and active galactic nuclei. Stars have their own life cycle in which their beginnings and endings are relatively short periods.

Astrophysicists in the 1970s studying the CMB immediately noticed that the locations of galaxies are not random: clusters containing thousands of galaxies up to 3 million light-years across are separated by voids almost devoid of galaxies 300 million light-years across.

Early gravitational contraction may be associated with black holes. 13.6 billion years ago the death of a massive star began the supermassive black hole (a hole in the fabric of the universe) that 13.5 billion years ago became Sagittarius A* (A-Star), with a mass 4 million times the mass of our sun (4.31 ± 0.38 million solar masses, according to Richard Genzel, who monitored the stellar orbits around Sagittarius A* for 16 years. As we have just seen, its gravitational energy may have contributed to the shape of the universe and all galaxies, including our galaxy, the Milky Way, in which it has a central place 26,000 light-years away from us. This was first seen by NASA in 2015 as a flash of X-rays (hot radiation) on the Event Horizon telescope that was launched in 2009 (the year *The New Philosophy of Universalism* was published). This telescope is situated outside the earth's atmosphere, and its first image of a black hole appeared in 2019.

A pre-1993 image of our galaxy, the Milky Way, on the front cover of Nicholas Hagger's *The Universe and the Light.*

Recent calculations have increased the number of galaxies and stars. There is a consensus that there are now 2 trillion galaxies, and (as we saw on p.47) 200 billion trillion (200 sextillion) stars, more than 10,000 stars for each grain of sand. As technology improves these figures can be expected to increase.

There are 400 billion stars in our galaxy, the Milky Way, and the two Gaia telescopes (launched in 2013) showed there had been collisions between our galaxy and smaller galaxies. A larger galaxy, Andromeda, has been found to be heading towards the Milky Way, a flattened disc of stars about a 100,000 light-years across, and may collide with it in 4.5 billion years' time.[77] But the Hubble telescope (launched in 1990) showed the universe is expanding at an accelerating rate and galaxies are receding, and in due course the Milky Way, perhaps having collided with the larger Andromeda and lost its structure and had its stars scattered, will be isolated and alone, with all other galaxies out of sight.[78]

The pillars of Creation, stars in the Eagle Nebula 7,000 light-years from earth in the Milky Way.

The James Webb Space Telescope has a view 15 times wider than the Hubble Telescope (which has been in orbit since 1990). It has revealed that most galaxies have a black hole at their centre and has captured the pillars of Creation, plumes of interstellar gas and dust forming stars in the Eagle Nebula 7,000 light-years from earth, which is on the cover of my book *The One and the Many* (1999). By the time light emitted by the earliest stars or galaxies reaches a telescope orbiting the earth it is stretched and arrives as infrared which is invisible to the naked eye. To observe infrared the James Webb Space Telescope will orbit round a point where gravitational forces will allow it to stay in the same place with a helium-powered refrigerator at –230°C so it can see light received as infrared, emitted 13.5 million years ago, just 300,000 years after the Big Bang when the cosmic microwave background radiation began to be detectable and the first stars outside the earliest galaxies were coming into being before evolving around 12 billion years ago. It is hoped that the James Webb Space Telescope will detect the first light after the Big Bang and the birth of the universe. The size of the universe is right for life:[79]

The size of the universe. There are at least 10^{23} stars in the observable universe, as many as the grains of sand on earth, and 10 million billion stars will be similar to our sun. There are 10^{24} planets; 10^{17} hypothetical black holes; 10^{11} galaxies; 10^7 clusters.[80] There are 10^{20} planets suitable for life forms[81] of which one, discovered in 2007, orbits the star 41 light-years from earth known as 55 Cancri,[82] and another, also discovered in 2007, orbits a small red star 20.5 light-years from earth known as Gliese 581. (Currently 228 extrasolar planets are known to orbit stars, some of which may be able to sustain life.)[83] It may seem a hugely wasteful scheme to have so many suns in the universe with life happening only on the earth. Yet those billions of potential homes for life may have to exist so that life can happen on earth. The linear size of the universe may be about 13.7 billion light-years (a figure based on the beginning of the universe 13.7 [since updated to 13.82] billion years ago) from the base point to the rim of our expanding "shuttlecock", assuming inflation is only going in one direction. We can in theory see as far as light has had time to travel during the last 13.7 billion years although we cannot see beyond the observable horizon, which is taken to be about 10 billion light-years, the farthest limit current telescopes can see. It takes a few billion years for a galaxy to form, for the first stars in it to process hydrogen and helium into heavy elements and for neutrinos to explode stars to scatter materials into space. It takes more time for new stars and planets to be formed from the *débris* and for life to evolve.[84] If the universe had just one galaxy the size of our Milky Way containing 200–400 [since updated to 400] billion stars, it would have expanded for only a month, according to mathematical formulae, and would not have been able to produce life.[85] The process from start to finish cannot be completed short of 13.7 billion years as it requires a good supply of *débris*-providers, and the billions of stars and billions of light-years may be necessary if life is to be produced on just one star, our planet.

All this follows a variation of +A + –A = 0: +A (expanding universe of hydrogen atoms after inflation) + –A (gravitational cooling and condensing into gases, clouds and galaxies of stars, then cores converted to helium) = 0 (the One universe with 200 billion trillion stars in all the galaxies in the observable universe, as distinct from the 2 trillion galaxies estimated as being in the entire universe, observable and unobservable, with perhaps 20 billion earth-like worlds).

(7) The formation of the solar system and earth

The sun was formed 4.6 billion years ago.[86] The earth was formed 4.6–4.55 billion years ago. Liquid water first flowed 4.4 billion years ago, perhaps created by the sun (as solar wind can make water in space dust), and the first rocks were 4.2–3.96/3.9 billion years ago.

In 2021 NASA reported that its Parker Solar Probe (launched in 2018) has dipped three times in five hours into the sun's corona (the solar atmosphere of extremely hot, radioactive particles) where temperatures can reach 3.6 million degrees F (2 million degrees C) and will fly through the corona to within 3.7 million miles of the sun at 430,000 mph, collecting data on the sun.

The New Philosophy of Universalism has this to say on the formation of the sun and the solar system:[87]

About 4.5 [now updated to 4.6] billion years ago our sun condensed from a cloud of gas and dust. We know it is a second-generation star as elements heavier than iron are present in its black body radiation spectrum. The sun gravitationally attracted a rotating gas cloud into a flat disc. The cloud rotated faster, larger particles captured smaller ones and most of the material was swept into the sun. Smaller entities became planets, comets and asteroids (rocks that fell to earth as meteors or shooting stars if they crossed the earth's path). Comets are dust embedded in ice, formed from water and methane. (Halley's Comet, seen in 1986, returns every 76 years.) The solar system's planets rotated round the sun in the same direction and roughly in the same plane. The inner planets, the earth, Venus and Mars are formed of small rocky high density materials. The outer planets, Jupiter and Saturn, are large, gaseous and low-density.

The sun grew larger and nuclear fusion took place. There were nuclear explosions, and material was blown outwards against gravitational forces. The sun radiated energy and a "solar wind" also blew smaller particles (charged particles such as hydrogen and helium, which are light elements) to the outer regions of the solar system. Sudden surges in the sun's energy have continued as solar flares and sunspots.

The planets were too small for nuclear fusion to take place and so they have gradually cooled and are only visible through reflected light from the sun. Their original atmospheres were lost when thermonuclear fusion took place in the sun. Their secondary atmospheres resulted from the release of gases from the planets' interiors. Earth's atmosphere has been modified by biological processes.

Galaxies are generally elliptical and spiral in shape. Our own galaxy, the Milky Way, is spiral with a central bulge and a disc of stars, dust and gas, which rotate round the galactic centre. In spiral galaxies new stars appear blue. In elliptical galaxies there is no disc and stars look red. Some galaxies have a luminous quasar or active galactic nucleus at their centre. Radio waves are used in mapping the stars at the centre of our galaxy as interstellar dust is transparent to radio wavelengths. The rotating stars' speed remains constant at all distances from the galactic centre, which does not have enough mass to exert a pull to draw them in. Stars are travelling too fast to remain gravitationally bound to the Milky Way.

The Swiss astrophysicist Fritz Zwicky proposed that there must be an extra gravitating material that contributes to the pull. This hypothetical, unevidenced material known as dark matter does not emit electromagnetic radiation like stars or galaxies. Dark matter could be present in MACHOs, massive compact halo objects such as brown dwarfs (hypothetical crosses between a planet and a star, failed stars with too little mass to ignite hydrogen), black dwarfs and black holes; and also WIMPs, weakly interacting massive particles such as neutrinos with few islands of atoms that pass through everything. We have seen that dark matter is hypothetically thought to make up 23 per cent of the universe according to the WMAP estimate.

The universe now contained 10^{18} (a billion billion) stars, one per cent of which (10 million billion) are probably similar to our sun. In the observable universe there were then 10^{10} (10 billion) galaxies. (There are now 10^{11}, a hundred billion [since updated to 2 trillion] galaxies.) The number of stars in all the galaxies in our observable universe is about 10^{23}.[88] The total number of grains of sand is about 10^{23}.[89] There is one star in our observable universe for every grain of sand [since updated to more than 10,000 stars for every grain of sand]....

The New Philosophy of Universalism says on the formation of the earth:[90]

The earth, which began about 4.6 or 4.55 billion years ago,[91] is one of the smallest planets grouped in our solar system round a central star (the sun). As all the planets revolve in the same plane round the sun, it is likely that they condensed from a single revolving disc of gaseous matter and that the earth originated as a mass of molten rock with no atmosphere. As it cooled, a crust formed. It was disturbed by volcanic activity (hot gases escaping from the molten interior). When the surface cooled to 100 degrees C, an atmosphere developed that consisted of water, ammonia, carbon dioxide, methane and hydrogen. Geological evidence shows that the early earth was virtually without oxygen – hence iron in reduced form is found in rocks. As the earth continued to cool, water vapour in the atmosphere condensed and fell as rain, forming rivers, lakes and oceans[92] about 4.4 billion years ago[93] while erosion formed rocks.

All this follows a variation of $+A + -A = 0$: $+A$ (expanding universe, gaseous matter) $+ -A$ (gravitational cooling into matter and planets revolving round the sun) $= 0$ (the One solar system).

(8) Acceleration of the expanding universe, the surfer and the infinite

The Big Bang exploded and flung energy in all directions at the speed of light (300 million metres per second, a million times faster than an H-bomb explosion), within the shape of an expanding shuttlecock (a cork with a ring of curved feathers). The ripples of the cosmic microwave background radiation are rippling in all directions, and all stars and galaxies are moving apart in all directions through space – within a universe shaped like a curved shuttlecock. Inflation (between 10^{-36} seconds to some time between 10^{-33} and 10^{-32} seconds after the Big Bang) was therefore in all directions, and the expansion of space is therefore in all directions – within the curved shuttlecock, and the acceleration of the expansion during the last 5 billion years means that the universe appears to be expanding faster than the speed of light – in its curved shuttlecock-like shape. The universe is not expanding out from a centre into space. The whole universe is expanding equally at all places – as from the cork base of an expanding curved shuttlecock.

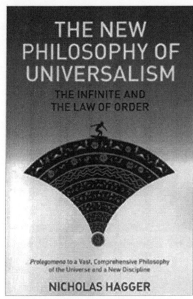

The front cover of *The New Philosophy of Universalism* showing the shuttlecock-shaped expanding universe with the surfer (see p.47) on its surge, breasting the infinite.

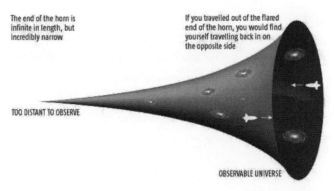

The shape of the universe as a funnel or medieval horn,
according to Frank Steiner, University of Ulm, Germany.

Big Bang and expanding universe shaped like a bell by Mikkel Juul Jensen.

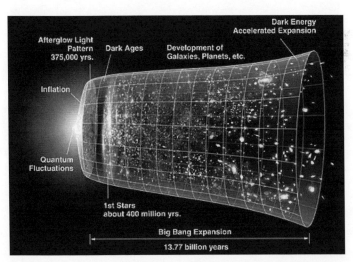

The shape of the expanding universe as a wide glass with the glow from the Big
Bang on the left surrounded by dark.[94] The figure of 13.77 has since been updated
to 13.82 (Also see the shape of the expanding universe as a narrow glass on p.307.)

The universe has expanded by a factor of about 1,000 from when it was 380,000 years old until today, and the expanding universe has been found to have been accelerating for the past 5 billion years. To put it another way, 5 billion years ago, the expanding universe began accelerating. The universe is shaped like a shuttlecock (see picture of the front cover of *The New Philosophy of Universalism* on p.54), and as its 'top' rushes into the infinite, where is a notional surfer surfing on the edge? Zeno stated paradoxes, and this question is Hagger's paradox. The answer: his feet are in space-time, the accelerating, expanding universe like a foaming tide, and the rest of him is in the infinite. Again, we can follow this development in *The New Philosophy of Universalism*, under a heading 'Acceleration, the Surfer and the Infinite', which describes what it would be like surfing on the edge of the accelerating, expanding universe, and attributes the acceleration to dark energy (Einstein's 1917 "cosmological constant"):[95]

In 1998 it was discovered that the universe's expansion has been accelerating during the last five billion years.[96] Having accelerated greatly immediately after the Big Bang, the expansion had become less rapid. After some 8 billion years of deceleration from inflation, renewed acceleration began and no more galaxies could form – the expansion is moving too fast for matter to resist and gather into galaxies. The mysterious speeding-up contradicted the second law of thermodynamics (which states that a system begins ordered and that its passage through time brings increasing disorder).[97]

What shape does this speeded-up universe have? In geometry there are three possible shapes the universe could have: spherical with curved space (called a closed universe); a flat space extending in all directions with zero curvature (known as a flat universe); and a hyperbolic ("plane curve") universe with no curved space where space expands forever (known as an open universe).[98]

Einstein's equations tell us that the universe may have no edge as space is gently curved around upon itself, making the equivalent of the surface of a sphere. The universe may appear a closed sphere even though it is open. Yet as it is accelerating, the curvature does not form a perfect sphere but rather a curved shuttlecock – curved as the feathers are curved on a shuttlecock, the base of which is the point or singularity and the top of which is open. The universe is not a closed sphere that can be walked around forever, like a fly walking around a snooker ball, and therefore that has no boundary.[99]

The universe is finite and shuttlecock-shaped (or trumpet-shaped, shaped like a trumpet swelling to its bowl, or shaped like a loudspeaker).[100] If the finite universe is accelerating it must be accelerating into something, and therefore must have an edge or boundary. What would happen if one could peer over the edge?[101]

Let us ask ourselves, what would it be like to stand on the edge or tip of the shuttlecock-shaped universe? The rim of the shuttlecock-shape is forever surging forward, so one would be like a surfer in a spacesuit crouching, arms out, knees bent on the crest of a surging wave. What is the surfer advancing into on the edge of our expanding, accelerating universe? It cannot be material space for space-time was created with the Big Bang and *is* the expanding, accelerating universe. There is no space or time outside the universe, yet the universe is surging forward into something, which is by definition not material space or time.

The something can only be the infinite which pre-existed the universe of space-time and from which the point or singularity emerged. It can only be the metaphysical Reality beyond space-time, and the spacesuited surfer, crouching on the crest of the wave that is forever thrusting forward like an endless *tsunami*, ever surges – hurtles – into the infinite, the Void. The infinite is metaphysical because it is outside the wave of expanding, accelerating space-time and outside our physical universe.

Hitherto, the infinite has been a theoretical concept in metaphysical philosophy. But, within my rational concept, what the surfer is *experiencing is* the infinite. The infinite is thus evidential, for the surfer has practical experience of it. This is a crucially important idea that advances the reconciliation between the metaphysical and scientific emphases. The infinite is both theoretical and, to the surfer, scientifically evidential and its existence would be confirmed by scrutinising scientists such as Aristotle and Bacon if they were to join the surfer on his endless *tsunami* wave.

So if the surfer throws a stone from surging space-time out into the infinite, across the edge or boundary, what will happen? Does it go into the infinite? Does it cease to exist?[102] The infinite is outside the boundary of space-time. And so the stone, and all space-time, must cease to exist outside space-time when three-dimensional space ceases to exist, where the edge ever surges forward into the infinite.

We are accelerating into Absolute infinity. From the start of radiation (10 seconds after the Big Bang) to around 380,000 years later, the generally accepted end of the radiation era, the expansion was controlled by the gravitational pull of radiation, which had a slowly decelerating effect. In fact, the expansion was still controlled by the gravitational pull of radiation 379,000 years after the Big Bang. For about 380,000 years the expanding gas de-ionized and nuclei and electrons combined into atoms.[103] The beginning of the matter era maintained this reduced expansion. We have seen that about 5 billion years ago it is thought that the hypothetical dark energy took over and that as deceleration changed to acceleration no more galaxies could form as the expansion was moving too fast for matter to resist.

Our universe is still, after 13.7 [since updated to 13.82] billion years, expanding. Hubble time ranges from 12 to 20 billion years, older than but corrected by the 13.7 billion years [since updated to 13.82] calculated as a result of NASA's Wilkinson Microwave Anisotropy Probe in 2003 (see illustration below]), which refined the 15 billion figure that was previously the perceived norm for the age of the universe.[104] Hubble time measures the expansion of the universe by a simple sum: the present distance of a galaxy from the earth divided by its present recession speed from the earth = the time since the Big Bang, with adjustments for cosmic deceleration. On this measurement our universe expands closer than 2 per cent to the dividing line that separates indefinite expansion from eventual contraction. Our universe is a finite universe that at present therefore looks as if it will expand forever.[105]

The mysterious force causing the acceleration is thought to be the hypothetical dark energy, which Einstein predicted in 1917. He called it the "cosmological constant". He saw this energy as enmeshed in the fabric of space-time and not having a source. It is thought that dark energy (assuming it exists) is responsible for making space-time flat.[106]

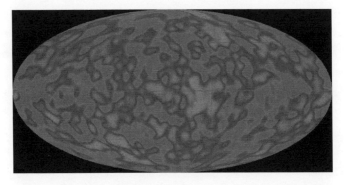

1992 COBE (Cosmic Background Explorer) map of the cosmic microwave background radiation, produced by NASA (above); and (below) 2003 WMAP (Wilkinson Microwave Anisotropy Probe) of the cosmic microwave background radiation, produced by NASA (with 35 times better resolution than the COBE map).

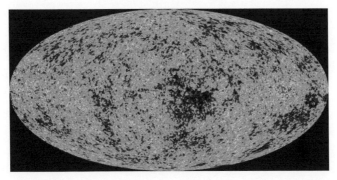

To put it another way, the total number of protons, the building blocks of matter, is 10^{78},[107] a factor of ten billion trillion times too small.[108] A proton's lifetime exceeds 10^{33} years.[109] It is conjectured that our universe must contain two-thirds of its energy density in the form of dark vacuum energy. What the universe's energy density should be is larger than what is observed by a factor of 10^{120}.[110] The hypothetical dark energy in the present vacuum is thought to have an energy scale that is 3 trillion quadrillion (3×10^{27}) times smaller than that of inflation.[111]

Einstein added his "cosmological constant", which for him was an anti-gravity constant, to his static universe to counter gravity, and later regarded the move as his worst mistake. For soon afterwards, in 1929,

Hubble showed that the universe was expanding and not static.

In 1998, Einstein's solution was back in vogue as his constant was regarded as supplying the force or energy within the vacuum that has speeded the expansion up. Of Friedmann's three models, the first one that showed the universe ending in disorder with a Big Crunch could now be discounted. On balance it looked as if the universe would expand forever but the hypothetical dark matter may act as a brake and slow it down towards flatness.

Flatness, we have seen on p.56, is a geometric concept. The period of rapid expansion, the inflation phase which began at 10^{35} seconds, flattened the geometry of the universe. This meant that the rules of Euclidean geometry are observed in the cosmos, that light not being bent by gravity travels in straight lines to infinity, not in curves, and the angles of a triangle still add up to 180 degrees. Space is expanding at such a rate that it is Euclidean rather than curved, giving an expanding but flat universe.[112] We have seen that Einstein held that space appears curved, but an insect crawling on the inside of a fully-inflated balloon would experience the surface as flat even though the expansion makes it appear curved.

We have seen that the critical density of the universe is 6 hydrogen atoms per cubic metre. The number of this critical density is Ω (omega). In a closed universe Ω is greater than 1, in a flat universe $\Omega = 1$ and in an open universe Ω is less than 1. In a closed universe the density of matter is greater than the critical density and causes gravitational forces to reverse the expansion and bring about a Big Crunch. Our observable universe is accelerating because of the hypothetical presence of dark energy, and because of the hypothetical braking dark matter Ω works out at 0.2 hydrogen atoms per cubic metre, or 0.3 of the density needed to halt expansion.[113] Our observable universe is flat and looks as if it will expand forever [although the sun will become a black dwarf, see below].

As our observable universe has a density of matter that is *less* than critical density (0.3 in relation to 1), space is not curved into the finiteness of a spherical ball but is an essentially flat surface curved by expansion like an inflated balloon. It obeys the laws of flatness but has been curved by inflation. The visible, observable universe is finite. To a multiversalist our universe could be in a bubble and there could be a multiverse outside our bubble; and our local conditions within our bubble may be different from the conditions in the totality of the multiverse outside our bubble. The universe could in theory be infinite. However, as it is in fact shaped like a shuttlecock (or trumpet or loudspeaker) as we have seen, it is finite and advancing into infinity (or Absolute infinity in Cantor's terms).

The universe, then, is now expanding faster than five billion years ago, and this can only be explained by the existence of the mysterious anti-gravity force that opposes gravity hypothetically known as dark energy, which, it is argued, transformed slowing-down into speeding-up. This anti-gravity force balances gravity. Einstein's constant was a universal repulsion or anti-gravity force which balances gravity, and we have seen [on p.41] that Newton also sought an anti-gravity force – in light – which would balance gravity. The 10^{78} atoms in the universe form only 4 per cent of matter, and it must be emphasized that there is no evidence other than imaginative mathematics for the existence of dark matter and dark energy, which WMAP estimated comprise 23 per cent and 73 per cent of the matter in the universe respectively.

The question of whether the universe is closed, flat or open is important as which of the three it is affects its ultimate fate, as we shall see in numbers (19)–(20). For now we need to note that I have found that the universe at present looks flat, as if it will expand forever, accelerating due to hypothetical dark energy, but that there is a likelihood that hypothetical dark matter will act as a gravitational brake. Things may change during the next 2,000 billion years, and it is forecast that the sun will be a black dwarf in 5 billion years' time. We must not get ahead of ourselves. *The New Philosophy of Universalism* has this to say on the subject of flatness:[114]

Flatness. In a flat universe expansion and gravitational pull are balanced so eventually expansion stops without collapsing back, making the relation between distance and angles the same as in Euclidean space. Matter is contained in the universe in such a density that it expands at the right speed to allow carbon-based life forms and human life to form. We have seen that our universe can be open (with continued expansion forever) or closed (with expansion until it collapses back) or balanced between these two and flat, with a critical density of matter in space of $\Omega = 1$, a universe which expands at an ever-slowing rate and comes to a stop at zero velocity. Today Ω has a value of 0.3 (after an adjustment for dark matter), and the solar system is expanding so fast that eventually it might fly apart. Today the density of the universe is within a factor of ten (between one-tenth and ten times) of the critical value corresponding to a flat universe. As the universe is now, 13.7 [since updated to 13.82] billion years later, only a factor of ten away from flatness, one second after the Big Bang the universe must have been flat to within a factor of 10^{15} (one part in a million billion) and so the amount by which the density differed from the critical value was 0.0000000000000001. Between time zero (the Big Bang) and 10^{-43} seconds after the Big Bang when the gravitational force separated, soon after which at 10^{-36} seconds after the Big Bang the universe began expanding, the flatness of the universe was precise to within one part in 10^{60} – which precisely gives all the conditions at the very outset that allow stars, galaxies and human life to form.[115]

All this shows a variation of $+A + -A = 0$: $+A$ (acceleration due to thrust from hypothetical dark energy of the expanding universe with a critical density within a factor of ten from $\Omega = 1$ and of no more than 5 or 6 hydrogen atoms per cubic metre, and a density of matter that is *less* than critical density (0.3 in relation to 1) $+ -A$ (gravity with hypothetical braking of dark matter) $= 0$ (the One universe).

(9) The bio-friendly universe and order

We saw in (4) that there is "implicate" order within the quantum vacuum, according to Bohm, who had discussions with Einstein and shared some of these discussions with me.

There is order within the universe, which therefore could not be random (despite what reductionists assert). We need to sum up the position regarding order in relation to the infinite and the finite so far, before we proceed to the bio-friendly universe. *The New Philosophy of Universalism* has this to say on the order behind the universe from the moment of the Big Bang, and the co-existence of the finite and the infinite:[116]

I have already formed some conclusions about the structure, context and future of our "boundless"-born, shuttlecock-shaped universe. The scientific evidence so far suggests that the expanding universe may not end in disorder, in a Big Crunch. It is between expanding forever and being a flat universe balanced between gravitation and electromagnetic forces which may eventually bring the expansion to a halt. At present it looks as if we are expanding forever provided the hypothetical, unevidenced dark matter keeps us accelerating and does not slow us down.

We have seen that what lies spatially ahead of our accelerating shuttlecock-shaped universe is an "edge", beyond which there is a Void which our astronaut surfer breasts as he rushes forward into it. Universalists hold that it contains within itself latent properties that include virtual pre-particles, one of which could have surfaced as a singularity and begun our universe.

The universe, space and time are finite in the sense that they are bounded by extremes: the curvature round the sides of our universe resembling the curve round a shuttlecock's feathers; and the hot beginning (the Big Bang) in our past and the conceptual end of the universe and Nature in the future. Time's arrow is between the singularity at its beginning and the singularity at its end. Time, we have just said, is a succession of material spatial events that began when spatial events began. Time is overlaid on timelessness, the Void or infinity which preceded the Big Bang to which the accelerating universe is rushing, and which would continue after a Big Crunch, the end of the moving universe when all motion ceases in infinity at a temperature of absolute zero or −273.16C. This Big Crunch would only happen if the pull of dark matter caused the universe to decelerate, and by then there would be no life on earth as the sun will burn up in five billion years' time.

Although the laws of Nature break down when being sucked back into a singularity or black hole, they emerge from a singularity as a working unity, having "revved up" from inactivity in a reverse of "breakdown". My view of the Big Bang has suggested that our universe came out of an infinite timelessness as one point or singularity in which the laws of Nature and four forces were united, to separate within a second in structures conducive to order; and that our universe co-exists with the infinite timelessness from which it emerged until it eventually returns to it. The shuttlecock-shaped universe is within finite boundaries and is surrounded by the boundless infinite into which the spacesuited surfer is continuously borne.

All this follows a variation of +A + −A = 0: +A (boundless infinite timelessness) + −A (finite boundaries in space-time) = 0 (the One). Also: +A (electromagnetic and orderly forces) + −A (gravitation) = 0 (the One flat universe balanced).

Physicists' theories about the structure of the universe identify six possible models.[117] The six models and my comments on them are:[118]

1. the absurd, accidental, random universe that cannot be explained, it just is and intelligence is an accident;
2. the unique physical universe unified by a Theory of Everything (perhaps string/M theory);
3. the multiverse (an infinite series of universes);
4. intelligent design by God or Aristotle's first cause, a theological explanation;
5. a bio-friendly life principle that works with the four fundamental forces to drive the universe towards life and intelligence – a more metaphysical model than 2; and
6. the self-explaining or participatory universe (Wheeler's "bootstrap" universe in which intelligent life can create the universe that gives rise to it by backward causation, a reversal of time, mind collapsing matter backwards in time to make the universe more bio-friendly forwards in time).

I omit obviously bizarre and freakish models such as one that asserts that the universe is a fake, a virtual-reality simulation in which we, too, must also be simulations.

Even before our investigation has begun, the Universalist can reject the absurd, accidental model (model 1) for enshrining randomness and ignoring order. The Universalist can also reject backward causation (model 6) for its bizarre idea that we are now creating the past universe. Suggesting that the present gives rise to the past – for example, that we are now creating the Second World War – is incredible both to philosophical reason and to common sense. Universalism focuses exclusively on *this* universe and is not interested in speculative other universes of an unproven multiverse (model 3). (The multiverse model shifts the problem of how the universe began into the multiverse – but the problem then becomes, how did the multiverse begin?) Our unique physical universe (model 2) is a disguised variation of model 3. It rejects the infinite and puts its trust in string theory and many dimensions, which are hypothetical and unevidenced. We have seen that a Theory of Everything has eluded theoreticians so far, and that there is no prospect of finding a purely physical solution as the temperatures required to reunite the four forces experimentally are so high. Universalism can reject the unique physical universe.

That leaves us focusing on a model created by an intelligent designer/God (model 4), which we shall consider shortly; and a bio-friendly life principle (model 5).

Of the six models of the universe, I have gone with the bio-friendly model. The bio-friendly model draws attention to the conditions that are ideal for human life – the tweaking of any one of which would make life impossible. *The New Philosophy of Universalism* has a section 'The Bio-Friendly Model', which includes the following on the conditions left by the Big Bang being ideal for human life:[119]

> One of the principal reasons for rejecting the idea of a chance Big Bang and a haphazard universe condensing, cooling and in a random way putting up layered atmosphere is that the conditions left by the Big Bang turned out to be ideal for human life and any tweaking of these conditions would make human life impossible. In the words of Fred Hoyle, the universe looks like "a put-up job".[120]...
>
> Universalism, focusing on this universe and not other possible universes, has a different angle on so-called anthropic data. It sees the

laws of physics and the initial cosmological conditions as providing a *context* for life that seems orderly rather than fortuitous and random. It would be truer to say that Universalism affirms a Contextual Principle, that the context for life is a system that is too precise to have formed by accident or chance or to be random and which is an Order Principle, a principle of order.

But the system is more extensive than the regularities we are aware of each day suggest.

The main conditions of this system can be extracted from work in classical cosmology, physics and astrophysics, quantum mechanics and biochemistry. There are at least 40 conditions that are just right for human life. I now present them, not in their particular disciplines, although the first 12 happen to come from cosmology, but arranged in relation to the importance and fundamental nature of their contribution to the structure of the universe. I broadly proceed from macrocosm to microcosm, but break up different emphases and avoid slavishly lumping together all the different treatments of gravity and protons while seeking to illuminate through juxtaposition. Thus, a force mentioned in one condition has a carry-over effect to the next condition, assisting understanding.

I then set out 40 conditions which have benefited life. These can be read in detail in *The New Philosophy of Universalism*, pp.102–122. Here I refer to each condition in a word or two, the topics that make it easier for life:[121]

1. The Big Bang.
2. Gravity and rest-mass energy.
3. Dimensions.
4. The rate of expansion.
5. Density.
6. Anti-gravity and acceleration.
7. Gravitational-electromagnetic ratio and constant.
8. Nuclear binding.
9. The heat of the universe.
10. Mathematical laws.

11. Flatness.

12. Carbon.

13. Molecules of life inside stars.

14. Temperature.

15. Weak and electromagnetic forces.

16. The strong nuclear force.

17. Gravitons.

18. Neutrinos.

19. Weak force and helium.

20. The nuclear forces, di-protons and stars.

21. Proton-electron ratio.

22. Vacuum energy.

23. The size of the universe.

24. The emptiness of the universe.

25. Proton-electron charges.

26. Electron-neutron mass ratio.

27. Quantization.

28. Freedom of particles.

29. A neutron-proton mass differential.

30. Sunlight and chlorophyll.

31. Bonding elements.

32. Atmosphere.

33. Water and rainfall.

34. Particles-antiparticles.

35. Bosons.

36. Cold dark matter.

37. Dark energy and zero vacuum.

38. Glacials and climate.

39. The speed of light.

40. Constants and ratios.

It is important to consider that all the following cosmic densities'
units have the value of unity or 1:[122]

the density of the universe   in protons per cubic metre
the density of galaxies   in protons per cubic centimetre

galactic dark matter density	in particles per cubic centimetre
density of stars within galaxies	in grams per cubic centimetre
the mean density of the sun	in grams per cubic centimetre
the typical density of planets	in grams per cubic centimetre
the typical density of life-forms	in grams per cubic centimetre
the density of a white dwarf	in metric tons per cubic centimetre[123]

The contents of the cosmos also have total unity or 1 in density, suggesting that inflation did indeed take place:

Component in the cosmos	Density as proportion of cosmos density (Ω x)
Radiation	0.00005
Stars	0.005–0.01
Neutrinos	≥ 0.003
All baryons	0.04
Dark matter	0.25
All matter	0.30
Dark energy	<u>0.70</u>
Total	1.00[124]

The detail of this table is slightly out of date as dark matter and dark energy are now thought to form 23 per cent and 73 per cent of the contents of the cosmos respectively (i.e. 0.23 and 0.73 as a proportion of 1), but the point is, the proportions that contribute to a density of 1 are broadly right.

Outside this range, human life would be impossible. If the constant of gravity were stronger – 10^{25} times less powerful than the strong nuclear force instead of 10^{38} times weaker – the universe would be small and swift, the average star would have 10^{-12} times the mass of the sun and would exist for only a year and no life would develop. If galaxies were to form, the early universe would have to be broadly homogeneous, meaning the same everywhere. This is only approximately true, but is an excellent approximation when averaging over the large regions of galactic space. The inhomogeneity ratio must not be greater than 10^{-2}, otherwise non-uniformities would condense into black holes before stars formed. It must not be less than

10^{-5}, otherwise inhomogeneities would be insufficient to condense the galaxies.[125] Classical cosmology permits a new parameter, S, the entropy (disorder) per baryon in the universe, which is about 10^9. Unless S is less than 10^{11}, galaxies would not be able to form, and planets and therefore life could not exist. S is a consequence of the baryon asymmetry in the universe, quarks and antiquarks being asymmetrical before 10^{-6} seconds after the Big Bang.[126] In short, no matter what constants or ratios we examine, all are within a correct band that permits galaxies, stars and human life to form. It is my contention that all these 326 constants and 75 ratios within them are evidence that the universe is controlled by the universal principle of order, or order principle.

It must be stressed that any change to just one of the 326 constants and 75 ratios (the 40th bio-friendly condition) would make human life impossible. To emphasise their collective bio-friendliness I now list them.

For the supporting detail of the 326 constants and 75 ratios, see *The New Philosophy of Universalism*, pp.372–395,[127] where tables give each constant's value, uncertainty and unit. Below are the 326 constants without the supporting detail:

1. alpha particle-electron mass ratio
2. alpha particle mass
3. alpha particle mass energy equivalent
4. alpha particle mass energy equivalent in MeV
5. alpha particle mass in u
6. alpha particle molar mass
7. alpha particle-proton mass ratio
8. Angstrom star
9. atomic mass constant
10. atomic mass constant energy equivalent
11. atomic mass constant energy equivalent in MeV
12. atomic mass unit-electron volt relationship
13. atomic mass unit-hartree relationship
14. atomic mass unit-hertz relationship
15. atomic mass unit-inverse meter relationship

16.   atomic mass unit-joule relationship

17.   atomic mass unit-kelvin relationship

18.   atomic mass unit-kilogram relationship

19.   atomic unit of 1st hyperpolarisability

20.   atomic unit of action

21.   atomic unit of 2nd hyperpolarisability

22.   atomic unit of charge

23.   atomic unit of charge density

24.   atomic unit of current

25.   atomic unit of electric dipole Mom.

26.   atomic unit of electric field

27.   atomic unit of electric field gradient

28.   atomic unit of electric polarisability

29.   atomic unit of electric potential

30.   atomic unit of electric quadrupole Mom.

31.   atomic unit of energy

32.   atomic unit of force

33.   atomic unit of length

34.   atomic unit of mag. Dipole Mom.

35.   atomic unit of mag. Flux density

36.   atomic unit of magnetisability

37.   atomic unit of mass

38.   atomic unit of momentum

39.   atomic unit of permittivity

40.   atomic unit of time

41.   atomic unit of velocity

42.   Avogadro constant

43.   Bohr magneton

44.   Bohr magneton in Ev/T

45.   Bohr magneton in Hz/T

46.   Bohr magneton in inverse meters per tesla

47.   Bohr magneton in K/T

48.   Bohr radius

49.   Boltzmann constant

50.   Boltzmann constant in Ev/K

51.   Boltzmann constant in Hz/K

52. Boltzmann constant in inverse meters per kelvin
53. characteristic impedance of vacuum
54. classical electron radius
55. Compton wavelength
56. Compton wavelength over 2 pi
57. conductance quantum
58. conventional value of Josephson constant
59. conventional value of von Klitzing constant
60. Cu x unit
61. deuteron-electron mag. Mom. ratio
62. deuteron-electron mass ratio
63. deuteron g factor
64. deuteron mag. Mom.
65. deuteron mag. Mom. to Bohr magneton ratio
66. deuteron mag. Mom. to nuclear magneton ratio
67. deuteron mass
68. deuteron mass energy equivalent
69. deuteron mass energy equivalent in MeV
70. deuteron mass in u
71. deuteron molar mass
72. deuteron-neutron mag. Mom. ratio
73. deuteron-proton mag. Mom. ratio
74. deuteron-proton mass ratio
75. deuteron rms charge radius
76. electric constant
77. electron charge to mass quotient
78. electron-deuteron mag. Mom. ratio
79. electron-deuteron mass ratio
80. electron g factor
81. electron gyromag. ratio
82. electron gyromag. ratio over 2 pi
83. electron mag. Mom.
84. electron mag. Mom. anomaly
85. electron mag. Mom. to Bohr magneton ratio
86. electron mag. Mom. to nuclear magneton ratio
87. electron mass

88. electron mass energy equivalent
89. electron mass energy equivalent in MeV
90. electron mass in u
91. electron molar mass
92. electron-muon mag. Mom. ratio
93. electron-muon mass ratio
94. electron-neutron mag. Mom. ratio
95. electron-neutron mass ratio
96. electron-proton mag. Mom. ratio
97. electron-proton mass ratio
98. electron-tau mass ratio
99. electron to alpha particle mass ratio
100. electron to shielded helion mag. Mom. ratio
101. electron to shielded proton mag. Mom. ratio
102. electron volt
103. electron volt-atomic mass unit relationship
104. electron volt-hartree relationship
105. electron volt-hertz relationship
106. electron volt-inverse meter relationship
107. electron volt-joule relationship
108. electron volt-kelvin relationship
109. electron volt-kilogram relationship
110. elementary charge
111. elementary charge over h
112. Faraday constant
113. Faraday constant for conventional electric current
114. Fermi coupling constant
115. fine-structure constant
116. first radiation constant
117. first radiation constant for spectral radiance
118. hartree-atomic mass unit relationship
119. hartree-electron volt relationship
120. Hartree energy
121. Hartree energy in Ev
122. hartree-hertz relationship
123. hartree-inverse meter relationship

124. hartree-joule relationship
125. hartree-kelvin relationship
126. hartree-kilogram relationship
127. helion-electron mass ratio
128. helion mass
129. helion mass energy equivalent
130. helion mass energy equivalent in MeV
131. helion mass in u
132. helion molar mass
133. helion-proton mass ratio
134. hertz-atomic mass unit relationship
135. hertz-electron volt relationship
136. hertz-hartree relationship
137. hertz-inverse meter relationship
138. hertz-joule relationship
139. hertz-kelvin relationship
140. hertz-kilogram relationship
141. inverse fine-structure constant
142. inverse meter-atomic mass unit relationship
143. inverse meter-electron volt relationship
144. inverse meter-hartree relationship
145. inverse meter-hertz relationship
146. inverse meter-joule relationship
147. inverse meter-kelvin relationship
148. inverse meter-kilogram relationship
149. inverse of conductance quantum
150. Josephson constant
151. joule-atomic mass unit relationship
152. joule-electron volt relationship
153. joule-hartree relationship
154. joule-hertz relationship
155. joule-inverse meter relationship
156. joule-kelvin relationship
157. joule-kilogram relationship
158. kelvin-atomic mass unit relationship
159. kelvin-electron volt relationship

160. kelvin-hartree relationship
161. kelvin-hertz relationship
162. kelvin-inverse meter relationship
163. kelvin-joule relationship
164. kelvin-kilogram relationship
165. kilogram-atomic mass unit relationship
166. kilogram-electron volt relationship
167. kilogram-hartree relationship
168. kilogram-hertz relationship
169. kilogram-inverse meter relationship
170. kilogram-joule relationship
171. kilogram-kelvin relationship
172. lattice parameter of silicon
173. lattice spacing of silicon
174. Loschmidt constant (273.15 K, 101.325 kPa)
175. mag. Constant
176. mag. Flux quantum
177. molar gas constant
178. molar mass constant
179. molar mass of carbon-12
180. molar Planck constant
181. molar Planck constant times c
182. molar volume of ideal gas (273.15 K, 100 kPa)
183. molar volume of ideal gas (273.15 K, 101.325 kPa)
184. molar volume of silicon
185. Mo x unit
186. muon Compton wavelength
187. muon Compton wavelength over 2 pi
188. muon-electron mass ratio
189. muon g factor
190. muon mag. Mom.
191. muon mag. Mom. anomaly
192. muon mag. Mom. to Bohr magneton ratio
193. muon mag. Mom. to nuclear magneton ratio
194. muon mass
195. muon mass energy equivalent

196. muon mass energy equivalent in MeV
197. muon mass in u
198. muon molar mass
199. muon-neutron mass ratio
200. muon-proton mag. Mom. ratio
201. muon-proton mass ratio
202. muon-tau mass ratio
203. natural unit of action
204. natural unit of action in Ev s
205. natural unit of energy
206. natural unit of energy in MeV
207. natural unit of length
208. natural unit of mass
209. natural unit of momentum
210. natural unit of momentum in MeV/c
211. natural unit of time
212. natural unit of velocity
213. neutron Compton wavelength
214. neutron Compton wavelength over 2 pi
215. neutron-electron mag. Mom. ratio
216. neutron-electron mass ratio
217. neutron g factor
218. neutron gyromag. ratio
219. neutron gyromag. ratio over 2 pi
220. neutron mag. Mom.
221. neutron mag. Mom. to Bohr magneton ratio
222. neutron mag. Mom. to nuclear magneton ratio
223. neutron mass
224. neutron mass energy equivalent
225. neutron mass energy equivalent in MeV
226. neutron mass in u
227. neutron molar mass
228. neutron-muon mass ratio
229. neutron-proton mag. Mom. ratio
230. neutron-proton mass ratio
231. neutron-tau mass ratio

232. neutron to shielded proton mag. Mom. ratio

233. Newtonian constant of gravitation

234. Newtonian constant of gravitation over h-bar c

235. nuclear magneton

236. nuclear magneton in Ev/T

237. nuclear magneton in inverse meters per tesla

238. nuclear magneton in K/T

239. nuclear magneton in MHz/T

240. Planck constant

241. Planck constant in Ev s

242. Planck constant over 2 pi

243. Planck constant over 2 pi in Ev s

244. Planck constant over 2 pi times c in MeV fm

245. Planck length

246. Planck mass

247. Planck mass energy equivalent in GeV

248. Planck temperature

249. Planck time

250. proton charge to mass quotient

251. proton Compton wavelength

252. proton Compton wavelength over 2 pi

253. proton-electron mass ratio

254. proton g factor

255. proton gyromag. ratio

256. proton gyromag. ratio over 2 pi

257. proton mag. Mom.

258. proton mag. Mom. to Bohr magneton ratio

259. proton mag. Mom. to nuclear magneton ratio

260. proton mag. Shielding correction

261. proton mass

262. proton mass energy equivalent

263. proton mass energy equivalent in MeV

264. proton mass in u

265. proton molar mass

266. proton-muon mass ratio

267. proton-neutron mag. Mom. ratio

268. proton-neutron mass ratio

269. proton rms charge radius

270. proton-tau mass ratio

271. quantum of circulation

272. quantum of circulation times 2

273. Rydberg constant

274. Rydberg constant times c in Hz

275. Rydberg constant times hc in Ev

276. Rydberg constant times hc in J

277. Sackur-Tetrode constant (1 K, 100 kPa)

278. Sackur-Tetrode constant (1 K, 101.325 kPa)

279. second radiation constant

280. shielded helion gyromag. ratio

281. shielded helion gyromag. ratio over 2 pi

282. shielded helion mag. Mom.

283. shielded hellion mag. Mom. to Bohr magneton ratio

284. shielded helion mag. Mom. to nuclear magneton ratio

285. shielded helion to proton mag. Mom. ratio

286. shielded helion to shielded proton mag. Mom. ratio

287. shielded proton gyromag. ratio

288. shielded proton gyromag. ratio over 2 pi

289. shielded proton mag. Mom.

290. shielded proton mag. Mom. to Bohr magneton ratio

291. shielded proton mag. Mom. to nuclear magneton ratio

292. speed of light in vacuum

293. standard acceleration of gravity

294. standard atmosphere

295. Stefan-Boltzmann constant

296. tau Compton wavelength

297. tau Compton wavelength over 2 pi

298. tau-electron mass ratio

299. tau mass

300. tau mass energy equivalent

301. tau mass energy equivalent in MeV

302. tau mass in u

303. tau molar mass

304. tau-muon mass ratio

305. tau-neutron mass ratio

306. tau-proton mass ratio

307. Thomson cross section

308. triton-electron mag. Mom. ratio

309. triton-electron mass ratio

310. triton g factor

311. triton mag. Mom.

312. triton mag. Mom. to Bohr magneton ratio

313. triton mag. Mom. to nuclear magneton ratio

314. triton mass

315. triton mass energy equivalent

316. triton mass energy equivalent in MeV

317. triton mass in u

318. triton molar mass

319. triton-neutron mag. Mom. ratio

320. triton-proton mag. Mom. ratio

321. triton-proton mass ratio

322. unified atomic mass unit

323. von Klitzing constant

324. weak mixing angle

325. Wien frequency displacement law constant

326. Wien wavelength displacement law constant

Below is a list of the 152 fundamental physical constants grouped under fields (dashes indicate that a constant is linked with the one above):[128]

(1) Universal

1. speed of light in vacuum

2. magnetic constant

3. electric constant

4. characteristic impedance of vacuum

5. Newtonian constant of gravitation

6. Planck constant

7. Planck mass

8.   Planck temperature

9.   Planck length

10. Planck time

(2) Electromagnetic

11. elementary charge

12. magnetic flux quantum

13. conductance quantum

14. – inverse of conductance quantum

15. Josephson constant

16. von Klitzing constant

17. Bohr magneton

18. nuclear magneton

(3) Atomic and Nuclear

General

19. fine-structure constant

20. – inverse fine-structure constant

21. Rydberg constant

22. Bohr radius

23. Hartree energy

24. quantum of circulation

Electroweak

25. Fermi coupling constant

26. weak mixing angle

Electron

27. electron mass

28. – energy equivalent

29. electron-muon mass ratio

30. electron-tau mass ratio

31. electron-proton mass ratio

32. electron-deuteron mass ratio

33. electron to alpha particle mass ratio

34. electron charge to mass quotient

35. electron molar mass

36. Compton wavelength

37. classical electron radius

38. Thomson cross section

39. electron magnetic moment

40. – to Bohr magneton ratio

41. – to nuclear magneton ratio

42. electron magnetic moment anomaly

43. electron g-factor

44. electron-muon magnetic moment ratio

45. electron-proton magnetic moment ratio

46. electron to shielded proton magnetic moment ratio

47. electron-neutron magnetic moment ratio

48. electron-deuteron magnetic moment ratio

49. electron to shielded helion magnetic moment ratio

50. electron gyromagnetic ratio

Muon

51. muon mass

52. – energy equivalent

53. muon-electron mass ratio

54. muon-tau mass ratio

55. muon-proton mass ratio

56. muon-neutron mass ratio

57. muon molar mass

58. muon Compton wavelength

59. muon magnetic moment

60. – to Bohr magneton ratio

61. – to nuclear magneton ratio

62. muon magnetic moment anomaly

63. muon g-factor

64. muon-proton magnetic moment ration

Tau

65. tau mass

66. – energy equivalent

67. tau-electron mass ratio

68. tau-muon mass ratio

69. tau-proton mass ratio

70. tau-neutron mass ratio

71. tau molar mass

72. tau Compton wavelength

Proton

73. proton mass
74. – energy equivalent
75. proton-electron mass ratio
76. proton-muon mass ratio
77. proton-tau mass ratio
78. proton-neutron mass ratio
79. proton charge to mass quotient
80. proton molar mass
81. proton Compton wavelength
82. proton rms charge radius
83. proton magnetic moment
84. – to Bohr magneton ratio
85. – to nuclear magneton ratio
86. proton g-factor
87. proton-neutron magnetic moment ratio
88. shielded proton magnetic moment
89. – to Bohr magneton ratio
90. – to nuclear magneton ratio
91. proton magnetic shielding
92. proton gyromagnetic ratio
93. shielded gyromagnetic ratio

Neutron

94. neutron mass
95. – energy equivalent
96. neutron-electron mass ratio
97. neutron-muon mass ratio
98. neutron-tau mass ratio
99. neutron-proton mass ratio
100. neutron molar mass
101. neutron Compton wavelength
102. neutron magnetic moment
103. – to Bohr magneton ratio
104. – to nuclear magneton ratio
105. neutron g-factor
106. neutron-electron magnetic moment ratio

107. neutron-proton magnetic moment ratio

108. neutron to shielded proton magnetic moment ratio

109. neutron gyromagnetic ratio

Deuteron

110. deuteron mass

111. – energy equivalent

112. deuteron-electron mass ratio

113. deuteron-proton mass ratio

114. deuteron molar mass

115. deuteron rms charge radius

116. deuteron magnetic moment

117. – to Bohr magneton ratio

118. – to nuclear magneton ratio

119. deuteron-electron magnetic moment ratio

120. deuteron-proton magnetic moment ratio

121. deuteron-neutron magnetic moment ratio

Helion

122. helion mass

123. – energy equivalent

124. helion-electron mass ratio

125. helion-proton mass ratio

126. helion molar mass

127. shielded helion magnetic moment

128. – to Bohr magneton ratio

129. – to nuclear magneton ratio

130. shielded helion to proton magnetic moment ratio

131. shielded helion to shielded proton magnetic moment ratio

132. shielded helion gyromagnetic ratio

Alpha particle

133. alpha particle mass

134. – energy equivalent

135. alpha particle to electron mass ratio

136. alpha particle to proton mass ratio

137. alpha particle molar mass

(4) Physico-Chemical

138. Avogadro constant

139. atomic mass constant
140. – energy equivalent
141. Faraday constant
142. molar Planck constant
143. molar gas constant
144. Boltzmann constant
145. molar volume of ideal gas
146. – Loschmidt constant
147. Sackur-Tetrode constant
148. Stefan-Boltzmann constant
149. first radiation constant
150. first radiation constant for spectral radiance
151. second radiation constant
152. Wien displacement law constant

The 75 ratios can be extracted from the list of 326 constants and are:[129]

1. alpha particle-electron mass ratio
2. alpha particle-proton mass ratio
3. deuteron magnetic moment to Bohr magneton ratio
4. deuteron magnetic moment to nuclear magneton ratio
5. deuteron-electron magnetic moment ratio
6. deuteron-electron mass ratio
7. deuteron-neutron magnetic moment ratio
8. deuteron-proton magnetic moment ratio
9. deuteron-proton mass ratio
10. electron gyromagnetic ratio
11. electron gyromagnetic ratio over 2 pi
12. electron magnetic moment to Bohr magneton ratio
13. electron magnetic moment to nuclear magneton ratio
14. electron to alpha particle mass ratio
15. electron to shielded helion magnetic moment ratio
16. electron to shielded proton magnetic moment ratio
17. electron-deuteron magnetic moment ratio
18. electron-deuteron mass ratio

19. electron-muon magnetic moment ratio

20. electron-muon mass ratio

21. electron-neutron magnetic moment ratio

22. electron-neutron mass ratio

23. electron-proton magnetic moment ratio

24. electron-proton mass ratio

25. electron-tau mass ratio

26. helion-electron mass ratio

27. helion-proton mass ratio

28. muon magnetic moment to Bohr magneton ratio

29. muon magnetic moment to nuclear magneton ratio

30. muon-electron mass ratio

31. muon-neutron mass ratio

32. muon-proton magnetic moment ratio

33. muon-proton mass ratio

34. muon-tau mass ratio

35. neutron gyromagnetic ratio

36. neutron gyromagnetic ratio over 2 pi

37. neutron magnetic moment to Bohr magneton ratio

38. neutron magnetic moment to nuclear magneton ratio

39. neutron to shielded proton magnetic moment ratio

40. neutron-electron magnetic moment ratio

41. neutron-electron mass ratio

42. neutron-muon mass ratio

43. neutron-proton magnetic moment ratio

44. neutron-proton mass ratio

45. neutron-tau mass ratio

46. proton gyromagnetic ratio

47. proton gyromagnetic ratio over 2 pi

48. proton magnetic moment to Bohr magneton ratio

49. proton magnetic moment to nuclear magneton ratio

50. proton-electron mass ratio

51. proton-muon mass ratio

52. proton-neutron magnetic moment ratio

53. proton-neutron mass ratio

54. proton-tau mass ratio

55. shielded helion gyromagnetic ratio
56. shielded helion gyromagnetic ratio over 2 pi
57. shielded helion magnetic moment to Bohr magneton ratio
58. shielded helion magnetic moment to nuclear magneton ratio
59. shielded helion to proton magnetic moment ratio
60. shielded helion to shielded proton magnetic moment ratio
61. shielded proton gyromagnetic ratio
62. shielded proton gyromagnetic ratio over 2 pi
63. shielded proton magnetic moment to Bohr magneton ratio
64. shielded proton magnetic moment to nuclear magneton ratio
65. tau-electron mass ratio
66. tau-muon mass ratio
67. tau-neutron mass ratio
68. tau-proton mass ratio
69. triton magnetic moment to Bohr magneton ratio
70. triton magnetic moment to nuclear magneton ratio
71. triton-electron magnetic moment ratio
72. triton-electron mass ratio
73. triton-neutron magnetic moment ratio
74. triton-proton magnetic moment ratio
75. triton-proton mass ratio

Any change in the value of just one of these 326 constants and 75 ratios would make human life impossible, and the fact that they are all in a perfect balance with each other is an overwhelmingly powerful argument for there being an order principle behind the creation of the universe rather than a random event.

All this follows a variation of +A + −A = 0: +A (8 densities of 1) + −A (density of the contents of the cosmos of 1, 326 constants and 75 ratios) = 0 (the One, total unity expressed in +A and −A by the densities figure 1 − +1 + −1 = 0).

To put what I have just said about order and bio-friendliness another way, in many cases the values of the 326 constants and 75 ratios are just right for life, and if a particular ratio were higher or lower, the rest would not be able to sustain life. Such a fine balance involving 326 constants and 75 ratios could not be a random accident.

To the 75 ratios should be added Fibonacci numbers which determine all growing things, including the eight clockwise and 13 counter-clockwise spirals of pine cones. The sequence of Fibonacci numbers is arrived at by adding the two previous numbers starting at 0 (0, 1, 1, 2, 3, 5, 8, 13, 21, etc.). Dividing each Fibonacci number by its preceding number gives the golden ratio of approximately 1.6180339. This determines the design proportion of the human frame (elbow to wrist: wrist to fingertip; belly button to the soles of the feet: top of the head to belly button; ideal height: shoulder to fingertip – each is in a ratio of 1.6180339, see *My Double Life 1: This Dark Wood*, p.xxviii).

All this follows a variation of +A + –A = 0: +A (40 orderly bio-friendly conditions, 326 constants and 75 ratios) + –A (random, accidental disorder) = 0 (the One universe that is creating the conditions for life).

(10) The origin of life: the first cell
The physical universe came out of one point, a singularity. All biological species came out of the same point, initially out of one cell. We need to see how that first cell emerged.

The first chemical evidence for the origin of life and the first cell was 4–3.8 billion years ago. The first prokaryotic (unicellular) cells appeared 3.8–3.5 billion years ago. The first chemical evidence of unicellular eukaryotic cells with a membrane-bound nucleus appeared 2.7 billion years ago, and multicellular eukaryotic cells (containing genetic material) appeared 2.5–2 billion years ago.

It can be said that the first living thing on earth appeared 4 billion years ago. The first fossils of primitive cyanobacteria (that obtain energy via photosynthesis) appeared 3.7–3.5 billion years ago, and of bacteria living on land appeared 2.6 billion years ago.

It is possible that the first living thing on earth appeared 4.28 billion years ago. It was announced in 2022 that a fist-sized rock in Quebec, Canada 4.28 billion years old may hold the earliest evidence for bacterial life on earth. It shows evidence of fossilised tubes claimed to have been made by microscopic life forms multiplying on the sea-floor only 300 million years after the earth formed, although some scientists have said that the shapes may have been formed by chemical reactions. However, no known chemical reaction could have produced them, and there is therefore still a possibility that bacteria appeared 4.28 billion years ago.

We have seen that (see (9)) the 40 bio-friendly conditions of the earth made biological life possible 4–3.8 billion years ago, and again we need to look at the process described in *The New Philosophy of Universalism*, which covers all the alternative theories behind the reductionist approach that runs counter to the Universalist approach. *The New Philosophy of Universalism* has a section 'The Origin of Life' that challenges the reductionist view of the origin of life that ignores the order principle:[130]

The earth, which began about 4.6 or 4.55 billion years ago,[131] is one of the smallest planets grouped in our solar system round a central star (the sun). As all the planets revolve in the same plane round the sun, it is likely that they condensed from a single revolving disc of gaseous matter and that the earth originated as a mass of molten rock with no atmosphere. As it cooled, a crust formed. It was disturbed by volcanic activity (hot gases escaping from the molten interior). When the surface cooled to 100 degrees C, an atmosphere developed that consisted of water, ammonia, carbon dioxide, methane and hydrogen. Geological evidence shows that the early earth was virtually without oxygen – hence iron in reduced form is found in rocks. As the earth continued to cool, water vapour in the atmosphere condensed and fell as rain, forming rivers, lakes and oceans[132] about 4.4 billion years ago[133] while erosion formed rocks.

Reductionism eliminates all features of a complex process that are not essential to its function to isolate its most important features. It then creates a simple model for a process that may otherwise be too complicated to study effectively. Reductionism subdivides complex situations into components which are simple enough to be investigated. According to reductionists, if evolution is a fact of life then the origin of life and the first cell (the first chemical evidence for which was 3.8 billion years ago)[134] should all be explainable within the evolutionary (neo-Darwinist) scheme, as should the origin of eukaryotic cells (cells containing genetic material), the origin of multicellular organisation, the rise of flowering plants, the origin of vertebrates from non-vertebrates and the origin of viruses.

According to the reductionist view, life began when it was formed from earthbound, chemically non-living matter, a process known as abiogenesis.[135] There are a number of biogenetic hypotheses, none of which have yet been proved.

The best known reductionist hypothesis is that life began in a "primordial soup".[136] Charles Darwin had suggested in a letter to Joseph Dalton Hooker on February 1, 1871 that the original spark of life may have begun in a "warm little pond, with all sorts of ammonia and phosphoric salts, light, heat, electricity, etc. present, that a protein compound was chemically formed ready to undergo still more complex changes".[137] He added that "at the present day such matter would be instantly devoured or absorbed, which would not have been the case before living creatures were formed".

In 1924 the Russian plant physiologist Aleksandr Ivanovich Oparin showed that atmospheric oxygen prevented the synthesis of organic molecules that are the building blocks of life. In 1936 in *The Origin of Life on Earth*, he suggested that life began in the sea, and that a primeval "soup" of organic molecules could be created in an oxygenless atmosphere through the action of sunlight. He argued that these molecules would combine and dissolve into droplets which would fuse with other droplets and reproduce through fission.[138] Around the same time the English geneticist and physiologist J.B.S. Haldane suggested that the earth's pre-biotic oceans would have formed a "hot dilute soup" in which organic compounds would have formed, the building blocks of life.[139] The formation of small molecules (such as amino acids and monosaccharides) would lead to the formation of larger polymers (proteins, lipids, nucleotides and polysaccharides) and in due course from polymers to the formation of organisms (simple prokaryote cells).[140] In 1953, taking his cue from Oparin and Haldane, the American chemist Stanley Miller, a Chicago graduate student working under Harold Urey, carried out an experiment on the primeval soup.[141]...

There were problems with the "soup" theory. Amino acids have to become proteins and one protein requires 100 amino acids of 20 varieties which requires 10^{130} combinations of amino acids. The amino acids are building blocks, not the assembled structure, and it would be hard to achieve the right protein by accident. Furthermore, the soup theory assumes a lack of any presence of oxygen in the earth's primeval atmosphere as that would poison early forms of life. However, geological evidence of oxides in minerals suggests that the early atmosphere included oxygen; oxides in minerals could not have formed without the presence of oxygen. Oxygen aside, the early atmosphere on earth was formed of different gases from those used by Miller and Urey. There was no methane or ammonia in the primeval atmosphere, and experiments with the atmospheric gases that would have been present such as carbon dioxide did not produce plentiful amino acids. Carbon in particular does not break out to make larger organic molecules. What is more, amino acids would have to form protein while the second law of thermodynamics applied. This says that a system becomes less and less organised over time, meaning that amino acids cannot be more and more organised over time and form protein. The primordial soup would have been too diluted to generate proteins and lacked a 'mechanism' – a self-organising principle – that could have created proteins.[142]

Repeat experiments based on speculations regarding varying earth conditions in early times have formed more biochemicals (all 20 amino acids, sugars, lipids, polynucleotides and ATP). Despite the experiments and work of Sidney Fox (1950s and 1960s), Juan Orowin (1961), Manfred Eigen and Peter Schuster (early 1970s), Günter Wächtershäuser (1980s) and Tom Scheck, who discovered riboenzyme (enzymes made of RNA, 1986), no one has succeeded in synthesising a protocell using basic components. Most reductionists nevertheless believe that prebiotic synthesis took place on the earth. The "primordial soup" hypothesis therefore remains an unproved hypothesis.[143]

There have been a number of alternative hypotheses. Miller's experiment created the field of exobiology, the study of life beyond the earth. This looked for the origin of life elsewhere in the solar system. The extraterrestrial hypothesis favoured by Sir Fred Hoyle, who criticised abiogenesis on the grounds of improbability, asserts that primitive life may have originally formed in space or on a nearby planet such as Mars or the satellites of Jupiter, or the outer solar systems.[144] Hoyle sees life including that of viruses and insects as coming to earth from space.[145]

He was supported in this view by the British biochemists Francis Crick and Leslie Orgel, both of whom thought that life began with spores and bacteria borne on meteorites. It is interesting that Orgel, having spent a lifetime researching into how life began on earth, concluded that molecules preceded RNA, which preceded DNA – which was too complex to be the first repository of genetic information, a view Crick, the discoverer of the structure of DNA, shared – and that these molecules did not arise from a primordial soup but from outer space. Meteors, dust and comets may have supplied a few per cent of organic compounds.[146]

Various hypothetical models have been offered to explain how organic molecules become protocells. There are theories connected with RNA further to the work of Tom Scheck, who discovered that enzymes made of RNA can replicate themselves. There are theories that look at metabolism rather than genes further to Wächtershäuser's iron-sulphur theory. There is a clay theory put forward by Graham Cairns-Smith in 1985 that holds that organic molecules arose from clay in silicate crystals. The same applies to honeycombs and seashells, which also have repetitive order with low information content. There is a view by Thomas Gold in the 1990s that life began some five kilometres below the earth's surface in deep rocks where nanobes (filomental structures that are smaller than bacteria but may contain DNA) have been found. A 2,600-feet-deep volcanic system of towering pinnacles and chimneys in a hydrothermal vent field on the mid-Atlantic ridge produces hydrocarbons – oil and gas and molecules, the building blocks of life – by the chemical interaction of sea water and rocks. Their source is non-living, and they are organic building blocks from a non-biological source. It has been suggested that the floor of the Atlantic Ocean contains the origins of life.[147]

There is a view connected with lipids. Phospholipids spontaneously form bilayers in water, the same structure as in cell membranes. These self-replicating molecules were not present in primeval times but may reveal early information storage which led to RNA and DNA. There are views connected with polyphosphates, ecopoesis and polycyclic aromatic hydrocarbons (PAHs) in nebulae. There is a view popularised by Richard Dawkins, the British ethologist, in his book *The Ancestor's*

Tale, citing the Scripps Research Institute in California (2004), that autocatalysis accounts for the origin of life. Autocatalysts are substances that catalyze the production of themselves and therefore are molecular replicators for Dawkins. All these experiments and theories are unproved hypotheses.[148]

The trouble is that for every theory and experiment there are difficulties and counter-arguments. The problem with RNA, for example, is that there are no known chemical pathways for the abiogenic synthesis of the pyromidine nucleobases cytosine and uracil under prebiotic conditions. There is also the difficulty of nucleoside synthesis (from ribose and nucleobase), ligating nucleosides with phosphate to form the RNA backbone and the short lifetime of nucleoside molecules, especially cytosine. There are doubts about the size of an RNA molecule capable of self-replication. Different hypotheses lack complete evidence, and cannot be related to the exact conditions in the prebiotic universe.[149]

So where does this leave the origin of life? In plain English, a Thermos flask with water, a heat source, a flash to simulate lightning, simple molecules and a coil demonstrated that two weeks after this soup was left cooking there was molecular activity. Simple liquids threw up enzymes like bubbles from dishwashing detergent. This suggested that the action of wind, heat and lightning on waves would have a similar effect and throw up molecules that might form a cell. However, an initial cell could not copy itself until the molecules were complicated enough to replicate themselves and become a lipid group with proteins trapped inside along with RNA. RNA could copy itself if a wave hit an enzyme and broke it into two small "bubbles", which became a cell and a printed copy.[150] The chances of our world coming out of this soup by accident are extremely slight.

To the reductionist, chemical life began when carbon combined with five other atoms: hydrogen, oxygen, nitrogen, sulphur and phosphorus. Life to a reductionist is an electrical gradient or imbalance of charges across a membrane. A membrane is like a sandwich with no filling: two layers of fat (the slices of bread) with water-loving heads of lipids on the outside of one layer and water-avoiding tails of lipids pointing inwards on the outside of the other layer. If three sodium ions are pumped in and two potassium ions are pumped out there is a charge. That charge, to a chemical reductionist, is life.[151]

All life and all species grew from a first cell, and there are now trillions of trillions of cells. It takes more than 10^{15} cell divisions to proceed from one cell to a human being and in each mitosis there is a one-in-a-million (10^{-6}) chance of error, of a mutation being made. For every time a cell divides it copies DNA, and every time there is a copy of DNA there is a one-in-a-million (10^{-6}) chance for an error to be made in human cells. The four building blocks of DNA are A, P, G and C. If a cell copies itself so P comes first, there is a mutation, and the two cells are different. One cell will grow better and take over the population of cells that stems from further mitosis from that cell. We cannot say that one cell is better than the other, just different. It will prove to be better if it survives and will prove not to be better if it is extinguished. If a cell receives too many instructions it commits suicide. If the lipid pump is poisoned, a cell will die. Mutations that are different and do not die are creative.[152]

There have been a huge number of cells created over the years. Multiply the number of cells in each human being, more than 10^{15},[153] by the number of human beings in the world at present, 6.7 billion. Making allowances for births and deaths and working backwards, taking account of the fact that there were one billion human beings alive in 1804 and two billion in 1927,[154] we can in theory work back to the first human – or at any rate the estimated number of early humans alive 40,000 years ago – and estimate the number of human beings who have ever lived: somewhere between 45 billion and 125 billion, most probably between 90 and 110 billion.[155] We can then calculate the number of human cells that there have ever been. Bearing in mind that life began on earth between 3.8 billion and 3.5 billion years ago[156] we can in theory calculate an estimate for the number of cells in all mammals, reptiles, birds, fish, insects, plants and bacteria since the very beginning.

All these cells came out of one cell, the first cell or protocell. A long time had to elapse during which cells learned to hold together and exchange information. We thus have a Tree of Life for the period and pattern from the first cell to the inflation of species (see p.93), which replicates the concept of the period and pattern from the Big Bang to the inflation of galaxies. But whereas the atoms that came into existence with the Big Bang are still in existence and have not died out, cells – and branches of species – *have* died out. Now all human beings have one thing in common: we are all cells.

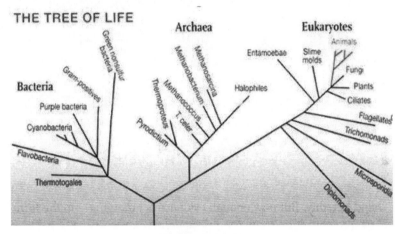

The Tree of Life.

We are made of atoms of hydrogen that comprise one-tenth of our weight, which came from the Big Bang, and of stardust, which forms 0.01 per cent of the universe and comprises: atoms of helium from the Big Bang and later stars, iron atoms carrying oxygen from exploding white dwarf stars, oxygen from exploding supernovas and carbon from planetary nebulas.[157]

The reductionist view of the origin of life is a string of hypotheses that are not proved. Reductionism identifies self-replication but has not identified a 'mechanism' by which self-replication takes place in a right way that leads from cell to organism. 'Mechanism' is a reductionist word whose mechanistic implications the Universalist does not accept, hence the inverted commas. What is missing is the 'mechanism' for a self-organising, ordering principle which can take organic molecules to the first cell, and the first cell to the many species of our world without being driven by blind, random accident, the chances of which happening are very highly improbable.

There is now evidence that before life began a meteorite landed on the earth with the core materials that make up DNA, transported from space. Only three of the five main chemical components (nucleobases, nitrogen-containing compounds) of DNA had been found in space rocks until recently (adenine, guanine and uracil), but a study published in the journal *Nature Communications* (26 April 2022) by astrochemist Yasuhiro Oba of Hokkaido University, Japan, Danny Glavin of NASA's Goddard Space Flight Center in Maryland and five others has found the other two chemical components (thymine and cytosine) when they re-examined meteorites with new techniques. See https://www.nature.com/articles/s41467-022-29612-x. The five chemical components make an individual's genome and provide individuals with a unique genetic code in DNA, the molecule that carries genetic instructions (as an algorithm conveying order) in living organisms, and RNA, the molecule crucial for controlling the action of genes. The five chemical components have been found in samples from the Murchison meteorite which landed in Victoria, Australia in 1969; from the Murray meteorite that landed in Kentucky, USA in 1950; and from the Tagish Lake meteorite that landed in British Columbia, Canada in 2000.

(All this follows a variation of +A + −A = 0: +A (self-organising, ordering of self-replication of cells and DNA into organism) + −A (molecules from water treated by lightning forming a cell and chemical life) = 0 (the One universe).

(11) Evolution and environment: Expansionism and Darwinism
Early evolving life from unicellular to multicellular organisms after the first cell was linked to the geological and climatic environment and conditions. The geological history of the earth spans nearly 4 billion years, as *The New Philosophy of Universalism* describes:[158]

> The oldest known rocks have an isotopic age of 3.96–3.9 billion years ago. The earliest fossils may be 3.8 billion years ago.[159] From the time the earth came into existence about 4.55 billion years ago according to geologists and astronomers there are about 650 million years (between 4.55 and 3.9 billion years ago) for which no geological record exists.[160]

The Precambrian period stretches from 3.9 billion years ago to 570 million years ago, 75 per cent of the earth's history which included the first photosynthesis about 1.5 billion years ago.[161]

Evidence for past cold climatic conditions has been found in rocks. There have been five Ice Ages, the oldest being the Huronian Ice Age 2.7–2.3 billion years ago:[162]

2.7 to 2.3 billion years ago, the Huronian Ice Age during the Proterozoic Era in the Precambrian times. This Ice Age is hypothetical and is not well documented by geological, chemical and palaeontological evidence. Some authorities date the beginning of the first Ice Age to 2.3 billion years ago.

The first evidence of bacteria living on land was 2.6 billion years ago. 2.5 billion years ago the Great Oxidation Event gave the earth an atmosphere in which all could breathe: cyanobacteria (blue-green algae) filled the atmosphere with oxygen, and multicellular life forms could exist. At that time there was one land mass on earth, Gondwana.

The earth and atmosphere requires the right level of oxygen, which first developed 1.9 billion years ago:[163]

Atmosphere. The earth's atmosphere requires gravity to pull strongly enough to prevent it from evaporating into space, as would happen on the moon if it had an atmosphere, for gravity is weaker there. The atmosphere requires the presence of water, and the earth's collision with an asteroid gave it a shield from solar wind and created the moon which, by gravity, pulls the oceans towards it, creating tides that assist evaporation into the atmosphere and aid life. The earth must not be too hot like Venus or too cold, or too near or too far from a long-living stable star such as the sun. Otherwise there would be no life. The earth is made of iron and during volcanic eruptions its molten core releases carbon dioxide, which regulates the earth's temperature. The earth's orbit must be stable. The atmosphere must be rich in oxygen, which means it must be transformed by bacteria in its early history.[164] The present atmosphere is 21-per-cent oxygen, the upper limit for safety of life as the risk of forest fires being started by lightning increases by 70 per cent for every 1-per-cent increase in the atmosphere's oxygen. If the oxygen were 25 per cent, all land vegetation would be at risk from being incinerated from lightning-strike fires. Oxygen forms ozone in the upper atmosphere, which screens out far ultraviolet radiation which would otherwise destroy carbon-based life. Until plants produced sufficient oxygen to sustain creatures which crawled out of oceans, life was only possible in water, which filters out far-ultraviolet radiation. The loss of hydrogen gas from the earth's atmosphere should be counterbalanced by an influx of hydrogen from the sun. As nitrogen forms nearly 80 per cent of the earth's atmosphere and the nitrogen in the atmosphere can be converted into ammonia, the source of nitrogen in biological compounds, nitrogen is essential for the building blocks of amino acids and therefore for the chemistry of life.[165] All these conditions can be found on earth, which make it just right for an atmosphere with air to allow life to form and survive.

The oldest multicellular forms appeared 1.8 billion ago. The first photosynthesis was 1.5 billion years go.

There was another Ice Age, the Cryogenian Ice Age, 850–630 million years ago, and the first multicellular life was 800 million years ago:[166]

850–630 million years ago, the Cryogenian Ice Age. This was the most severe Ice Age in the last billion years and may have resulted in the entire earth's being covered in sheet ice, making the earth resemble a snowball. This is the earliest well-documented Ice Age, and it is possible that its end may be responsible for the Ediacaran and Cambrian explosions which saw progress towards the emergence of *Homo sapiens*.

There were two more Ice Ages: the Andean-Saharan Ice Age of 460–430 million years ago, and the Karoo Ice Age of 350–260 million years ago. 450 million years ago, the late Ordovician mass extinction happened and a climate shift left Gondwana covered with glaciers. Later there were plate movements that created the continents out of one landmass (Gondwana) some 200 million years ago. These changes assisted evolution:[167]

There is a consensus that these Ice Ages were caused by a combination of conditions: the composition of the atmosphere, and in particular the concentrations of carbon dioxide, methane, sulphur dioxide and other gases; changes in the earth's orbit around the sun and possibly the sun's orbit around the galaxy; the movement of tectonic plates which affected the continental and oceanic crust; variations in solar energy; the orbits of the earth and moon; the impact of large meteorites; and the eruption of super volcanoes. Of these the most significant were likely to be: the changing of continental positions and the lifting of continental blocks; the reduction of carbon dioxide in the atmosphere; and changes in the earth's orbit. In our own time, greenhouse gases may affect the composition of the earth's atmosphere and bring about climate change.[168] On the other hand, climate change may have been caused by the glacial-interglacial pattern which can affect the earth's atmosphere by changing the rate at which the weather removes carbon dioxide.

The fifth Ice Age began 40 million years ago, and did not end 11,750–10,000 years ago. We are still in it because there are still ice sheets in Greenland and Antarctica.

Sponges are 760 million years old; jellyfish 500 million; horseshoe crabs 445 million; coelacanths and lampreys 360 million; and horseshoe shrimps and sturgeon 200 million. Evolution so far looks as follows:[169]

Million years ago	Event
800	First multicellular life
700/575	First animals
543	First shelled animals
533–525	Cambrian explosion
520	First land plants
500–450	First fish, insects and other invertebrates move on land
	First colonisation of earth by plants and animals
460–430	Andean-Saharan Ice Age
400	First vascular and seed plants, and insects
365	Tiktaalik moves onto land

360	Four-limbed vertebrates move on land
355	First reptiles
350–260	Karoo Ice Age
320	First amphibians
290	First conifers
250/225	First mammals and dinosaurs
200	Dinosaurs dominate
160	First birds
135	First beaked birds, first flowering plants
65	Large dinosaurs extinct, mammals dominate

All this follows a variation of +A + −A = 0: +A (expansionist self-organising, purposive, ordering of self-replicating cells of organisms with purposive coded messages in DNA in bio-friendly environment) + −A (random, accidental natural selection in harsh environment of rocks, Ice Ages and limited oxygen) = 0 (the One universe).

Darwin's theory of natural selection has been controversial but has developed into neo-Darwinism:[170]

Early biologists believed that species were created by God and had continued without change. Charles Darwin's grandfather Erasmus, a doctor, believed in "organic evolution" (the evolution of the living world) and became the first of five generations of Darwins to become a Fellow of the Royal Society in 1761. He read Buffon, the 18th-century scientist who realised that species were not fixed and unchanging.

Erasmus Darwin came to believe in evolution by variation and "improvement" of species by their own activities. He believed that characteristics acquired by a parent can be transmitted to their offspring. He wrote about the evolution of life from "tiny specks in the primeval sea", which led to the forming of fish, amphibians, reptiles and human beings.[171]...

Neo-Darwinism restates the concept of evolution by natural selection in terms of Mendelian and post-Mendelian genetics. Genetic variation involves chromosome and gene mutations, apparently random assortment of maternal and paternal homologous chromosomes (pairs with the same structure) during crossing over and the recombination of segments of parental homologous chromosomes. Genetic variation is expressed in phenotypes (observable characteristics), some of which are better able to survive than others. Natural selection causes changes in the proportions of particular genes that may lead to new species.[172]

In short, natural selection involves:

- a variation in a population;
- an advantage over others depending on local environment (for example the advantage of one bird over other birds); and
- a variation being passed on to offspring.[173]

Selection is in accordance with local conditions. It starts with a small mutation in DNA that makes for diversity which can be physically inherited. Thus a seed-eating finch encountering no seeds on a new island in the Galapagos, but plenty of cacti, adapts and develops a beak that can eat cactus. Local conditions were determined by the weather, which affects population growth.

The first primates appeared 60 million years ago, and the first apes 35 million years ago. Apes diverged from other primates 15 million years ago. Orangutans diverged from other apes 14 million years ago. Gorillas diverged from chimpanzees, bonobos and ancestors of humans 8–6 million years ago.

Some biologists claim that our human ancestors appeared 7–5 million years ago. Ancestors of humans diverged from chimpanzees and bonobos 5.3–5 million years ago. The first hominids appeared 4.43 million years ago, early *omo*s 2.3–2 million years ago. Humans are primates and share common ancestors with chimpanzees but do not descend from monkeys, which are another primate. Our kind of humans are 200,000 years old: ancestors of modern Cro-Magnon *Homo sapiens*. Humans are still evolving.

The New Philosophy of Universalism has a section 'Apes and Humans', which details the descent of modern humans, *Homo sapiens sapiens*:[174]

In linking all species including humans to the first cell the new philosophy of Universalism affirms the oneness of all living creatures and is open to, and prepared to be convinced by, the most controversial aspect of Darwin's theory of evolution, humans' descent from apes, though in a context of bio-friendly order rather than one of random accident.

A more immediate common ancestor for apes and humans than the tiktaalik may have been the ape/monkey *Aegyptopithecu* (whose fossils date to 35 million years ago, or mya, and were first found in Egypt). It evolved into the true ape *Proconsul* (whose fossils date from 20 million years ago) and the human-like ape *Ramapithecus* (whose fossils date from 10 million years ago first in India and Pakistan, and later in Kenya and the Middle East, long after Africa joined up with Eurasia and ceased to be an island 16 or 17 million years ago).[175] Nearer our time there were more ape-like hominids:[176]

Sahelanthropus tchadensis	7–6 mya, West Africa, Chad
Orrorin tugenensis	6 mya
Ardipithecus kadabba	5.5–4.4 mya
Ar. Ramidus	over 4.4 mya, Africa

There were 22 intermediate extinct species between apes and, or alongside, man (*Homo sapiens sapiens* being the twenty-third [species, not yet extinct]). The intermediate levels began with *Australopithecus*, a *genus* of fossil hominids, particularly *Australopithecus Afarensis* (a family of three of which left fossilised footprints in volcanic ash in Laetoli 3.6 million years ago, proving the species walked upright on two feet). The first partial skeleton of this hominid, named Lucy, discovered in Afar, Ethopia in 1974, was 3.5 million years old.[177] It is possible that modern *Homo sapiens* descended from her, and that from her evolved *Homo habilis* and *Homo erectus* before *Homo sapiens neanderthalensis* and modern *Homo sapiens* (ourselves) shared a common ancestor about 400,000 years ago.[178]

The full list is as follows:[179]

Australopithecus	4.2–2 mya, fossils beyond 4 mya
1. *Au anamensis*	4.2–3.9 mya, North Kenya
2. *Au afarensis*	3.8–3 mya, East Africa
3. *Au africanus*	3 mya, found in 1924
4. *Au bahrelghazali*	3.5–3 mya, Chad

5. *Au garhi*	2.5 mya, Ethopia
Kenyanthropus	
6. *Kenyanthropus platyops*	3–2.7 mya, Kenya
Paranthropus	3–1.2 mya
7. *P. aethiopicus*	2.5–1.2 mya, East and South Africa
8. *P. boisei*	2.3–1.2 mya, East Africa
9. *P. robustus*	2.0–1.0 mya, South Africa
Species of the *genus Homo*	
10. *Homo habilis*	2.5–1.5 mya, Olduvai, Tanzania/Africa
11. *Homo rudolfensis*	1.9–1.7 mya, Kenya
12. *Homo georgicus*	1.8–1.6 mya, Georgia

One theory holds that archaic ancestors of *Homo sapiens* belonging to the genus *Homo* arose simultaneously in Africa, Europe and Asia some 2 million years ago.[180]

13. *Homo ergaster*	1.9–1.4 mya, East and South Africa
14. *Homo erectus*	2–0.03 mya, Africa, Eurasia, Java, China
15. *Homo cepranensis*	0.8? mya, Italy
16. *Homo antecessor*	previously thought to be 0.8–0.35 mya, now 1.2–1.1 mya, Spain, England
17. *Homo heidelbergensis*	0.6–0.25 mya, Europe, Africa, China
18. *Homo sapiens neanderthalensis* (chronologically descended from *Homo erectus*)	0.35–0.03 mya, Europe, West Asia
19. *Homo rhodesiensis*	0.3–0.12 mya, Zambia

Another theory holds that archaic ancestors of Cro-Magnon *Homo sapiens* arose in Africa about 0.4 million years ago, and certainly from 0.25 or 0.2 million years ago; and that modern *Homo sapiens* emerged from Africa 0.2 million years ago (200,000 years ago) and replaced all other species of *Homo*.[181]

20. *Homo sapiens,* ancestors 0.4/0.25 or 0.2–0.01 mya, Africa
of *Cro-Magnon man*
(The bones at
Cro-Magnon, France,
date to 35,000–10,000BC.)

21. *Homo sapiens idaltu* 0.16–0.15 mya, Ethiopia, France

22. *Homo sapiens sapiens* 0.1/0.07/0.05 mya, Africa/ worldwide
(i.e. modern *Homo
sapiens*)

23. *Homo floresiensis* (This 0.10–0.012 mya, Indonesia
may in fact be a dwarf
version of *Homo sapiens*
rather than a new species)

Humans began with the *genus Homo* (a name chosen by the 18th-century Swedish botanist Carolus Linnaeus for his classification scheme) some 2 million years ago, but studies of human evolution include earlier hominins (a specialist term used by palaeontologists to denote members of the subfamily or tribe of *Homininae* or "humans", i.e. chimpanzees and humans). Within the genus *Homo* of archaic and modern humans there are species, such as *Homo erectus* and *Homo sapiens*.[182]

According to geneticists, modern humans are hybrids created by interbreeding between early hominids (a general term used by archaeologists for the family of ape-men and humans) and chimpanzees/ bonobos, which began 6.3 million years ago.[183] Hominids and chimpanzees/ bonobos are thought to have diverged from gorillas between 6 and 8 million years ago, perhaps 7.5 million years ago (though the recent discovery of *Chororapithecus abyssinicus* suggests the split might have been 10–11 million years ago). Hominids are thought to have diverged from chimpanzees and bonobos about 5.5–5.3 million years ago, before which they had a common ancestor.[184] (See diagram below.) For a million years hominids acquired chromosomes from chimpanzees/bonobos until hominids finally broke free from them 5.3 million years ago.[185] (Chimpanzees and bonobos are our closest living relatives and if they became extinct then gorillas would become our closest living relatives.)[186]

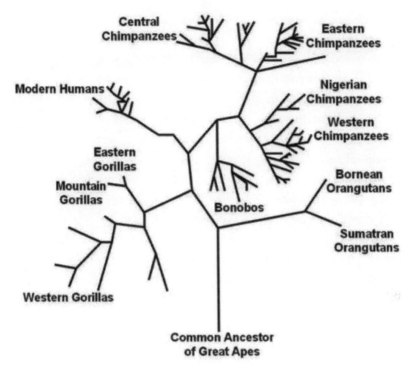

Divergence of gorillas, chimpanzees and bonobos and modern humans.

There is uncertainty regarding the line of descent:[187]

A. afarensis is widely recognised as the ancestor of A. africanus, but A. africanus is no longer recognised as the ancestral hominid for modern Homo sapiens. The descent from Australopithecus to Homo is now believed to be more indirect.[188] The diagram on p.104 allows for alternative evolutionary pathways for the descent.

There are two competing theories as to how modern humans spread across the globe. One, the "multi-regional" idea, holds that modern humans appeared simultaneously in Africa, Europe and Asia, descended from Homo erectus who left Africa some 2 million years ago. This theory sees Homo erectus as evolving and producing modern Homo sapiens in Africa.[189] Recent research suggests that Homo ergaster, who lived between 1.9 and 1.4 million years ago, rather than Homo erectus is the link between apes and humans.[190] Evidence for this came from the discovery of the bones of over 30 people 780,000 years old at

Gran Dolina on Atapuerca Hill in Spain in 1994.[191] On this view, *Homo ergaster* passed through *Homo heidelbergensis* and then branched into *Homo erectus* in China and S-E. Asia 1 million years ago, and into the ancestors of *Homo neanderthalensis* 0.5 million years ago and of modern *Homo sapiens* 0.4 million years ago.[192] (See map on p.107.)

We have seen that a species is defined by breeding and breeds within itself, and that modern *Homo sapiens*, belonging to one species, mates within its species but not outside it. Books say that the predecessors of modern *Homo sapiens*, Neanderthal and Cro-Magnon man, did not interbreed with modern *Homo sapiens* as they were separate species. However, they had a common ancestor, archaic *Homo sapiens*,[193] and there is some evidence that the two siblings did interbreed with modern *Homo sapiens*.[194] When they died out, only modern *Homo sapiens* remained. By 27,000 years ago only one hominid species survived – our own.

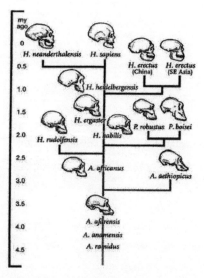

Family tree or bush: descent of modern
humans from 4.5 million years ago.

The other theory holds that modern humans, known as Cro-Magnons, emerged from Africa some 0.2 million years ago and spread across the globe, replacing all other species of *Homo* including *Homo erectus*.[195] It must be said that there is no clear link between the appearance of modern *Homo sapiens* in Africa 200,000 years ago and older human species.[196] Climate change may have been responsible for modern *Homo sapiens'* move out of Africa, our true "Garden of Eden", into Eurasia. We have seen that a Milankovitch 100,000-year advance of mountain glaciers peaked around 200,000 years ago, and this may have impacted on Africa by bringing colder weather, driving modern *Homo sapiens* northwards. There seems to have been a new migration of modern humans, *Homo sapiens sapiens* (our species), from Africa about 100,000 years ago, and certainly between c.70,000 and c.50,000 years ago, perhaps also on account of climate change. This migration replaced all existing hominid species in Europe and Asia. There is fossil evidence in Israel that modern *Homo sapiens* had reached the Near East 100,000 years ago.[197] The "out-of-Africa" idea sees a more recent African origin for modern *Homo sapiens*, and there is more of a consensus for this African siting.

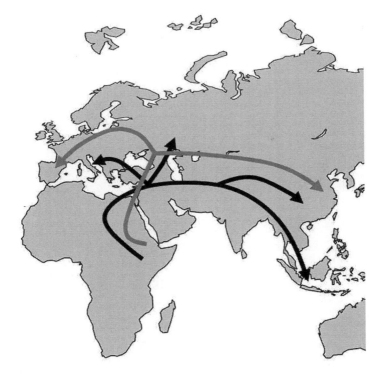

▶ *Homo erectus* expansion beginning about 1.8 million years ago

▶ *Homo heidelbergensis* expansion beginning about 0.6 million years ago

There is a compromise between the multi-regional theory – which proposes that there was no single geographical origin for modern humans and that there were independent transitions from *Homo erectus* (or *Homo ergaster*) to *Homo sapiens* in regional populations – and the "recent African origin" theory that all non-African populations descend from a *Homo sapiens* ancestor who evolved in Africa just over 100,000 years ago. It is the assimilation (or hybridisation) theory that gene flow between the early human populations was not equal. Modern humans could therefore have blended modern characteristics from African populations with local characteristics in archaic Eurasian populations, so that genes are from archaic African *and* non-African populations.[198]

To sum up, the descent of *Homo sapiens* looks as follows:

Thousand years ago	
400–250	First archaic *Homo sapiens*
250	First *Homo neanderthalensis*
200	Glacier advance peaks
	Ancestors of modern Cro-Magnon *Homo sapiens* leave Africa
150–130	Anatomically modern humans (*Homo sapiens sapiens*) arise in Africa
115	Glacier advance peaks
100–50	Modern *Homo sapiens sapiens* leave Africa
40/35–10	Cro-Magnon man in France
29/27	*Homo neanderthalensis* extinct

All Creation came out of one cell, and so there is a oneness between all living things and all humankind, as DNA shows. *The New Philosophy of Universalism* has a section 'DNA and the Oneness of all Living Things', which details the interrelatedness of humankind (the unity of which is a key plank of Universalism):[199]

The Universalist philosopher is open-minded about "the new scientific interpretation" of the origin of *Homo sapiens*, but it has to be said that there is no widespread agreement on the exact links in the descent from ape to human. Darwin wrote in *The Descent of Man*: "In a series of forms graduating insensibly from some ape-like creatures to man as he now exists it would be impossible to fix on any definite point when the term man ought to be used." Today modern *Homo sapiens sapiens* looks to be less than 100,000, perhaps between 70,000 and 50,000, years old despite 13 previous species of *Homo* and a previous 9 intermediate forms – 22 previous species in all. There is therefore a weakness in Darwin's view that humans are descended from apes.

Nevertheless, modern humans have 98.4 per cent of genes in common with chimpanzees – 99.4 per cent if we omit bases that could be changed without affecting amino acid[200] – and evidence of our descent can be found in anatomical traces of our tails and in the different levels of the human brain (some instinctive, some rational). There are moves to include chimpanzees in the same *genus* rather than the same taxonomic family as humans.[201] As humans share at least 98.4 per cent of their genes with chimpanzees the links are likely to be permutations of the intermediate species in the family tree.

Emergence of modern humans from Africa c.50,000 years ago.

The discovery of DNA, the universal genetic code, has confirmed that all humankind within modern *Homo sapiens* is interrelated.[202] (DNA was first discovered in 1869 but its role in genetic inheritance was not demonstrated until 1943 and its structure was not determined until 1953.)[203]...

Human DNA is thought to go back to around 74,000 years ago at the most. Yet, the species of modern *Homo sapiens*, the ancestors of Cro-Magnon man, is about 200,000 years old. So, why does DNA make modern *Homo sapiens sapiens* look younger than he appears from the fossils? Is it because *Homo sapiens sapiens* did not emerge until just over 70,000 years ago, with different DNA from the ancestors of Cro-Magnon man and other *Homo* species, whom he replaced? Perhaps. But the answer may lie in ancient volcanic activity. Some 75,000 years ago, there was a huge eruption of a volcano at Toba in the Indonesian island of Sumatra. This released energy equivalent to about one gigaton of TNT (three thousand times greater than the 1980 eruption of Mount St Helens), and the ash-cloud is thought to have reduced the average global temperature by 5 degrees C for several years and to have intensified the Ice Age. The fall-out and the ensuing climate change are thought to have wiped out most of our species, reducing us to between 1,000 and 10,000 breeding pairs[204] and perhaps damaging our DNA, so that in effect we had to start again. The Toba catastrophe theory is just a theory, but there *is* a dearth of genetic material before 75,000 years ago.

We can all now experience the relatedness of humankind according to DNA. In 1994 Professor Bryan Sykes, a leading world authority on DNA and human evolution, was asked to examine the 5,000-year-old Ice Man, who had been found trapped in glacial ice in northern Italy. By examining the Ice Man's DNA he was able to locate a living relative of the man. His research located a gene which passes undiluted from generation to generation through the maternal line. Having plotted thousands of DNA sequences from all over the world he found that they clustered around 36 groups or clans, of which there are only seven in Europe. Therefore everyone of native European descent can trace their ancestry back to one of seven women. He found that there are 17 paternal clans, five of which are found in Europe. (The Sykes technique traced my maternal line back 10,000 years and my paternal line back 40,000 years.) Thus modern *Homo sapiens* – the 6.7 billion currently alive – can be tracked back to one woman, our "Eve", who lived about 140,000–160,000 years ago as one of only one or two thousand (or as many as ten thousand) individuals whose lines did not survive.[205] This woman is the mother from whom all living human beings are descended, and in terms of her all human beings are literally related as distant cousins many times removed.

We all look back to one "mitochondrial Eve" and "one Y-chromosome Adam", who is thought to have lived some 60,000 years ago. Both, in turn, had a long line behind them. In the "multi-regional" theory, both were at one time thought to have been ultimately descended from *Homo erectus* via the Neanderthals. (We have just seen that *Homo ergaster* may be a better link between apes and humans than *Homo erectus*). Fossil and genetic evidence suggest that anatomically modern humans emerged from Africa around 50,000 years ago or at the most 100,000 years ago (as distinct from the migration of their predecessors that began 200,000 million years ago),[206] and in a few thousand years replaced all other species of humans across the Old World.

It has been suggested that our "Eve" might have been Lucy, but she lived 3.5 million, not 140,000–150,000, years ago. In the "out-of-Africa" theory, both Eve and Lucy were descended from the early humans who may have left Africa 200,000 years ago and certainly did leave 50,000 years ago. We have seen that the compromise assimilation theory sees humans as blending African and non-African genes.

In seeing all human beings of the modern *Homo sapiens* species as coming from 36 verifiable clans with one clan mother and clan father, DNA research dramatically reveals the fundamental oneness of the modern *Homo sapiens* species and humankind.[207] The evidence is even more startling,[208] for it just as dramatically reveals that all living things have DNA and their genetic material has a universal genetic code which can be read. Biochemical analysis suggests a common origin for all living things. Thus the evidence suggests that just as biology came out of one point, *all* biological Creation is ultimately one and interconnected via measurable DNA, and that all species came out of one cell. DNA brings us face to face with the fundamental interconnectedness between every living thing and with the hypothesis or theory we have just considered, that, as Darwin believed, life itself – the first cell – evolved from non-living, inorganic resources of the earth.

There have been doubts about Darwinism:[209]

The Universalist philosopher affirms the oneness of living things and is open to the theory of evolution but dubious that our complexity could have evolved by random accident.

The evidence for evolution has accumulated since Darwin's day, when the adaptive radiation of the Galapagos Islands' birds was the strongest evidence for natural selection. There is now considerable evidence from palaeontology – notably from fossils and rock-dating, but also pollen and spore microfossil analysis. (Spores date back more than 450 million years ago, and the oldest pollen to 330 million years ago.)[210] Although no exact links have been found in the descent from ape to human, comparative biochemistry has now established that a chimpanzee is at least 98.4 per cent related to humans in terms of its DNA. Immunology detects differences in proteins and also establishes that chimpanzees and gorillas (African apes) are 97–98 per cent related to humans. More evidence for evolution has been provided by geographical distribution, comparative anatomy, comparative embryology, artificial selection and systematics (the classification of living things within a phylogenetic scheme).[211]...

There are many doubts about Darwinism.

In biology, Darwin's theory requires complex structures to work immediately if they are to be preserved in the struggle for survival. How are such structures built up without being extinguished? How can a stick insect evolve safely from a structure that does not look like a stick?[212]

In palaeontology, new species develop suddenly. In the Cambrian explosion, biology's Big Bang in Burgess Shale, Canada 530–525 million years ago, at least 19 and perhaps as many as 35 out of 40 phyla first appeared on earth. (Phyla are the highest biological categories in the animal kingdom, for example corals and jellyfish, squids and shellfish, crustaceans, insects and trilobites, sea stars and sea urchins.)[213] There was also a Big Bang of birds and an explosion of flowering plants.[214] Darwin thought new species would develop gradually, not suddenly.

The earth was thought to be no more than 100 million years old in Darwin's time.[215] We have seen that it is now thought to be 4.55 billion years old, and there have been at least 3.5 billion years of the evolutionary history of life on earth.[216] Life forms are vastly more complex than Darwin thought and 100 million years was not enough for such complexity to develop. 4.55 billion years is not enough when the first 3.5 billion produced only bacteria, leaving too little time for such complex organs as the human eye to evolve. Can random fluctuations in accordance with the theory of natural selection produce such complex and organised patterns which seem to be purposive?

Some species of hydra hunt and protect themselves with poisonous guns, tiny stinging cells that fire a poisoned hair. The flatworm generally will not touch a hydra, but occasionally it seeks one out and swallows the poison guns without digesting them and then places the guns on its own body for its own protection and fires them at attackers. When it runs low on poison ammunition it seeks out another hydra and swallows its poison guns.[217] According to Darwinism chance, random natural selection and the instinct to survive account for this purposive and intelligent, co-ordinated behaviour. This sense of purpose seems to be present in even the lowest forms of life, contrary to Darwin's theory.

Darwin's theory of sexual selection ignores the fact that some species (for example, the whiptail lizard whose young are produced from unfertilized eggs) are only females; that some animals change sex to maintain their numbers and that over 300 species indulge in genetically useless homosexual behaviour, contrary to Darwin's theory.[218] If apes, fish and dinosaurs evolved or mutated into humans, mammals and birds, why have no convincing and undisputed intermediate forms been found, either in fossils or bones?

Other doubts include:[219]

1. Species that evolve the least last longest.
2. Antibiotic resistance comes from an orderly, organising 'mechanism' (a reductionist word Universalists dislike) within an organism when evolution and natural selection are supposed to be chance, random, blind.

3. Co-operation is important in evolution, contrary to Darwin's survival of the fittest.

4. Nature has patterns that exceed chance – for example, marsupial mammals occupy the same or similar niches as eutherian mammals. The two diverged in the Cretaceous period and show parallel adaptive radiation.

5. New species appear without a long succession of ancestors whereas Darwinism requires gradual changes from one species to another.

6. Some creatures (such as cockroaches and coelacanths) survive without evolving, even though they mutate.

7. A bird's lung, a rock lobster's eye and a flagellum of bacteria are too complex to be produced by chance.

8. There is little evidence for large changes as a result of natural selection.

9. There is not enough time for random evolution, as we have seen, let alone random evolution that is continually impeded by natural selection.

10. There is no universally accepted descent from apes to humans, as we have seen, and although there are 22 intermediate extinct species, the descent is subject to academic dispute.

Doubters often put forward their doubts in considered good faith while admiring some aspects of Darwin's generalising theory. Such people have thought deeply and would be shocked by the tone adopted by some Christian dissenters, one of whose websites[220] claims: "I believe that Darwin was dishonest and lied to help get his theory accepted." He claims that the Galapagos finches had almost nothing to do with the formulation of Darwin's theory; that Darwin's claims about the role of the finches in evolution were retrospective in the second (1845) edition of his *Journal*; that Darwin only belatedly came to believe that "one species had been taken and modified for different ends"; and that he revised his *Journal* in 1845 to inject into his pre-1839 text what he had pieced together between 1839 and 1845. "Dishonesty" and "lies" are grave charges to be levelled at Darwin and do not do justice to the process of the evolution of the idea of natural selection in Darwin's mind in the 1830s and 1840s, a slow musing familiar to all creative people and involving revisiting and reinterpreting half-formed ideas.

Darwin's theory of natural selection is a hypothesis, just as much of a theory as the origin of life. It is not finally provable because mutations over millennia cannot be replicated in a laboratory.

Neo-Darwinists are reductionists, and there is another way of regarding evolution, the Expansionist way, which is set out in *The New Philosophy of Universalism*'s section 'Expansionism: "Mechanism" of Order':[221]

The Universalist philosopher thus says "Yes" to Darwinism as an imperfect but unbettered description of a process, what happens in evolution, but "No" to Darwinism when it claims that why it happens is always a chance, random accident.

To reductionist cell biologists, biochemists or Darwinists, there is no reason why we are alive beyond accidental combinations of atoms into a cell, and accidental mitoses and DNA copies that have led to mutations and new species. To reductionists the Darwinist process of adaptation, variation and the struggle for survival by accident and chance explain the drive from the first cell to all the species in our universe.

But there is another way of thinking, the Expansionist (or integrationist) way. Whereas reductionism reduces from the very complex to the most simple and from the Whole, which it takes to bits, to its parts, on whose functions it dwells, Expansionism expands from very complex individual cells, individual structures, organisms and species to the whole universe and sees humans as products of the whole universe, the One. Whereas reductionism explores the functions of parts broken off from the Whole and reduced or simplified, Expansionism explores the functions of parts in relation to the Whole, integrates and acts integrationistically,[222] putting back together what has been broken off.

Such thinking is what Coleridge meant by the "esemplastic power of the imagination",[223] "esemplastic" coming from the Greek *"eis en plattein"*, "to make into One". It is the opposite to the faculty of the reason which analyses, makes distinctions and reduces into bits. It puts the bits back together into a whole. It is like finding a shattered Roman vase in potsherds in the ground and sticking it together to restore its original shape.

Such thinking is also Universalist thinking. It has found fragments and attempted to unite them. In science it has found mechanism and vitalism, reductionism and Intelligent Design, and has attempted to put them together. Whereas reductionism works by simplifying – some say, oversimplifying – and leaving out crucial aspects such as order, Expansionist Universalism relates the functions of broken-off parts back to the function of the Whole from which they were fragmented, and through this relation brings newer meaning to the parts.

The question now is how to put all atoms and all cells back together into an Expansionist view of the universe that, besides uniting relativity and quantum theory in cosmology in an underlying unified field, includes both the thinking of reductionism and of Intelligent Design.

What is the Expansionist view of the origin of life? It acknowledges reductionist Darwinism but adds a sense that we are products of the universe, of the 40 "coincidences". It brings to Darwinism a sense of the complex order in the universe that made life possible: a self-organising drive. It proposes a 'mechanism' within cells that has driven evolution by natural selection so that all species have an instinctive place in a vast ecosystem, the whole bio-friendly universe in which, through 40 "coincidences", humankind can sustain itself. It proposes that, as Newton surmised, sunlight has a role in driving the evolution of plants whose seeds have chlorophyll and carotene pigments that trap light energy, and also of insects, fish, birds, reptiles, mammals and humans. It studies the nutrients in leaf and skin. In short, Universalism reforms Darwinism by introducing a 'mechanism' of self-organising order which is unfortunately not biology's pressing need to identify.

Biological holism has tried to establish such a 'mechanism'. Holism is a 1920s idea that originated with Jan Smuts, the South African philosopher who later became prime minister of South Africa from 1939 to 1948. In *Holism and Evolution* (1926) he argued that everywhere we look at Nature we see wholes which are hierarchical, whole fields at ascending levels. Smuts held that the universe is dynamic and produces ever higher-and-higher-level wholes which are ever more organised and complex. Evolution is the driver to ever higher unities, and Expansionism proposes that a principle of order drives evolution. The holistic view of organisms holds that irreducible wholes are the determining factor in biology, and that control-and-regulation 'mechanisms' operating at molecular level make these organism systems cybernetic or automatic-control systems which follow self-determining and self-organising principles.

The concept of holism has become contaminated by vague, New-Age accreted meanings. One of these is the anti-reductionist Gaia hypothesis put forward by the scientist James Lovelock, who took up a suggestion by his friend William Golding, the Nobel prize-winning novelist, that the holistic universe should be named after the Greek goddess Gaia. The Gaia hypothesis regards the earth, the atmosphere, the oceans and organisms as one living process, a single evolutionary entity whose environment regulates itself as a whole and maintains its self-regulation by homeostasis so that the processes of life remain comfortable.

That the earth is an equilibrium regulating itself by homeostasis is an idea that is being taken increasingly seriously. The idea that the earth is alive as a single entity is as old as the human race and is in tune with Wordsworth's view of the Lake District's mountains. However, it has not convinced scientists, and including a goddess in a holistic label has not helped the scientific acceptance of the holistic view.

The holistic principle is under attack by rationalist reductionists[224] who claim that holists assert that the whole is more than the arithmetical sum of its interconnected (or associated) parts, and that it is a falsity to abstract the whole from the parts and then reify this abstraction, implying that the whole possesses a kind of existence over and above that of its parts in their actual state of interconnectedness. There is also a circularity as it explains the parts in relation to the whole, which is explained in terms of the parts.

Rational Universalists can counter this. It is known that plants have sensory apparatuses as they grope for water and move in the direction of the sun, but can one see the atmosphere, oceans and mountains as being sensory and self-regulatory? Clearly a principle present within atmosphere, oceans and mountains is doing the regulating, and I shall have more to say on this in chapter 9. Without this principle within the Whole, holism is essentially an intuitional insight, a perception. In the last analysis, Gaia's form of holism without a principle within the Whole remains a theory.

To Expansionism, the survival of the fittest plants and animals is not controlled solely by external forces of natural selection but by internal forces as well, which may be triggered by sunlight. Mutations are driven by an organism's response to the environment. Thus plants set apart from each other produce more seeds. Snails grow thicker shells if crabs are in the vicinity. They select thicker shells before the crabs appear. In this they display intelligence and are ahead of natural selection. Barnacles grow thicker shells when predatory snails are about. Survival is not a matter of the weeding out of thin-shelled forms by natural selection. Changes in the phenotype ("observable characteristics") are purposeful rather than a random mutation. Evolution takes place at the level of an organism rather than the population. Plants, snails and barnacles are evidence of a Non-Random Evolutionary Hypothesis (NREH).[225] This hypothesis states that the capacity of organisms to adapt to a new environment is built into their adaptive genes. As the capacity to adapt is within each organism, each adaptation can be made purposefully, raising notions of meaning and design. Cell biologists speak of "molecular vitalism", which evokes Goethe's view of plants.

I return to the idea that the drive for evolution may be a matter of inner will, a persistent drive for self-improvement, an inner inspiration to self-betterment which leads to self-transcendence. Evolution may be a thrust to realise better possibilities in relation to the environment, a persistent drive in the imagination to fulfil an image of the organism's next stage of development. This may be contained in coded instructions in the cells via DNA, and suggest the workings of the order principle – and also of Expansionism.

Universalism, seeking to reconcile conflicting views, would prefer to steer clear of the term "holism" and start afresh with "Expansionism". Expansionism covers the ground of biological holism and order's purposive drive to ever more complex hierarchical wholes. In this respect it looks back to the philosopher Whitehead's "philosophy of organism" in *Process and Reality*, which was delivered as lectures in 1927–1928 and published in 1929. Whitehead, lecturing a year after Smuts' book, held that there is a definite property of wholeness that enables organisms to develop. Universalism, taking the baton from biological holism and its philosophical cue from Whitehead, identifies a biological 'mechanism' within the neo-Darwinist model which spearheads an organism's teleological drive at molecular level to greater and greater wholes. In 1953 Crick and Watson elucidated the structure of the DNA molecule and found that strings of nucleotides, precisely sequenced chemicals in DNA, store and transmit assembly instructions for building protein molecules the cell needs to survive. The human genome of one human being consists of six billion "letters" of DNA[226] within 10^{360} possible strand combinations within 10^{23} bits of information[227] and, given the complexity of one individual's genetic make-up – the DNA recipe book that he or she inherited from his or her parents – it is quite possible that among the six billion letters there are coded instructions for the organism to implement the principle of order and to be purposive in driving towards more and more complex hierarchical wholes. The instructions are in four-character digital code-like letters in a written language or symbols in a computer code. Richard Dawkins, the British ethologist (or behavioural zoologist) has pointed out, "The machine code of the genes is uncannily computer-like." The letters appear to have been designed and no theory of undirected chemical evolution has explained the design in these letters. There is too much information in the cell to be explained by chance.

Perhaps this 'mechanism' is within the coded information within the complex order of the double helix of DNA which instructs cells to make specific proteins – that perhaps contain purposive drives to more and more complex wholes? Perhaps this 'mechanism' is stimulated by sunlight, which also stimulates chlorophyll in plants through photosynthesis, as Newton believed?

Cosmology has found reasons to suggest that the Big Bang is bio-friendly. The same bio-friendly considerations apply to the emergence of the first cell from the soup, which may have been helped by interstellar bacteria. The bio-friendly universe and first cell suggest a life principle and principle of order. Where do they come from? If they did not just emerge from the soup by accident in a random way, a view I have rejected, then they developed from a predisposition to order with laws that contain within them the drive to develop towards self-improvement in a certain direction. The action of ultraviolet light on the cosmic "soup" may have been an aspect of this drive to order.

We have seen that there is a return to Einstein's cosmological constant that [in the unlikely event of there being no gravitational braking from hypothetical dark matter] avoids a Big Crunch, the eventual collapse of the universe. Universalism links the universal predisposition to order to this constant and to the development into higher consciousness of modern *Homo sapiens* which is a self-organising drive. This higher consciousness has become higher and higher over millions of years and can appreciate Beethoven and Shakespeare. It has been able to create sophisticated religions far beyond the powers of the ape.

The stage modern *Homo sapiens* has reached today does not represent the final stage in evolution. We are an intermediary stage, and beyond us is a far more conscious being than we are, a modern self-improved and self-developed *Homo sapiens* with a far higher consciousness that will be able to open to forces that we cannot open to, just as in relation to us the ape could not appreciate beautiful art or open to the mystic Light. For *Homo sapiens* higher and higher levels of consciousness are ahead.

As regards the human genome of 6 billion letters (see p.117), it was announced in 2022 that the human genome has at last been fully decoded by scientists at the Telomere-to-Telomere (T2T) consortium in Washington, following a quest that began in 1990 and was completed 32 years later in 2022. The 3-billion-letter code (not now 6 billion, as earlier thought) has all the blanks filled in, and one of the newly-deciphered sections may provide clues as to how humans have developed bigger brains than other apes. There is now a full blueprint for what makes humans distinct, and within the coded instructions may be found instructions from the algorithm of Creation to implement the principle of order in many ways, and to be purposive. There are exciting decades ahead as the decoding comes to be fully understood.

All this follows a variation of +A + −A = 0: +A (Expansionist, self-organising ordering in self-replicating cells of organisms with coded messages in DNA in bio-friendly environment, in relation to the One) + −A (random, accidental, reductionist natural selection in harsh environment) = 0 (the oneness of all living things that evolved from the first cell in the One universe).

(12) Ecology: Nature's self-running, self-organising system
Order within a self-organising system is found throughout Nature, a 'mechanism' in the DNA of all species (mammals, insects, trees and plants) that thrusts organisms forward to self-improvement and self-betterment and advances evolution. *The New Philosophy of Universalism* has a section 'Instinct: Inherited Reflex or Transmitted Order?' that describes how order governs instinctive behaviour:[228]

> Universalist philosophy, seeking the 'mechanism' that thrusts evolution forward to self-improvement and self-betterment, needs to take account of the instincts of all the species on which Darwin focused. Their behaviour patterns are amazingly ordered from the moment of their birth, and they go to elaborate lengths to renew their species and keep the system running. All the species are born with instinctive knowledge of how they are to operate. It seems that this instinctive knowledge has been inherited.
>
> From about 1800 Jean Baptiste de Monet de Lamarck, the French biologist, discussed the idea that species were not "fixed" (or "immutable" in Darwin's word, unchangeable). Lamarck's mechanism of change was his "law of use and disuse" in which organs used a lot become well-developed whereas organs falling into disuse become atrophied − a change that he claimed was transmitted to offspring over several generations or even within a single generation as the inheritance of acquired characteristics. Thus the giraffe, faced with too little vegetation at ground level, stretched its neck to feed from trees and after stretching its neck for several generations acquired the characteristic of a long neck which it passed on to its offspring. Some biologists scoff at the concept of neck-stretching being passed on. August Weisman was largely responsible for the rejection of

Larmarckian inheritance of acquired characteristics. He claimed that the gametes (cells able to unite in reproduction) passed from parents to offspring were not changed by the surrounding body cells. The 'mechanism' of organic evolution was then thought to be natural selection, which is followed by neo-Darwinism.[229]

DNA has had a profound effect on how we regard inheritance of characteristics. To reductionists, inheritance of such acquired characteristics seems to be bipolar: it moves between a biological tendency (or substrate) passed into DNA by inheritance, and an environment to which it reacts.[230]

Thus, a young bird that has undergone a variation, such as the Galapagos cactus finch, has a general tendency to look for something prickly. The young bird searching for its food has to be guided by its parents to an actual cactus in the environment, which it soon remembers through its high-order memory system. Particular cultural or behavioural inheritance then comes into play via the persistence of memory.

A young robin is born with a tendency to pull at string or grass. It sees a parent pulling a worm, and learns to apply its tendency to particular worms in the environment. Again, a young cat is born with a tendency to tug string. It sees a parent pull at a mouse's tail, and learns to apply its tendency to particular mice in the environment. In the same way a sea lion's pup is born with a tendency to suck and has to be guided to its mother's nipples in the environment, which it drains as the mother may be absent for five or six days fishing.[231] A similar tendency can be observed in young human babies who mouth a sucking action within minutes of being born, and have to be guided to their mother's nipples in the environment.

Reductionist biologists explain this instinct in terms of skills developed at random by the struggle to survive. To reductionists instinctive behaviour is more often innate (a tendency) than learned, a complex system of conditioned reflex – of stimulus and reflex to survive – learned in the struggle to survive and passed on genetically into DNA by inheritance.

The discovery of the genetic and structural aspects of DNA in 1943 and 1953 has focused anew on the inheritance of acquired characteristics, but with a time span of several generations rather than a single generation. Thus Darwin's cactus-eating finch transmitted the tendency to look for something prickly over several generations. The first cactus finch to transmit this knowledge created a mutation in the species that has continued to be transmitted through DNA.

It should be pointed out that sportsmen cannot transmit their sporting prowess, though they can pass on their physical athleticism through DNA and teach their children or grandchildren to emulate their achievements through cultural and behavioural example. It should be stressed that this is not automatic. Einstein's children did not become geniuses even though they had the tendency to study and were in contact with the cultural and behavioural example of Einstein.

We can see this interpretation if we look at animal behaviour. The observation of animals advanced through the work of Karl von Fritsch (on bees), Konrad Lorenz (on the fixating of jackdaws and greylag geese, which he found to be innate or instinctive, and genetically determined) and Nikolaas Timbergen (on gull chicks pecking a red spot on the side of their parents' bill to make them regurgitate food). They collectively won the Nobel Prize for physiology or medicine in 1973 for establishing ethology (the study of zoological behaviour which is related to ecology, genetics and physiology) as a modern science. Reductionists assert that we must avoid attributing human feelings (anthropomorphism) and planning (teleological outcomes) to animals, whose behaviour is determined by their need to seek out a favourable environment in which to survive. They do this by receiving sense data by using feedback, the 'mechanisms' for response and co-ordination in the nervous system and the effecter organs. Their behaviour is therefore controlled by homeostasis.[232]

Instinctive behaviour can be a reflex response to danger or pain (for example, the earthworm's escape) or orientation (as when woodlice move quickly in dry conditions but slow down in greater humidity). Darwin defined instinctive behaviour as a series of complex reflex actions that are inherited – today we would say "genetically determined" – and therefore subject to natural selection. Nest building, courtship and food selection must be completed in the right order if the animal is to survive, and offspring without parental care have to learn to know how to survive very early on. Ready-made behaviour patterns can be found in animals that have no time for trial-and-error learning, like the new-born turtles.[233]

But there is another view, that instinct conveys the principle of order through a law that is transmitted in the DNA of all organisms (the same mechanism reductionists cite). To a Universalist the universal principle of order through DNA transmits innate tendencies or instinctive behaviour to organisms so they can integrate themselves into the limited ecosystem which orders their lives. The many limited ecosystems make up one gigantic whole ecosystem, within which there is a predisposition to order.

The Universalist explanation for this innate, instinctive behaviour – that it is genetically determined through DNA by the universal principle of order which is driving all members of all species to survive and play their part within an orderly system – can be seen to operate in many species. Thus, a cactus finch is driven by the principle of order within its DNA to look for something prickly and eat the fruit of a cactus, for in doing so it takes its place in the order of the ecosystem of the Whole. A young robin, a kitten and a sea lion's pup are driven by the "order principle" transmitted within their DNA to eat a worm, a mouse and to suck nipples.

There are many examples of Nature's self-running, self-organising system in food chains, food webs and the order in Nature's ecosystems:[234]

Further evidence for an order principle can be found in the food chains which form part of wider food webs. Every creature eats or is eaten in such proportions that by and large the balance of every species' population in relation to that of all other species is preserved in a harmonious way, so that there is enough food to go round for all. In the eating habits which all creatures instinctively know from birth the Universalist philosopher detects a self-running, self-organising, self-perpetuating system of bewildering and amazing complexity.

An ecosystem contains many plants and animals living together in a community of organisms and different species. Ecology is the study of interactions among organisms and between organisms and their environment, and plants are fundamental as they derive energy from sunlight.[235] On land and in water, the two life-sustaining processes are photosynthesis as sunlight combines with water to sustain plants, and respiration – as oxygen, the waste product of photosynthesis – adds to the oxygen in the atmosphere. Animals need nutrients which they eat and digest. These nutrients are dependent on plant nutrition.

In every ecosystem there is a food chain during which matter and energy are transferred from organism to organism as food.

Plants convert solar energy to food by photosynthesis. In a predator chain the plants are then eaten by an animal, which is then eaten by a larger animal. In a parasite chain a small organism eats part of a larger host and is then fed on by smaller parasite organisms. In a saprophytic chain, micro-organisms feed on dead or decayed organic matter. As energy in the form of heat is lost at each step, there are usually only four or five eating. When humans eat cereal grains directly without eating animals that have fed on cereal grains, they take in increased energy.

There are four principal ecosystems featuring ponds, woodlands, lakes and oceans. In each of these, there are herbivores, which eat plants; carnivores, which eat animals; insectivores, which eat insects; decomposers, which eat the surface of mud; and detritivores, which eat dead or decayed organic matter and the detritus below the mud. In the food chains of all ecosystems herbivores eat plants and then are eaten by carnivores. There are some 10^{18} insects on earth on which insectivores can feed. In the many food chains there are producers (green plants, leaves, algae, seaweed) and consumers (fish, birds).[236]

The New Philosophy of Universalism gives details of the four principal ecosystems.

Self-running, self-organising and self-perpetuating are found in co-operation between species and in symbiosis:[237]

The self-running, self-organising, self-perpetuating system operates in a co-operative way as well as in a competitive way. All species regard each other as allies against prey when they are not hunting or being hunted. To the Universalist philosopher the self-running, self-organising, self-perpetuating system appears to advance the interests of the Whole, the One, through the massive display of co-operative behaviour which keeps the species healthy between the hunting culls which keep the populations of species stable, and therefore contributes greatly to the order in Nature. It looks as if maintaining the Whole's checks and balances, which require limits on species' populations, is the important thing, and that preying on individuals within the species is unimportant, a sacrifice to achieve the overall aim of maintaining the self-running, self-organising system harmoniously. The workings of the order principle can therefore be detected in such co-operative behaviour.

The place in the ecosystem occupied by organisms feeding on each other and competing for food is called a niche. To be more exact, a niche is the smallest unit of a habitat an organism occupies, and its role in the community of organisms found in the habitat. An organism's niche is how it feeds and where it lives.[238]

Competition is at its most intense between species occupying identical or similar niches (or habitats). For example, aphids and caterpillars compete for leaves; competing caterpillars intimidate each other; and competing caterpillars and beetles intimidate each other.

All organisms also avoid direct competition and share resources so that food goes round. Species that do not inhabit identical niches may co-exist. Organisms instinctively operate within their ecosystem to utilise its resources efficiently. The organisms know at birth instinctively what they have to do and slot into a system, feeding in the right part that is "allocated" to them and guarding themselves against predators.[239]

The New Philosophy of Universalism gives details of three forms of co-operative relationships – 70 examples of mutualism, 40 examples of parasitism and 25 examples of commensalism, a total of 135 co-operative relationships:[240]

1. Mutualism.[241] In mutualism, two organisms of different species live in an association that provides advantages or benefits to both.

This co-operation between species makes common cause within the ecosystem where there are always predators and prey within a food web. Some mutualisms are defensive (involving defending a plant or herbivore), and others are dispersive (involving the dispersal of seeds).

There are many instances of humans, birds, insects, sea fish and plants associating with another species to the benefit of both.

2. Parasitism.[242] Parasitism is an association in which one organism lives on or in another organism, its host, for much of its life cycle and depends on the host for food. In other words, one organism benefits at the expense of the other, usually without killing it. Some flowering plants lack chlorophyll and depend on a host plant for water and nutrients. Some depend on a host for water alone. Some live in the host's body, some on the outside. The host apparently receives no benefit. There may be a reason for the negative influence parasites have on their host, one that has not been understood until now. Do parasites contribute to the running of a balanced system by keeping the host's population down?

3. Commensalism.[243] Commensalism is an association between two species in which one organism benefits, the commensal, which is usually the smaller. Its benefits can include nutrients, shelter, support, locomotion or transportation (which is called phoresy). The host receives no benefit but is not harmed.

There are also many examples of co-operation between healing plants and human ailments, of which *The New Philosophy of Universalism* gives 70 examples:[244]

Another aspect of the self-running, self-organising, self-perpetuating system is the way Nature grows cures for all ailments, illnesses and diseases. Humans have always known which herbs or plants help their digestion when they are unwell. It could be that plant-eating animals also know instinctively from birth which specific plants to seek out by smell when they are ill. More research needs to be done on this. Nature's provision of growing cures is further evidence of the workings of an order principle which has created an orderly environment for all living species and has anticipated every eventuality and contingency by providing an antidote.

We have seen that the universe is bio-friendly. The medieval monks knew and homeopaths still know of the bio-friendly connection between humans and plants.[245] Nature's system which permits ailments to arise also provides their cure. Nature acts as a primitive pharmacy. It may be eye-opening to list 70 examples[246] of the main diseases and ailments humans suffered from in the Middle Ages before the opening of pharmacies and provision of pills and tablets, and which they still suffer from (by no means an exhaustive list), and the various plants which (according to traditional belief and lore) were, and still are, held to provide natural remedies within a complex order....

It is as if every conceivable ailment a human "ex-ape" suffers from is taken care of by a plant in the natural world. In bygone times, these were intimately known by everyone, and in giving out plants for ailments (rosemary, pansies, fennel, columbine, rue and daisy) mad Ophelia acted out a country lore that was widely shared and a profound part of popular culture. The point here is that the complex orderly system in Nature has arranged for plants to grow which contain balms from which humans can benefit.

In conclusion, the vast ecosystem which includes all local ecosystems, the ecowhole, its food chains and food webs, of which we are part, its symbiotic relationships (mutualism, parasitism, commensalism) and the interacting between plants and humans who benefit from the healing properties of plants all provide countless instances of unrandom, seemingly bio-friendly order in Nature's ecosystems. If these instances illustrate complex order rather than randomness now, could they really have begun in a random way? It is the same question I asked of the regular rising of the sun, moon and stars: the laws of Nature seem so orderly, could they have come into being by random accident? And do our mathematics, which describe the running of the universe so precisely and with such unrandom complexity, describe a random structure?

All this follows a variation of $+A + -A = 0$: $+A$ (Nature's self-organising system that advances through self-improvement and self-betterment) + −A (random, accidental natural selection in harsh environment) = 0 (the oneness of all Nature that evolved from the first cell in the One universe).

Also, in the environment can be found an ordering, self-regulating, self-organising unity which gives the earth its own equilibrium amid the random destructive forces of humans and climate, as *The New Philosophy of Universalism*'s section 'Environmental Universalism' sets out:[247]

As the earth and our environment emerged from one point that preceded the Big Bang, environmental Universalism sees all environmental developments as being aspects of one process and interconnected. Perceiving the earth's environment as one, Universalists affirm a one-world, or to put it in its wider context, a one-universe environmental movement.

An astronaut in space sees our planet as one. We have seen from the 40 bio-friendly conditions that the earth cannot be said to have an exclusively accidental, chance environment. I have already referred to the idea that the earth might have an equilibrium and an environment that regulates itself by homeostasis. Such a process would be an environmental equivalent of the homeostasis that regulates the body and brain.

This equilibrium is hard to detect or corroborate in a time when the ozone layer has been depleted by CFCs, when greenhouse gases have caused global warming, when there has been a rapid growth in human population and when short-term warming conceals long-term cooling. Much of this obfuscation is due to human contamination.

The depletion of the ozone layer by CFCs was discovered in 1985,[248] when the British base at Halley in the Antarctic Peninsula discovered that the ozone layer above the Antarctic had a hole in it the size of the United States, Canada and Mexico.[249] This hole has continued to grow. The depletion in the stratosphere has been found to have been caused by artificial chemicals such as chlorofluorocarbons (CFCs) and halons in the northern hemisphere.

CFCs were first marketed by the DuPont corporation in refrigerators in 1929,[250] and they were used as aerosol propellants for hair, paint, insect sprays, air fresheners, foam, packing and insulation after the Second World War. Although forming less than one per cent of the atmosphere,[251] the ozone in the stratosphere filters out and restricts the amount of ultraviolet-B (UV-B) radiation reaching the surface of the earth. High levels of UV-B have been saturating Antarctica and the Southern Ocean in spring and early summer, threatening phytoplankton on which the Antarctic marine ecosystem depends. A band of shorter wavelength UV-C has also appeared.[252]

Just under one quarter of the earth's atmosphere is oxygen.[253] The CFC molecules rose to the stratosphere and dislodged ozone (which is three atoms of oxygen bound together). The atmospheric radiation separates chlorine from a CFC molecule. Tiny ice crystals form in the stratosphere. These crystals provide places where chlorine is separated to destroy ozone. The chlorine captures one of the three oxygen atoms, turning ozone into oxygen. The chlorine-oxygen alliance is swiftly broken when another single oxygen atom takes the oxygen atom and forms another oxygen molecule. The chlorine then looks to destroy another oxygen molecule and this pattern is repeated over and over again.[254]

Under the Montreal Protocol of 1987, an international treaty, the production of CFCs has been eliminated.[255] However, as a result of the backlog of CFCs in the atmosphere about 50 per cent of ozone is missing in the Antarctic each spring whereas only 5 per cent is missing in the Arctic. Scientists have found that the protective ozone over the South Pole will take until 2068 to recover.[256] It is expected that there will be an ozone hole every spring for the next hundred years, and that the global effects of the ozone hole will remain significant in a thousand years' time. The ozone problem in the Antarctic may cause the southern hemisphere to have a new Ice Age – or rather glacial.

That greenhouse gases cause global warming has been suspected since the 1940s. The sun's radiance and radiation warms the earth, and the earth's temperature allows it to radiate away the radiant energy it receives from the sun. About 50 per cent of solar energy – light and heat – reaches the ground, of which 70 per cent is absorbed by rocks, the vegetation on continents and the oceans, warming the earth, while 30 per cent is reflected back into space by bipolar ice.[257]

When water vapour, carbon dioxide, methane and other gases which together make up under 0.1 per cent of the atmosphere are exposed to infra-red radiation, they form greenhouse gases, trapping the earth's radiating-out of part of the sun's energy. About 60 per cent of global warming is caused by the greenhouse effect and 16 per cent by methane. Carbon dioxide, which formed 98 per cent of the early earth when the air temperature was 290°C, only forms 0.038 per cent today when the air temperature is 16°C. The missing carbon dioxide has been pulled out of water during the making of reefs and limestone.[258]

The Intergovernmental Panel on Climate Change (IPCC) has said that greenhouse gases have caused global warming, which is greatest round the north and south polar regions. Since the beginning of the 20th century the average temperature just above the earth's surface has increased by about 0.6°C. According to British data since the 1940s, there has been a sustained temperature increase in the earth's atmosphere of around 2.5°C in the Antarctic Peninsula. The National Climatic Data Centre and NASA (both in the US) and the Climate Research Group of the University of East Anglia (in the UK) concur.[259]

A report for the European Commission on climate change published in 2007 made dire predictions. By 2071 climate change will cause droughts and floods that will kill 90,000 people a year. Droughts in southern Europe will cause fires and reduce soil fertility and crop yields will fall by a fifth. There will be floods in Hungary's Danube basin, and sea levels will rise by up to a metre, causing damage that will cost tens of billions of euros. Oceans will acidify, and fish will migrate northwards.[260]

In Antarctica ice shelves are breaking up and glaciers are cracking, blocks of broken-away ice are falling into the sea and drifting as icebergs. Since 1995 the Larsen Ice Shelf, the Filchner-Ronne Shelf and the Ross Ice Shelf have all lost ice the size of Rhode Island (2002), larger than Delaware (1998) and a third of the size of Switzerland (2000) respectively. Winter ice seems to be contracting in West Antarctica but not in East Antarctica, where small shelves are stable.

Across the world, mountain glaciers are in retreat: in the Alps, Andes, Himalayas and Rocky Mountains; and in New Zealand, East Africa and Greenland. The US Glacier National Park may have no glaciers after 2050. There is later freezing and earlier melting from Alaska to the UK, where spring is coming earlier with flowers blooming and birds breeding earlier. As a result of melting glacier ice, the sea level has risen 4 inches since 1900.[261]

In Antarctica the annual air temperature averages −4.5°C, and in the winter when the sun disappears for six months the temperature can reach −80°C. The water temperature is warmer. About 55 million years ago the surface temperature of the Southern Ocean was 15°C. Now it is a maximum of 3°C, because surface waters have been carried down into deeper waters.[262] About 8 per cent of Antarctic fish only live in Antarctica. Having no blood cells as there is oxygen in sea water, they live in a water temperature of −1.8°C and if it rises above +1.5°C (a low-temperature tolerance of 3 per cent) they die. Coral reefs live at a temperature of 1°C and die if the water rises above 3°C. Melting ice and warmer waters affect krill, the staple food of whales, seals, dolphins, penguins and albatrosses, and there has been a reduction of 6 per cent to 12 per cent in marine stocks. Warmer-water species now invade the warmer cold waters, threatening the habitat of colder-water species and endangering their survival. As ice melts the sea level rises, and mangroves, coral reefs and sea-grass meadows will not survive. If, as has been proposed, global warming increases by 1.5°C to 5.5°C during the 21st century sea levels will rise, South Florida may be inundated and cities near the sea may be flooded, such as New York, New Orleans, Amsterdam, Copenhagen, Tokyo and Buenos Aires. Water vapour will increase and there may be a reduction in water levels of lakes, though increased rain may keep levels stable.[263]

Natural phenomena may have contributed to these atmospheric-based changes. The sun has cycles when small dark sunspots appear on its face, the last two sunspot cycles falling between 1979 and 2001.[264] Radiation from the sun produces chemical isotopes such as carbon-14 and beryllium-10. An estimate can be made of solar variability for the last few thousand years by examining these two in rocks and fossils. Measuring greenhouse gases began in the International Geophysical Year program of 1957–1958. In the 40 years since then the carbon dioxide in the atmosphere has gone up by 20 per cent, and is now up 0.5 per cent each year on what it was in 1750. Russian drilling in Antarctica has found cores that show the concentration of carbon dioxide exceeds that at any time since 420 million years ago. During this time high carbon-dioxide levels coincide with high temperatures, and low carbon-dioxide levels with low temperatures. We can assume there will be high temperatures – and high carbon-dioxide levels – for the rest of the 21st century.[265] Methane has doubled since 1750. These gases will decline as a result of the Montreal Protocol, but not yet. Microscopic particles and droplets known as aerosols caused by fossil fuels are also in the atmosphere but these do reflect some of the earth's sunshine back into space before it reaches the earth's surface. Volcanic eruptions have also spewed volcanic aerosols and ash into the atmosphere.

Despite these natural contributions to the atmosphere, in 2001 the IPCC scientists concluded that the increase in temperature cannot be explained by sunspots and volcanic eruptions alone, and that global warming was mainly due to human activity. The consequences may be that there is more plant growth due to the fertilisation of carbon dioxide, but soil will become more desiccated.[266]

Universalist thinking on global warming is very much in tune with the environmental movement which brings a global perspective to conserving energy, developing alternatives to fossil fuels and preventing greenhouse gases from escaping to the atmosphere. The globalist Kyoto Protocol of 1997 began the reduction of carbon-dioxide emissions.

The rapid growth of the human population is an essential part of the problem, as Universalist thinking realises. The growth of the human population was a flat line from 160,000 years ago to shortly before AD1800, after which it began rising dramatically. The human population reached 1 billion in 1804, 2 billion in 1927 (123 years later), 3 billion in 1960 (33 years later), 4 billion in 1974 (14 years later), 5 billion in 1987 (13 years later) and 6 billion in 1999 (12 years later). It [was] projected to reach 7 billion in 2013 (15 years later), 8 billion in 2028 (15 years later [subsequently revised to 2023]) and 9 billion in 2054 (26 years later).[267]

Man has had a huge impact on his environment, covering green lands with roofs and roads and filling the night sky with electric lights. Man has exceeded the earth's carrying capacity. It has been estimated[268] that humans are consuming 1,000 times more than other animals of our size; are producing 100,000 times more carbon dioxide than other animals of our size; and have 1,000 times the human population than other animals of our size. (We should be 6 million on this comparison and in 2008 we were well over 6 billion.) At the present rate of expansion, 750 years from now humans will cover the entire earth, and in 2,000 years' time the volume of human flesh will equal the volume of the earth.[269] Animals forage in 2 per cent of their area, man forages in 80 per cent and eats meals whose ingredients have come from many countries.

The human population has nearly trebled in my lifetime and resources have not kept up with this increase. There is a problem with water in some regions, for example in Indonesia which is over-populated and where human faeces are fertilising rice fields and polluting Djakarta bay. The human population is the problem behind global warming, with its expectations of an ever higher and higher life style, demand for water and production of effluents.

How should mankind deal with the problem of human population? There has already been an attempt to control it through the *Global 2000 Report to the President*.[270] Accepted by President Carter in 1980 when the human population was just over 4 billion, this project has failed in its attempt to cap and reduce the human population. Now, despite the project's having persuaded China to legislate and confine families to one child per family, the world's population is 6.7 billion and approaching 7 [since updated to 7.91 in October 2021] billion. Solutions that reduce the population back towards the 1980 figure are fundamentally neo-Hitlerian and racist. Many of the countries whose populations have dramatically increased recently are in the Third World, and the *Global 2000 Report*'s commentators referred to these as "useless eaters", meaning people who do not work but eat the earth's food. To put the genie back into the 1980 bottle would require the human population to be reduced by nearly 3 billion, a thought that cannot be countenanced.

Humans will resist attempts to reduce the human population. Many (including the Pope) would say that modern *Homo sapiens* should not be measured biologically by being given a status equivalent to the status of penguins, seals and whales; that modern *Homo sapiens* dominates

Nature as the only creature with self-awareness, self-consciousness, reason, intellect, and a perhaps immortal soul and spirit; and that modern *Homo sapiens* should have a privileged position and not be required to reduce his population. Some are scornful about global warming, seeing it as an excuse to make us buy new cars, televisions and kitchen goods that comply with anti-warming criteria and strengthen demand, which the capitalist system needs to remain healthy.

Many would say that Nature will solve the problem of human population by adjusting from overpopulation to underpopulation – perhaps by nuclear warfare or perhaps through disease. Modern *Homo sapiens* is, as we have observed, one species which can interbreed and is therefore susceptible to a new strain of a virus which could decimate our species, indeed wipe humanity out.

At this point Universalist thinking demands that we should stand back and take a long-term view for long-term cooling is the context of our short-term warming. We have found that within our Ice Age which began 40/33.7 million years ago, there are meteorological, geological and man-made influences which may accelerate or reverse this trend in the short term or in the long term.

From one point of view, there is a prospect of a long-term period of global cooling as greenhouse gases shield the earth from the sun. Within our interglacial period, a lesser, recent "mini"-glacial stage known (in terms of our definition, wrongly) as the "Little Ice Age" began c.1500 and reached its maximum advance c.1750. It receded until 1850 when warming began with the Industrial Revolution and lasted until 1950. Global warming in our time may therefore be to some extent part of a cyclical pattern, a continuation and intensification of the warming of 1850–1950.[271]

Predictions based on Milankovitch's work in palaeoclimatology suggest that, as we have seen, eccentric orbiting is now the most dominant influence, and that this and the axis wobble combined suggest that the ice sheet is in retreat, whereas the tilt of the axis suggests that it is advancing. From 20,000 years ago North America was under ice, and the ice sheet has been in retreat for the last 11,500–10,000 years. The sea level around Europe has risen, and there are predictions that if Greenland's ice melts, the sea will rise 6 metres, which would mean that Florida and Bangladesh would be flooded. Some predictions say this could happen in our lifetime, others that it will not happen until long after our time. If the Antarctic continent were to melt it has been estimated that the sea level would rise 66 metres – which, with the 6 metres from Greenland would make 72 metres in all – but no one is suggesting that this is imminent.[272] There has been a temperature change in Europe, which is now warmer than it was in the Middle Ages. It seems that the Age of Aquarius, the water-pourer in astrology, may indeed be ahead.

However, from this point of view we are still in an Ice Age, and though we must act to prevent global warming to ameliorate the conditions which our generation and our children's and grandchildren's generations will experience, it could be that in as little as a thousand years' time people will lament that what we have done to reduce global warming and to avoid enlarging the hole in the ozone layer will have contributed to the global *cooling* future generations will experience when the ice sheet advances again.

From another point of view, there is a prospect of a short-term period of global warming as our carbon emissions harm the earth's atmosphere and cause the earth to grow warmer, further melting ice sheets and raising sea levels. It is a measure of this global warming in our interglacial period that the Arctic's North-West Passage, which traverses the polar ice in a 3,200-mile waterway connecting the Atlantic and Pacific Oceans across the top of Canada and Alaska and which has defeated virtually all previous attempts to sail it until now, is now navigable. It may be regularly open by 2027, even by 2012. This is because satellite images have shown that the ice sheet has shrunk 38,600 square miles (100,000 sq kms) per year over the past decade, and ten times as much in 2006 alone.[273] Polar bears may soon be threatened with extinction as the ice sheet is breaking up, sometimes leaving them stranded on floating bits of ice. Some are drowning, some are dying of starvation due to climate change.

The same pattern of breaking ice can be detected in the Antarctic. Everywhere in Antarctica the glaciers are retreating. The large glacier, Risting glacier, filled the end of the Drygalski fjord in South Georgia in 1957, showing that in 50 years a huge amount has melted. The entire fjord was once the glacier, and as the glacier melts, so the fjord increases. Losses of ice due to melting in West Antarctica increased 59 per cent from 83 billion tons in 1996 to 132 billion tons in 2006, and 60 billion tons of ice melted from the Antarctic Peninsula in 2006, an increase of 140 per cent since 1996. In November 2007 I saw huge cracks in the glaciers on either side of the disused Argentinean station in Paradise Bay on the Antarctic Peninsula. Within the next hour part of one of the glaciers broke off and left several hundred yards of the bay under pack ice, which our Zodiac crunched through. I was told that calving glaciers often cause waves on the stony beach where I had landed. I encountered similar fracturing near the British Base A in Port Lockroy, so my own observation confirmed that the retreat of glaciers in these two places is typical of Antarctica. It looks as if more parts of glaciers may calve, and it was no surprise that in early 2008 a chunk of the Wilkins Ice Shelf 25.5 by 1.5 miles broke off, triggering the disintegration of 405 square kilometres of the shelf's interior.[274]

Short-term phenomena, then, such as the emission of greenhouse gases may affect the earth's atmosphere and climate, but these have to be seen against the background of long-term cycles, and against the prospect of a new glacial within our continuing Ice Age in perhaps 1,000–1,500 or 16,500–18,000 years' time. As the earth has a continent over the South Pole (Antarctica) and an almost landlocked polar sea over the North Pole (the Arctic Ocean), geologists believe that the earth will continue to endure glacial periods for the foreseeable future. To generations a thousand years from now I say, "I am sorry our generation compounded your problems, but we had to act in the best interests of *our* time to prevent the deaths of many millions from starvation and flooding."

The environmental movement has championed global warming but is caught within this glacial-interglacial cycle. Its findings that greenhouse gases contribute to changes in the atmosphere which can change climate and environmental conditions have to be acted on for our generation and for the coming generations. These findings about gases have to be balanced with the underlying pattern that climate changes may be consequences of the glacial-interglacial cycles of our Ice Age, which can change the earth's atmosphere.

In its approach, the environmental movement looks at human contributions to a deterioration in the earth's atmosphere and seeks to restrict and rectify the damage caused by emissions. In doing this it sees with a global perspective that focuses on the whole planet rather than regions and national boundaries. Whether it realises this or not, the environmental movement has adopted Universalist thinking.

Has the earth a self-regulating principle? If we forget about the label Gaia, is it a living self-organising unity? Are we on one self-regulating planet? If the earth has an equilibrium, its own version of the homeostasis we have found in the body and brain, its own version of negative feedback, it can be expected to adjust the imbalances of the depleted ozone layer, the greenhouse effect, the growth in human population, the damage to our oceans and atmosphere. Universalist thinking expects solutions to these problems.

All this follows a variation of +A + −A = 0: +A (self-regulating, self-ordering earth) + −A (random but consequential effects of climate change) = 0 (the One universe in relation to environment).

(13) Physiology: order within self-regulating body, brain and consciousness

Like Nature, the bodies of many species within it show an order principle within a self-organising and self-regulating system:[275]

We should expect the workings of the order principle in the vast whole ecosystem of Nature to be paralleled in the physiological make-up of each individual of the many species that operate within it, but nothing prepares us for the amazing degree of self-organisation and self-regulation in each organism of the many species within the hierarchical "great chain of being". It would be instructive to pursue this idea in a selection of minerals, vegetables, insects, reptiles, birds and mammals, but space is short and, bearing all the other species in mind, the Universalist philosopher must confine his attention to the body, brain and consciousness of humans as he concludes his "new scientific interpretation of the universe" by revealing the order principle in the physical workings of the human body and mind.

The species that came out of one point contain an astonishing intricacy and complexity. Physiology is "the science of the functions of living organisms and their parts" (*Concise Oxford Dictionary*). There are a number of self-regulating 'mechanisms' – a reductionist word, as I have said, whose mechanistic implications the Universalist does not accept, hence the inverted commas – "in the living organisms" (mammals, birds and reptiles as well as humans) which indicate self-organising systems. They suggest a highly complex degree of order that cannot have been arrived at by random chance.

The body's complexity can be judged from the fact that there are 10^{360} possible combinations of DNA strands;[276] there are 10^{23} bits of information in a human being;[277] and there are 10^{29} particles and (7×10^{27}) atoms in the whole human body. As we saw there are 10^{15} cells in each human body. There are 10^{12} bacteria on the surface of the human body and 10^{15} bacteria inside it.[278]

The body's self-regulating system is known as homeostasis:[279]

The Universalist philosopher, seeking the 'mechanism' that keeps organisms functioning within Nature's vast ecosystem, needs to focus on the body's homeostasis which enables all organisms to take part in Nature's ecosystem, and on the organisation, self-organisation and self-regulation by which this is achieved.

Physiologists divide the organism into seven layers of organisation:[280]

1. Chemical – many chemicals, or atoms combined into molecules, combining to form organelles (parts of cells such as the nucleus, mitochondria);
2. Organelle – many organelles combining to form cells (for example, heart-muscle cells);
3. Cell – many cells combining to form a single type of tissue (for example, heart muscle);
4. Tissue – one or more tissues combining to form an organ (for example, the heart);
5. Organ – one or more organs separately or jointly forming a system (for example, the cardiovascular system);
6. System – all the systems combining to form the organism (for

example, the cardiovascular, renal, respiratory, digestive, nervous, endocrine systems);

7. Organism (for example, a human being)....

Many scientists state that organisms combine to form an eighth organisational layer, one involving interaction with society, which acts as a yet more complex living entity as many organisms combine to form a society (or societies). Organisms interact with the population of their local community, habitat, ecosystem and biosphere.

In mammals, birds and the higher vertebrates, and in particular human organisms, body temperature, blood-sugar level, the oxygen and carbon-dioxide concentrations in blood, water potential and blood pressure are all regulated or held nearly constant within the 'mechanism' known as homeostasis (Greek for "staying similar").[281] Sweating, urine formation, heart rate and breathing rate vary but are controlled and regulated to maintain the variables nearly constant.

An organism self-regulates to maintain an internal environment of constant conditions. The self-regulatory system eliminates any deviation from the normal by negative feedback.[282] In negative feedback, a stimulus or input – a deviation from the normal or a need – is detected, responded to, fed back or put forward to a control centre or regulator, and then to an effector or output which responds to and eliminates the deviation from the normal or need and, having coped with it, then returns the system to the detector.

In a negative feedback loop, then, there is a challenge to homeostasis, a change to the variable being regulated, which is picked up by a sensor that alerts an integrator (often located in the brain), which compares input from the sensor to the normal condition or set point. It instructs an effector (a skeletal muscle) to produce a response that eliminates the problem and return the system to the sensor.

For example, a decrease in body temperature challenges homeostasis. The sensor, a group of temperature-sensitive nerve cells in the skin, detects the stimulus and sends a message to the integrator in the brain, which compares input (less than 37°C) to the normal condition or set point (37°C). The integrator passes the temperature drop on to the effector (the skeletal muscle), which produces a response (shivering, which increases heat production). This response compensates for the change induced by the challenge to homeostasis and the organism is returned to its normal homeostasis, under the care of its sensor.

Alternatively, if blood pressure falls due to blood loss sustained in a fight, blood-pressure arteries detect the drop in pressure and send a signal to the brain that activates pathways leading to the secretion of hormones. The effector (heart, blood vessels) produces a response (raising blood pressure back towards normal) that removes the stimulus to the sensor and returns the system to the sensor's watching care.[283]

Homeostasis is a fundamental characteristic of living organisms. It is the maintenance of the organism's internal environment within narrow, life-conducive tolerances (allowable variations). These tolerances vary only minutely between individuals, and then only rarely.

The ability of the internal environment of organisms to sustain life is affected by temperature, acidity and nutrient, ion, oxygen and carbon dioxide levels. As these directly influence the chemical processes essential for life there are numerous complex 'mechanisms' to keep them at effective levels. Blood pressure, heart rate, absorption, excretion and breathing rate are all controlled, and the thyroid regulates metabolic rate and bone remodelling. A 'mechanism' responsible for governing function in one organ or system may have components that work in other organs or systems: for example, the removal of excess acid from the blood (the circulatory system) by expulsion via the lungs (the respiratory system) under direction of the medulla (the central nervous system). Virtually nothing in physiology occurs in isolation, and the organism is dependent on this interaction between systems for survival.

Homeostasis is not the cause of these perpetually fluctuating changes; rather, it is an umbrella term given to the many independent but interacting processes at the chemical, organelle, cell, tissue, organ, and system levels that, in concert, keep the organism alive and healthy.

Below is a representative collection of 62 examples[284] of homeostasis covering each human organ system. The examples do not by any means provide a complete picture of human physiology, but they provide evidence for the ordered systems of the body's homeostasis. Some examples are of feedback 'mechanisms' in which a stimulus, detecting and responding to a deviation from the normal or a need, is conveyed to a regulator, which then responds. Other examples are simple "features" of physiology, where a simple condition facilitates continued function and health. Both demonstrate that the 'mechanisms' favour the survival of the organism and indicate a high degree of "order" rather than a chance set-up or random "accident". In fact, the feedback 'mechanisms' reveal order too complex and sophisticated to be a random accident.

The narrow tolerances (variations) of these self-regulating, self-organising systems and processes are interdependent. This is demonstrated by the rapid onset of illness that results if these tolerances are transgressed and/or if there are communication errors within or between physiological layers.

The New Philosophy of Universalism gives 62 examples of homeostasis in the body's temperature or blood loss, the embryological system, the pregnancy, birth and post-natal system, and in the immunological, neurological, haematological, cardiovascular, respiratory, gastro-intestinal, renal and reproductive systems.

The brain's self-regulating system has its own homeostasis:[285]

Further evidence for an order principle controlling the self-regulation of bodily homeostasis can be found in the homeostasis of the brain. There are different levels of brain function, some of them autonomic or automatic, others more "conscious", more associated with thought. The Universalist philosopher needs to approach the higher level of activity from the lower, more autonomous brain functions which involve the brain's self-regulation of homeostasis.

All species came out of one point, and there are self-regulating 'mechanisms' in the brains of all mammals, birds, reptiles and humans which suggest order. The complexity of brains' 'mechanisms' can be judged from the fact that the human brain has 10^{11} (100 billion) neurons.[286] There may be 10^{12} neurons and as many non-excitable neuroglial cells.[287]

These 'mechanisms' are connected with: the organisation of the central nervous system (i.e. the brain and spinal cord); the brain's self-regulation of its own health (for example, the production and reabsorption of cerebro-spinal fluid which nourishes the brain, and the protection offered by the blood-brain barrier; the brain's plasticity which makes possible learning, experience and habit and its ability to transfer function (for example, sensation, movement or memory) from an injured area of brain tissue to another area of the brain – after a stroke or head injury; the survival value of the link between memory and emotion; and the brain's control over the immune system, and the effects of cognition and mood on health and disease.[288]

The brain has a physical, structural organisation that is described by the way in which its physical constituents are arranged to form a whole. It also has a functional organisation that determines how specific areas of the brain provide particular functions such as sight, memory, sensation and emotion etc.

The brain's structural organisation has seven levels which correlate to seven levels of organisation within the body:

1. Chemical – many chemicals combining to form organelles;
2. Organelle – many organelles combining to form neurons (nerves, particular types of cell);
3. Neuron – many neurons combining to form a neuron circuit or tissue (such as a reflex, for example the knee jerk reflex);
4. Circuit – many circuits combining to form a structure such as the hypothalamus (hormone control) or the medulla oblongata (control of vital functions such as breathing, heart rate);
5. Structure – many interconnected structures combining to form a function (for example, memory, emotion, cognition);
6. Functions – all the functions combining to form the brain, a system;
7. Brain – not a separate organism from the body, but head of its own system.

As in the case of the body, there is an eighth organisational layer, one involving interaction with other brains to form a yet more complex living entity as many brains combine to form a society (or societies). The brain interacts with the population of the local community, country, region and humankind.

The brain's functional organisation is now recognised to be more complex than was first thought. In the past the brain was regarded as a single "organ". It is now more accurate to view it as a collection of many systems and structures, all of which carry out many different specialised functions.[289] At the same time all are interconnected with, and regulated by, each other to work together towards the optimal functioning of the brain's organism: a specific mammal, bird, reptile or human being. Thus, an area of the brain may have a role pertaining to memory when working with another area, but when working with yet another area may also have a reasoning function or perhaps a sensory one such as hearing.

Our current understanding of the brain's functional organisation is based on this model of the brain as a network of multi-functional areas. We no longer see it as having specific functional areas, each performing a specific task in isolation. This multi-functional approach[290] has important ramifications for the protection of brain function when the brain suffers injury, and for the organism's survival. The complex order of the brain's functioning can best be seen in action, in examples of how the brain regulates its own environment and thereby its health; stores information (facts, memories); transfers function from damaged areas to other areas; codes emotion (particularly fear) into some memories; and influences the immune system and thereby our health. These 'mechanisms' enable the brain and the organism to survive, survival being necessary for the perpetuation and evolution of the species.

The brains of mammals, birds, reptiles and humans seem to work in a broadly similar way, but I am now focusing primarily on the human brain. The human brain regulates its own homeostasis and environment via the feedback 'mechanism' of the "blood/brain barrier".[291]...

We have seen that feedback 'mechanisms' control the body. A feedback 'mechanism' also maintains adequate blood-flow to the brain. Although the brain only comprises around 2 per cent of the body's weight, it uses around 20 per cent of the glucose used by the body as energy.[292] Given the sensitivity of the brain to drops in glucose levels and the fact that oxygen deprivation causes brain death within four minutes, a stringently controlled blood supply to the brain is vital. Nutrient and oxygen requirements of the brain are relatively constant, even when the body is exercising strenuously, and so the brain must also be shielded from the rise in systemic blood-flow and pressure that occurs during exercise.

A feedback 'mechanism' regulates the brain's temperature as well as the body's. Brain function is very sensitive to changes, especially increases in temperature. The temperature-control 'mechanism' that applies to temperature drops in the body applies to raised temperature in the brain. It is significant that the body's thermostat, the hypothalamus, is sited centrally at the base of the brain, where it can respond very quickly to a change of brain temperature....

The higher levels of mental activity – thinking, choice, willpower – have been associated with the cerebral cortex but have not been located in one area. Do different parts of the brain also combine to give rise to these more sophisticated brain functions? We will return to this shortly when we consider consciousness.

The New Philosophy of Universalism gives 15 examples of homeostasis of the brain's blood/brain barrier and CSF (cerebral-spinal fluid), blood flow, temperature, neural circuits (learning through experience and habit), memory, damage and stress response.

The self-regulating homeostasis in body and brain admits photons, which are continuously converted into chemical energy in both body and brain in a process of light energy, as *The New Philosophy of Universalism*'s section 'How Photons Control Body and Brain' describes:[293]

The self-regulating homeostasis within body and brain has a surprising input from outside, which the Universalist philosopher needs to probe. Light energy is continuously being converted into chemical energy within the brain and body, and I can go so far as to say that this light energy makes homeostasis possible.

So far we have examined evidence for the workings of homeostasis in body and brain without considering how its feedback 'mechanisms' could be controlled by the order principle. Seeing the order principle at work in so many instances of homeostasis in body and brain requires some consideration of the entirely new science of nanostructures; the information-bearing potential of photons; and the way light is received in the eyes.

Because the new technology is being applied by scientists we must approach the workings of photons in body and brain via the technological and industrial application of photons. It has only relatively recently been discovered that photons can bear information. Leaving the question of what information existing photons bear to one side for the time being, we must grasp that a new science of photonics is being developed. This is the optical equivalent of electronics.

Instead of using electrons to transmit and process information as in telephone and computer technology, photonics use photons or living units of light. Advanced optics such as laser beams converging inside crystals the size of sugar cubes can form holographic images that work with quantum theory to process huge amounts of information.[294] Advances in fibreoptics have made it possible to transmit data on several wavelengths of light simultaneously along the same fibre through a technology called wavelength division multiplexing. Information can be sent a thousand times faster than current internet technologies can handle. Existing telecommunications networks use classical light to transmit information through optical fibres, and have to boost signals by repeater stations. Rapid progress is being made towards quantum networking.[295]

Fibreoptic lines can transmit a thousand colours at the same time alongside masses of photonic devices such as crystal switches and processors in which light controls light inside transparent crystals. All this is happening at present in laboratories, but as yet not on the internet. Photonics will not replace electronics but will work alongside them to transmit data over long distances, via network lines. Photons have thus made possible a future technology in quantum computing. Using quantum theory in which atoms or electrons can be in two places at the same time, computers based on quantum physics would have quantum bits ("qubits") that exist in both on and off states simultaneously, making it possible to process information much faster than conventional computers. A future internet of quantum computers may self-organise itself into an intelligent network of modes, and resemble a quantum brain.[296]

Optical microchips that can store light for short periods of time before sending it onwards have been constructed by IBM researchers in the US. Light is good at transmitting data at high rates as its signal can be switched on and off quickly, but it is difficult to store as photons interact weakly with each other. Light connections require optical buffers that can prevent packets of data from collecting at switching points. Already, light can be delayed by sending it through a length of optical fibre. A light pulse must pass through 21 centimetres of fibre to be delayed by one nanosecond.[297]

Using photons, electrons and quantum particles to carry information works as follows. The sender encodes the data as the quantum state of a particle, and the recipient measures the particle to infer the original quantum state. The laws of quantum mechanics specify the maximum amount of information that can be extracted. This theoretical maximum for transmitting information via photons and other quantum particles can be approached by choosing how information is encoded, transmitted and decoded.[298]

The amount of information that photons can carry can be increased by distinguishing between photons that have incrementally different amounts of momentum. A single photon with 32 possible orbital angular momentum values could transmit data five times as fast as binary-digit computers. Combining techniques including the polarization of light can multiply the information-carrying capacity of each photon.[299]...

The properties of photons that are currently being applied by scientists to machines can also be applied to the body and brain through the eye. Photons bombard the human retina, which is a filmy bit of tissue barely half a millimetre thick that lines the inside of the eyeball and develops from a pouch of the embryonic forebrain. The retina is therefore considered to be part of the brain.[300]

Each retina possesses about 200 million neurons and contains 120 million rod and 6 million cone photoreceptors.[301] The photoreceptors' rods are for low-light vision, the cones for daylight or bright-coloured vision. Nocturnal animals such as owls are dominated by rods. These sensory neurons respond to light and lead to intricate neural circuits that perform the first stage of image processing.[302] Photons collapse into the retina and are subsequently processed as information at the level of neural membranes. The information reaches the sensory cortical regions where it is converted into patterns of microtubule subunits and specific quantum states.[303]

The photoreceptors lie at the back of the retina against the back of the eyeball. They are in layers. In the second of three cell layers, called the inner nuclear layer, lie 1–4 types of horizontal cells, 11 types of bipolar cells and 22–30 types of amacrine cells. The numbers vary in each species.

The surface layer of the retina contains about 20 types of ganglion cells. Impulses from these ganglion cells travel to the brain via over 1 million optic nerve fibres. Synapses link the photoreceptors with bipolar and horizontal cell dendrites. This region is known as the outer plexiform layer. Where the bipolar and amacrine cells connect to the ganglion cells is known as the inner plexiform layer.[304]

Light rays – photons – must pass through the retina before reaching pigment molecules. This is because the pigment-bearing membranes of the photoreceptors must be in contact with the eye's pigment epithelial layer, which provides the vital molecule retinal, or vitamin A. Retinal becomes attached to the photoreceptors' opsin proteins, where it changes its form in response to photons, or packets of light. Retinal molecules are then recycled back into the pigment epithelium. This tissue behind the retina is very dark because its cells are full of melanin granules. The pigment granules absorb stray photons to prevent them from being reflected back into the photoreceptors, and causing images to blur.[305]

The retina is thus amazingly complex and plays an active part in perception. We do not understand the neural code that the ganglion-cell axons send as trains of spikes into the brain, but we are beginning to understand how ensembles of ganglion cells respond differently to aspects of the visual scene the retina sees. Ganglion cells construct aspects of what is seen, but the greater part of the construction of visual images occurs in the retina while the final perception of sight takes place in the brain.[306]

Much of the information transfer from photons to the retina depends on electrical connections among cells rather than on standard chemical synapses. Thus, the major neural pathway from the rods depends on electrical connections. Some fast-acting signals pass from amacrine cells into ganglion cells at gap junctions. Neuromodulators change the environment of the neuron circuits but act at a distance by diffusion rather than at synapses. A previously unknown ganglion cell type acts as a photoreceptor without input from rods or cones. The cell membrane of this ganglion contains light-reactive molecules known as melanopsins.[307]

All photoreceptors in humans are found in the outer nuclear layer in the retina at the back of each eye, and the bipolar and ganglion cells that transmit information from the photoreceptors to the brain are *in front of* them. Light must travel through the axons and cell bodies of other neurons before reaching the photoreceptors. A region in the centre of the retina, the fovea, which contains only photoreceptors, supplies high visual acuity or sharpness to compensate for the slow reception of photons. Each retina contains a blind spot, where axons from the ganglion cells go back through the retina to the brain. This also compensates for the way photons are received and information is dispersed.[308]...

We have seen that a photon can carry a thousand colours and many bits of information, and so it is reasonable to suppose that one of its thousands – perhaps millions – of bits of information may contain practical instructions for the order principle to photosynthesise plants and to influence the bodies and brains of all creatures via their eyes and visual system: their retinas on which photons collapse and the neural circuits along which they are fed as information to form an image of the world and, in all probability, to stimulate and direct the various feedback 'mechanisms' of the body and brain.

The bodily absorption of photons has been used in the medical measuring of blood flow, in a technique focusing on bio-photon emission. Either a laser Doppler blood flowmeter or an ultrasonic blood flowmeter can be used. A glass fibre is inserted into a blood vessel and irradiated by a laser beam. Blood flow is measured in terms of a variation of the wavelength of a reflected light or of an ultrasonic wave. The point is that blood vessels emit bio-photons, and these are measured in terms of units of time by converting them into an amplified electric signal. This method is used in combating cardiovascular diseases, cancers, strokes, hardening of the arteries and transient ischemic attacks (TIAs, dysfunctional blood circulation in the brain).[309]

As this medical technique suggests, human blood is full of bio-photons which have been received as photons in the retinas of the eyes. I submit that all the many instances of the body's and brain's homeostasis and negative feedback we have considered – each of the 62 examples of the body's homeostasis and the 15 examples of the homeostasis of the brain – can be seen as being serviced by information-bearing photons which continuously bombard the retinas, collapse and are dispersed throughout the brain's neural circuits and the body's blood as bio-photons which bear information containing instructions that direct and maintain the functioning of body and brain.

Much work remains to be done in this new area, in what will surely one day be a heavily researched, data-based and documented new science of photonology and biophotonology, but I submit that there is enough evidence within current optical and medical knowledge to substantiate the claim that photons trigger the directing and nourishment of the physical body and brain. I submit that Newton was right and that the photons of light control our bodies. I further submit that, like DNA, photons contain coded instructions through which the order principle controls our bodies and brains. I submit that the 62 examples of the homeostasis of body and the 15 examples of the homeostasis of brain are evidence for the workings of the self-improving order principle which controls all creatures' bodies and brains – and our consciousness, the high point of the evolution of body and brain.

Consciousness is also a self-regulating system that controls the brain as *The New Philosophy of Universalism*'s section 'Interaction of Brain and Levels of Consciousness' describes:[310]

Having established a link between photons and brains, the Universalist philosopher can now turn to consciousness, which seems self-regulatory in both its lower, autonomic and its higher, more thoughtful aspects and activities. The Universalist philosopher looks to science to determine whether consciousness is an effect of the brain like homeostasis, or whether it controls the brain as homeostasis controls both body and lower brain – and whether consciousness is implicated in controlling homeostasis.

Consciousness is a psychological condition, defined as "a state of perceptual awareness" (*Concise Oxford Dictionary*). It is also an awareness of self, and awareness of being aware.

The history of the study of consciousness has led to no satisfactory outcome. In the early 19th century some scientists, perpetuating Descartes' dualism of mind and body, saw it as a substance, "mental stuff" that differed from material substance. Others saw it as an attribute characterised by sensation and voluntary movement that separated animals and man from the lower forms of life and also separated their normal waking state from sleep, coma or anaesthesia ("unconsciousness"). It was also seen as a relationship or act of the mind towards Nature or objects, which provided a stream of mental sense data. (I once asked Sir John Eccles, a physiologist and Nobel prizewinner for his discovery of the chemical behaviour of nerve cells and a Cartesian dualist, where the mind is when the eye looks at a flower, and he replied, "Between the eye and the flower," a view of perception similar to that in Donne's 'The Exstasie': "Our soules... hung 'twixt her, and mee.")

For a long while, the main method of examining consciousness was introspection, "the perception of what passes in a man's own mind" (John Locke), looking within one's own mind to discover the laws of its operation. However, different trained observers disagreed on their observations and failed to reveal consistent laws. As a result, in the early 20th century mental states – and indeed, the concept of consciousness – were rejected by scientists, who switched to behaviourist psychology as described by John B. Watson. They approached consciousness from the outside, through behaviour....

Brain physiologists thought that human consciousness depended on the function of the brain, and that states of consciousness are correlated to brain functions. Levels of alertness and responsiveness were found to be correlated to patterns of electrical activity in the brain (brainwaves) which could be recorded on an EEG or electroencephalograph. In wide-awake consciousness, the brainwaves are irregular, of low amplitude or voltage. During sleep, the brainwaves are slower and of greater amplitude, coming in bursts. Beta and alpha rhythms are between 30 to 8 cycles per second. Theta and delta are from 8 to 0.5 cycles per second.

The gateway to higher-frequency waves seems to be the meditative state of 4 cycles per second.[311]

All this electrical and behavioural activity comes from part of the brainstem known as the reticular formation. This comprises the ascending reticular system, connections between the networks of each of more than 100 billion brain cells or neurons (10^{11}). There may even be a trillion (10^{12}) neurons, but if each of 10^{11} connects with an average of 3,000 other brain cells there are over 300 trillion connections (3 times 10^{14}). If each connection is capable of 10 levels of activity, there are 3 quadrillion potential brain states per second (1 quadrillion being 10^{15}).[312] Molecules within the brain cells may act as miniature computers within each cell and take the brain's computing power to 10^{27} operations per second.[313] Stimulating the ascending reticular systems with an electrical impulse rouses a sleeping cat to consciousness and activates the waking pattern of its brainwaves.

There are 12 levels of consciousness:[314]

Just as we saw that the brain has seven levels of organisation like the body – eight if society is included – so there are as many levels of consciousness. There are, in fact, at least twelve levels if we include the two unconscious ones at the beginning and admit two higher consciousnesses, one for rational philosophical and scientific thinking and the other for intuition and imagination, all of which between them cope with every known experience:

1. sleep – unconsciousness;
2. dream – dreaming while asleep;
3. drowsiness – the state of being half-awake when ideas can be received from the unconscious immediately after sleep, readiness for sleep, drunken or drugged consciousness (though some drugs temporarily induce higher levels of consciousness);
4. waking sleep – the robotic ordinary state of everyday automatic working consciousness, everyday passive social consciousness, reacting to external events, writing cheques, performing, travelling automatically back from work, tired

and drifting in a daydream, reactive choices, negative emotions (envy, jealousy, hate), attachment to objects, mindlessness while exercising or gardening, relaxing, listening to undemanding music, passive watching of films on TV (a primarily physical state controlled by the sense experience of the rational, social ego or temporal self);

5. awakening consciousness – more alert everyday consciousness, self-assertion, more considered choices, awareness of the world, noticing every detail of what one observes, alertness when active shopping, business deals, more positive emotions, sexual activity, the poet's observational view of the world (a more psychological state of awareness and observation);

6. self-consciousness – awareness of being aware, awareness of the rational, social ego that is doing the seeing, looking out from a centre that is not the ego, a mixture of positive and negative feelings, conscience, self-restraint and self-reproach, feeling anxious and discontented (a meditative, reflective state drawing on memory in a "wise passiveness");

7. objective consciousness – selflessness, losing oneself in concentration, thinking, active working on papers or computer, everyday active social consciousness, active choices, dictating, active reading of books and newspapers, doing research, listening to classical or spiritual music, being absorbed in art, calligraphy or pottery, lecturing, performing, writing, concentrating on the speed of the ball while playing sport, loving, deeply involved sexual activity, being completely absorbed, responding to religious experience, positive feelings, feeling contented and happy, seeing the universe with clarity through a centre that is below one's ego and known to poets as "the soul", and is often accompanied by a feeling of serenity or contentment (an intense soul state);

8. higher (or deeper) rational consciousness – detachment from senses, thinking very deeply, reflecting, discussing issues, debating, profound choices, being sceptical, a more intense

state of consciousness than 7, writing at a philosophical or scientific, rational level (a contemplative state);

9. higher (or deeper) intuitional and imaginative consciousness – detachment from senses, feeling very deeply, being immersed in a creative artistic project, imagining oneself into a character in a play or a novel, being transported into a different world and living a character's detailed experience (an inspirational-imaginative, visionary state);

10. superconscious or subconscious monitoring – an ordering, organising functioning that overnight finds and then identifies things left undone (a superconscious or subconscious editing state);

11. transpersonal consciousness – psychic and paranormal consciousness (telepathy, precognition, prophecy, healing), awareness of memories from the womb, past lives or another world seen in the near-death experience, mediumistic contact with departed spirits, rare glimpses seen through the centre known as the spirit which traditions claim has lived before (a spiritual state); and

12. universal or cosmic (or unity) consciousness – awareness of oneness with the universe and the infinite when time and space cease to exist, the mystic oneness that is perceived through the centre known as universal being and is often accompanied by a feeling of ecstasy, contemplative experience of the metaphysical Light (a mystical, contemplative state approaching the One Reality, traditionally known as the divine, the highest spiritual state and the highest-known state of consciousness).

All these hierarchical levels (in ascending order from numbers 1 to 12) are attested in accounts in literature, philosophy and mysticism. They may not all have been corroborated by science but these states have been reported on by those who have experienced them. They include all the experiences that a "system of general ideas" needs to interpret, including the experiences Whitehead listed.

Levels 8 and 9 could arguably be the other way round, elevating philosophy and science above intuition and imagination. Literary champions of the imagination are sometimes scathing about science's view of consciousness. Blake wrote of Newton (who as we have seen was actually very perceptive regarding the expanding force of light) that he had one level of consciousness: "May God us keep/From Single vision and Newton's sleep!"[315] Coleridge wrote in a letter "Newton was a 'mere Materialist' – Mind in his system is always passive – a lazy looker-on on an external world."[316] Blake and Coleridge were being more than a little unfair to Newton for his thinking about the universe was done in level 8. I could just as easily have designated Newton's level number 9. This would have challenged their belief in the superiority of the Romantic imagination over science. In fact, the consciousness of Blake and Coleridge was different in kind from Newton's consciousness rather than superior, and my levels 8 and 9 could both have been designated 8=.

The monitor of level 10 may be an extension of levels 8 and 9. It combs through recent work and throws up omissions. Only this morning I woke early and, when the images of a dream had receded, my consciousness pushed forward – or my brain threw up – three amendments to the text of a particular chapter I neglected to make last night: a need to change the order of some words, to insert a phrase I thought of yesterday and a detail involving a scientific fact. I duly wrote these into my text. What "I" do is being monitored by a part of my consciousness or brain which seeks out and finds my omissions and imperfections, and literally acts as a monitor. The monitor is an ordering function, and is perhaps the universal principle of order in operation.

There could be a thirteenth level, group consciousness, one involving interaction with society at different levels (State, country, region, globe), losing oneself in crowd emotion at a sporting occasion, but it could be argued that this is not hierarchically higher than level 12 as the social reality of a group is arguably less real than the mystical experience of level 12.

It could be argued that just as the brain controls and regulates the body with the help of photons and interacts with it so that at times the body also controls and regulates the brain, so the brain controls consciousness with the help of photons and interacts with it so that at times consciousness controls and regulates the brain with the help of photons. In this interaction, at times consciousness controls and regulates the brain and organises itself between these 12 levels. We are the quality of our consciousness. Those who spend the greater part of each day and evening in level 4 are completely different people from those who spend the greater part of each day and evening in levels 5–12. Those who spend their lives in level 4 will drift passively, whereas those who spend their lives in levels 5–12 are more likely to have drive and purpose. To a very large extent the quality of our lives is measured by the quality of our consciousness.

The 7th level of consciousness is the consciousness of the soul, and it is linked to the 8th and 9th levels of consciousness, higher rational and deeper intuitional consciousness.

Bodies, brains and consciousnesses show an order principle with self-organising, self-regulating systems:[317]

All bodies, brains and consciousnesses began in one point, the singularity that began the universe with the Big Bang, and later on in one cell, the first cell. The seed for all bodies, brains and consciousness emerged, like bio-photons and DNA, biologically along with the self-organising principle of order. In the course of the 3.8 billion years since its appearance, the first cell has evolved via the ape and self-improvement towards higher (or deeper) consciousness, which has only manifested itself during the last 2 million/200,000 years, and more significantly during the last 50,000 years.

All this follows a variation of $+A + -A = 0$: $+A$ (body's, brain's and consciousness's ordering self-organising system and homeostasis that advances and organises through self-improvement and self-betterment) + $-A$ (random, accidental body, brain and consciousness reacting to harsh environment) = 0 (the oneness of all body, brain and consciousness that evolved from the first cell in the One universe).

(14) Civilisations of history based on a vision of metaphysical Reality
As I showed in *The Fire and the Stones*, part of which was later updated as *The Rise and Fall of Civilizations*, history should be treated as a

whole. Civilisations rise round a metaphysical vision which a mystic has, that turns into a religion as surrounding peoples follow it (as happened after Mohammed's vision of the Fire in a cave), and decline when they turn secular. To enlarge, *The Secret American Destiny* (2018) says of civilisations' origin round a vision of metaphysical Reality and secular decline:[318]

> As we saw in the Prologue my study of history shows that for 5,000 years 25 civilisations have undergone a rise-and-fall pattern. They rise following experiences of the One (Reality, the metaphysical Light or Fire that is beyond the world of physics), which pass into their religion;[319] and they fall when the metaphysical vision ceases to be strong and turns secular. Civilisations have the structure of a curved rainbow which rises and falls as the metaphysical vision turns secular and social.
>
> Within each of the 25 civilisations there is a pair of opposites, a thesis and antithesis, a '+A + –A': the metaphysical and the social. The +A, the civilisation's metaphysical central idea, is strong during its growth; the –A, focus on the secular and social at the expense of the civilisation's original metaphysical vision, is strong during its decline and decay. When a civilisation has run its course it passes into another civilisation as did the Egyptian civilisation and its god Ra, which passed into the new Arab civilisation and worshipped Allah in 642.
>
> All civilisations from their outset are a tension between the '+A + –A' of their metaphysical/religious vision and their secular/social approach. Although during their growth the metaphysical vision is in the ascendancy and during their decline the secular mindset predominates, in every decade and century of a civilisation's life there is an underlying tension between the two, which is reflected in the seven disciplines....
>
> For 5,000 years history has had two conflicting approaches.
>
> Attempts to describe and interpret the events of history by philosophical reflection belong to speculative philosophy of history. Speculative philosophy assumed that history is linear, that it has a direction, order or design and is not a random flux without a pattern. In my work to prepare the ground for my study of civilisations[320] I reviewed the traditional approach of describing world history in terms of metaphysical Reality (as

expressed in religions) and the pattern of all civilisations.

In the religious texts of the ancient cultures there was always an attempt to explain the direction of human events and to interpret them. In the classical world the *Bible* interpreted human events in relation to Providence. St Augustine of Hippo saw history as being influenced by Providence in *City of God*. His approach was metaphysical and theological. Aquinas synthesized Aristotelian philosophy and Christianity, and saw historical events in relation to the eventual significance of history. Bossuet's *Discours sur l'histoire universelle* (*Discourse on Universal History*, 1681) attributed the rise and fall of empires to Providence. He tried to reconcile the existence of God with the existence of evil, and saw history as having a progressive direction. Leibniz raised the question of why a good God permits evil – he called the attempts to answer this question 'theodicy' in 1710 – and he and Bossuet grappled with the involvement of God in evil historical events just as I did in my poetic epic *Overlord*, which in book 5 asks why a good God permitted Auschwitz. The involvement of Providence was not considered deterministic as leaders had free choice which either furthered the Providential purpose or caused a reaction against their own policies – which furthered the Providential purpose.

Alongside this speculative linear tradition was another tradition of history, that it is a study of past events connected with individuals, individual countries, specific empires and social movements. In the classical world Herodotus, Thucydides and Livy narrated historical events and tried to assess their significance in a linear way, but were secular evidence-gatherers, realistic students of human nature who regarded human intelligence and judgment as contributing to cause and effect rather than the will of the gods. Herodotus investigated events, and the Greek word for 'inquiry' ('*historia*') passed into Latin and became our subject of 'history'. He is today considered to have been largely reliable in reporting what he was told, even though that was sometimes fanciful. He narrated the events of the Persian invasion of Greece in the 5th century BC, Thucydides the events of the Peloponnesian War and Livy the expansion of the Roman Empire. Machiavelli, in his *Discourses on Livy* (c.1517) about the expansion of Rome to the end of the Third Samnite War, saw political greatness such as Rome's as appearing in cycles....

I have continued both the metaphysical/Providential and secular/ speculative traditions in my own study of history, *The Fire and the Stones*, Part Two updated as *The Rise and Fall of Civilizations* (1991, 2008). I taught Gibbon, Spengler and Toynbee to postgraduate students in Japan and to Emperor Hirohito's second son from 1965 to 1967, and saw a fourth way, of seeing world history as a 'whole history', a common historical process in different cultures, which I describe as Universalism. As I have said, I see 25 civilisations going through 61 stages. Each civilisation then passes into a younger civilisation. The growth takes place through the importing of a metaphysical vision, as I have shown,[321] and when this vision begins to fail the civilisation turns secular and declines. I see history as both linear and cyclical:[322] cyclical because the Light recurs in different rising and falling civilisations, but linear because it spirals in a linear direction like a spiral staircase towards one world-wide civilisation, the coming World State. My rise- and-fall patterns are cyclical, but the direction of each civilisation's metaphysical vision is linear – progressing towards a World State.

Also, *The New Philosophy of Universalism* has a section 'Historical Universalism' that describes the rise of civilisations round a metaphysical vision and their decline when their citizens cease to believe in the vision and turn secular:[323]

As all history emerged from one point that preceded the Big Bang and then from the first cell and the first human historical event, historical Universalism perceives history as one, an interconnected unity, a whole.

The history of a particular country cannot be studied in isolation from the history of the rest of the world as that would be partial history. History is the combined events that have befallen humankind as a whole from the recorded beginnings to the present. And so, history has a worldwide pattern.

To his great credit, Arnold Toynbee writing in the tradition of Edward Gibbon, who charted the decline of the Roman Empire, and of Oswald Spengler, who wrote of the decline of the West, saw this and wrote of world history as a whole. He was a giant who was before his time, perhaps because he worked for the British Royal Institute of International Affairs, an advocate of world government, and was a leading figure at Chatham House from 1925 to 1955.[324] He saw rising and falling civilisations in a very regular pattern, and attributed their motive force to "challenge and response". In 1954 he completely changed his 1934 scheme of civilisations, having lost confidence in his original scheme. He wrote: "I have been searching for the positive factor which within the last 5,000 years, has shaken part of Mankind... into the 'differentiation of civilization'.... These manoeuvres have ended, one after another, in my drawing a blank."[325]

In my historical work,[326] I have examined the rise-and-fall pattern of 25 civilisations, and have found what I consider to be the law of history: that they go through 61 similar stages. Each begins as a culture in which there is a vision of Reality as the "ever-living Fire" or metaphysical Light, a vision which is received by a mystic (such as Mohammed, who saw the first page of the Koran written in Fire). People gather round the vision, and a religion forms round it. A new civilisation forms as the vision is taken abroad and the people expand. The civilisation grows as the vision is renewed (as in the quotations of Augustine, Hildegard and Dante). So long as the civilisation follows its metaphysical central idea, its growth is strong. A civilisation is healthy so long as the vision of the metaphysical Light is seen and is central.

When the civilisation ceases to believe in its Reality, it turns secular and goes into a long decline which ends in conquest and occupation by foreign powers, or other civilisations. Eventually it passes within a successor civilisation, whose gods replace its own gods. The civilisation ends when it has passed into another civilisation (as the ancient Mesopotamian and Egyptian civilisations passed into the Arab civilisation).

In my study of history, living civilisations can be overlaid on the dead ones that have been through 61 stages. Overlaying reveals where the living civilisations are now. The European civilisation is found to be two-thirds through, about to enter a Union (the European Union). The Byzantine-Russian civilisation has just moved out of this stage (Communism) and is now in federalism. The North-American civilisation is a quarter-through in a globalist stage – the equivalent stage of the Roman Empire when it ruled the world.

In all secular stages of all civilisations, cultures and times, human beings have encountered similar problems. In the early stages of civilisations, metaphysical visions of Being have been numerous (as in the monastic Dark Ages of the European civilisation). In the later stages of civilisations, Being tends not to be seen: the prevailing climate is one of scepticism, secular materialism. In such times, thinkers keep their civilisation alive by reminding their fellow-citizens of the "ever-living Fire" or metaphysical Light that inspired it. They are against the trend of their day, but they ensure that the civilisation's fundamental vision is handed on to the next generation. In the long term, they are therefore fundamental to the health of their civilisation. There is no space here to develop this view, which can be read in other works.[327]

My philosophy of history is Universalist as it shows the vision of Being, or "ever-living Fire" or metaphysical Light, as being central to history. It is the motive force which Toynbee sought in vain. He looked for it within history and did not find it because it is outside history: the Reality perceived by intuitional lone mystics which is later followed by kings, generals and their economists, whose struggles for succession, wars and material prosperity form the basis of most history books. My philosophy of history is based on a movement away from partial history to "whole history". Its approach is both rational (in setting out the pattern in history) and intuitional (in sensing the whole, esemplastically piecing the fragments of civilisations back into a whole "urn"). In short, in my philosophy of history I have charted a historical approach to Being, which I see as central to religions and therefore to civilisations. On this view, history is the consequence of the vision of Being.

Civilisation's intellectual metaphysical vision becomes corrupted and turns secular. This is paralleled by a more debased vision of Utopias which begins revolutions. In revolutions one class or part of a civilisation tries to eliminate another class – as the French bourgeoisie and proletariat tried to eliminate the aristocracy. Revolutions end in massacres (purges and guillotinings).[328]

The rational progress of the Enlightenment, which came to an end, shocked by the barbarity of the French Revolution, was opposed to revelation, preferring reason – which is another way of saying that it was rational and not intuitional. The rational Enlightenment saw history as a rational progress towards perfection, a view neo-Darwinists took up and applied to evolution, seeing history as driven by natural selection and therefore accidental, chance, chaotic and purposeless.

The view that visions of Being cause civilisations to rise and that absence of this vision causes them to decline is anti-Enlightenment. This view sees history as having an underlying order, a rise-and-fall pattern in which can be detected the workings of the universal principle of order and pattern – and perhaps an underlying purpose and meaning: to embody the idea of Being and metaphysical Light in religious buildings such as temples, cathedrals and mosques.

Universalist history, then, focuses on all humankind's history and therefore on all civilisations, their patterns of rise-and-fall and how their stages affect human beings in their lives. It is history in which the vision of the "ever-living Fire" or metaphysical Light is central to all cultures.

Being is seen in all cultures and ages, and historical Universalism focuses on the impact such visions of Being have had on stages in their civilisation's pattern. Each civilisation has local customs, traditions and differences while reflecting a universal global theme, and the whole of history has a richly unified and varied pattern that can reunify the fragmented discipline of history. In view of the coming emphasis on globalism, historical Universalism's time is coming.

The New Philosophy of Universalism also has a section 'Cultural Universalism', which sees a civilisation as a tree full of sap that later turns brittle:[329]

In each civilisation, during the growth phase when its vision of Reality or metaphysical Light is strong, its culture is unified round the central metaphysical idea like branches (to adapt Descartes' image) growing from a central trunk, which is the civilisation's religion – in the European civilisation, Christianity. During this early stage the trunk of the religion is fed with metaphysical sap and its branches – art, sculpture, music, literature and philosophy – all express the metaphysical vision of Being that is found in religion. Its leaves, individual works, express sap. There is unity of culture and "unity of Being".

In the European civilisation, during the Renaissance the philosopher

160

Marsilio Ficino and the artist Sandro Botticelli shared Dante's metaphysical and religious vision of Being. The central idea of our European civilisation and culture, and of all civilisations and cultures, is the metaphysical vision of Reality as the "ever-living Fire" or Light (expressed in European art as the halo) which is beyond the world of the senses but knowable within the universal Whole.

When the metaphysical sap stops, the branches grow brittle and the culture is fractured and fragmented. Art, religion and philosophy cease to be filled with metaphysical sap and, deprived of natural vigour, start falling apart. When the sap fails, the branches turn dry, sere leaves fall and the civilisation and its culture decline....

When civilisations lose contact with the One, with the metaphysical Light, there is cultural decline, and art, religion and philosophy turn secular, deal with surfaces and no longer embody the sap in the civilisation's religious trunk. The culture becomes shallow and unhealthy, and declines.

All this follows a variation of $+A + -A = 0$: $+A$ (history's civilisations based on a vision of the metaphysical Reality, the Light) $+ -A$ (secular accidental history based on historical events) $= 0$ (the One universe with one world history).

(15) Metaphysical and secular approaches in seven disciplines

All seven of the major disciplines I have focused on in my writings have the same '$+A + -A$' pair of opposites: a traditional metaphysical approach which is challenged by, and co-exists with, a secular, social approach. My work *The Secret American Destiny* covers each of these disciplines and opposite approaches:[330]

The growing North-American civilisation's metaphysical outlook is signalled by the confident 'In God we trust', which the US currency proclaims. Although the North-American civilisation is younger and more metaphysical than the more secular European civilisation, in both civilisations both the metaphysical vision and social approach co-exist within the seven disciplines of world culture, as we have just seen.

We can now see that in each of the seven disciplines the original metaphysical approach has been challenged by, and now co-exists with, a secular, social approach. In each of the seven disciplines there is a '+A + –A' or pair of opposites that make for disunity in world culture.

As we look at each of the seven disciplines, our focus is on the last 4,500–5,000 years and each discipline has a strong European contribution. America has inherited each discipline and added to it during the last century or more, often with distinguished and impressive results.

In mysticism, the traditional metaphysical approach involves the Light:[331]

The practical Mystic Way, which Underhill charts, begins with the awakening of the self to consciousness of Reality, and proceeds to purgation (which purifies the self) and detachment from the senses that leads to illumination and visions. The self is enabled ('now it looks upon the sun')[332] and is then plunged into a Dark Night in which the sun of Reality is absent. After recollection, quiet, contemplation, ecstasy and rapture, the self achieves union with the Light and experiences unitive life as a permanent condition.... The traditional mystic has a profound inner experience of Light – as I had on 10 September 1971[333] – in which the soul mirrors the sun.

I set out the inner, 'introvertive' tradition of the Light or Fire in *The Fire and the Stones* (1991), the first part of which was updated as *The Light of Civilization*. In Part Two of *The Light of Civilization* I collect all the mystic experiences of the world's mystics civilisation by civilisation. I state the Light within each of my 25 civilisations[334] and recount: the Indo-European Kurgan Light; the Mesopotamian Light; the Egyptian Light; the Aegean-Greek Light; the Roman Light; the Anatolian Light; the Syrian Light; the Israelite Light; the Celtic Light; the Iranian Light; the European Light; the North-American Light; the Byzantine-Russian Light; the Germanic-Scandinavian Light; the Andean Light; the Meso-American Light; the Arab Light; the African Light; the Indian Light; the South-East Asian Light; the Japanese Light; the Oceanian Light; the Chinese Light; the Tibetan Light; and the Central Asian Light....

Elsewhere[335] I have listed some of those who have known the Light in all cultures:

> Patanjali, Zoroaster, the Buddha, Mahavira, Lao-Tze, Jesus,
> St Paul, St Clement of Alexandria, Plotinus, Mani, Cassian, St
> Augustine, Pope Gregory the Great, Mohammed, Bayazid, Al-
> Hallaj, Omar Khayyam, Suhrawardi, Hafez, Symeon the New
> Theologian, Hildegarde of Bingen, Mechthild of Magdeburg,
> Moses de Léon, Dante, Angela of Foligno, Meister Eckhart,
> Tauler, Suso, Ruysbroeck, Kempis, Rolle, Hilton, Julian of
> Norwich, St Catherine of Siena, St Catherine of Genoa, St
> Gregory Palamas, Padmasambhava, Sankara, *Guru* Nanak, Hui-
> neng, Eisai, Dogen, Michelangelo, St Teresa of Avila, St John
> of the Cross, Boehme, Herbert, Vaughan, Crashaw, Traherne,
> Norris, Law, Cromwell, Marvell, Milton, Bunyan, Fox, Penn,
> Naylor, Mme Acarie, Baker, Pascal, St Francis of Sales, Mme
> Guyon, John Wesley, Blake, Swedenborg, Shelley, Emerson,
> Tennyson, Browning, Arnold, Newman, Mme Blavatsky, Trine,
> Jung and T.S. Eliot.

To this list could be added a host of others who enshrine the best of Western and Eastern culture.

The Light is a metaphysical experience.[336] It is 'beyond' or 'behind' (*'meta'*) the physical world of the five senses. We can see its 'beyondness' if we look at a couple of examples of experiences mystics on this list have had, starting with Augustine's 'eye-witness' view of the Light, c.400:

> I entered (within myself). I saw with the eye of my soul, above
> (or beyond) my mind, the Light Unchangeable.... It shone above
> my mind.... All who know this Light know eternity.[337]

Hildegarde of Bingen (who died in 1179) gives a similar 'eye-witness' account:

> From my infancy up to the present time, I now being more than
> seventy years of age, I have always seen this light, in my spirit
> (or soul, Jung's translation) and not with external eyes.... The
> light which I see is not located, but yet is more brilliant than the
> sun.... I name it 'the cloud of the living light'.... But sometimes
> I behold within this light another light which I name 'the living
> light itself'.[338]

Traditional inner mysticism is the tradition Underhill and I have described of the individual's inner, 'introvertive' quest to experience the metaphysical Reality that orders the universe. Alongside has always been present an alternative outer, 'extrovertive' mysticism that presents an awareness of the unity of the natural, social and outer universe without any inner vision. This is a form of pantheism in which the mystic experience of union is of an outer union with Nature in which stars, a heath, a lake or a pond are seen as part of the One as I experienced in my four experiences of oneness in 1946, 1954, 1959 and 1963.[339] Wordsworth described the oneness behind his Nature mysticism:

> ... A motion and a spirit, that impels
>
> All thinking things, all objects of all thought,
>
> And rolls through all things.[340]

This pantheistic mysticism is of the visible rather than the invisible (or seen shining behind closed eyes), of the seen rather than the unseen.

The secular, social view of the universe includes outer glimpses that all is a oneness in a system recognized by scientific Materialism. In this outer mysticism ultimate Reality is only implicitly present. There is no contemplation or spiritual vision, just a sense that Nature is one.

In mysticism +A (the traditional vision of the metaphysical Light) + −A (secular social love of all humankind) = 0 (the oneness of world mysticism).

In literature, the traditional metaphysical approach involves the quest for the One:[341]

For 4,600 years world literature has displayed two conflicting literary approaches.

In *A New Philosophy of Literature: The Fundamental Theme and Unity of World Literature* I showed that ever since literature appeared in the Mesopotamian civilisation c.4600BC and threw up, over several centuries, successive versions of *The Epic of Gilgamesh* there has been a tradition describing the hero's quest for metaphysical Reality. The literary Gilgamesh (based on a historical king who ruled Uruk in modern Iraq in the 27th century BC) wants immortality and the theme of the questing hero in literature originated in works associated with him....

I cover ten historical periods in ten chapters, and at the end of each
chapter I have a literary summary which shows how these characteristics
of the Universalist metaphysical aspect of the fundamental theme
of world literature are reflected in the literature of each of the ten
periods: the ancient world; the classical world; the Middle Ages;
the Renaissance; the Baroque; the Neoclassical, Romantic, Victorian
and Modernist periods; and the 20th-century anarchy. The literary
summaries track each characteristic of Universalism through each of
these ten periods. We can form a bird's-eye view of the metaphysical
aspect of the fundamental theme of world literature by following 8 of
these characteristics through the ten periods and noting the religious
texts, poets and literary authors in which they are expressed:

> The quest of the purified soul to confront death by journeying
> to the Underworld, and to receive the secret Light of infinite
> Reality – Gilgamesh's quest for immortality, the rituals of the
> Egyptian *Book of the Dead*, the *Avesta*'s 'place of Everlasting
> Light', the quest of the *rsis* (inspired poets), Arjuna, Buddha
> and Mahavira, the quest for the *Tao*, the quest for the Light of
> Yahweh in the Tabernacle; the Greek mysteries, Odysseus's
> visit to the Underworld, Plato's assertion that earthly life is one
> episode of a long journey, Aeneas's visit to the Underworld,
> the rites of Roman religion, and the *New Testament*; *Beowulf*, Sir
> Galahad's quest for the Grail, Dante and Zen; Ficino, More and
> Spenser's *The Fairie Queene*; Donne, Milton and Dryden; Goethe;
> Blake, Coleridge, Wordsworth and Shelley; Tennyson's *Idylls of
> the King*, Arnold's 'The Scholar-Gipsy'; Joyce's *A Portrait of the
> Artist as a Young Man* and Eliot; and Durrell's *Alexandria Quartet*
> and Golding.

Alongside this traditional, metaphysical aspect of the fundamental
theme of world literature from early times co-existed a social aspect:
condemnation of social follies and vices in terms of an implied
universal virtue. We can form a bird's-eye view of this social aspect of
the fundamental theme of world literature by following another of the
Universalist characteristics through the ten periods:

> universal virtue, a standard by which to measure human follies,
> vices, blindness, corruption, hypocrisy, self-love and egotism in

relation to implied universal virtue (when human interaction is considered from a secular perspective, separated from its context of Reality) – the stories of Gilgamesh, Dumuzi/Tammuz, the people in the Egyptian *Book of the Dead*, Mithras, Arjuna, Lao-Tze and the *Old Testament* prophets; Homer's criticism of Achilles, Aeschylus, Sophocles, Euripides and Menander, Plato, Plautus, Terence and Seneca, Horace, and the *New Testament*; Dante, *Everyman* and *Sir Gawayne and the Grene Knight*; Donne, Milton, Corneille, Racine, Molière and Restoration comedy; Pope, Swift, Johnson, Jane Austen and Goethe; Blake, Shelley's *Prometheus Unbound* and Byron's *Don Juan*; Pushkin and English, French and Russian novelists; Ibsen, Shaw and Forster; and Greene, Waugh, Orwell and Solzhenitsyn.

The two aspects of the fundamental theme are in a dialectic and are present throughout the history of world literature. My study shows that the conflicting metaphysical and social aspects – the quest for Reality and immortality and condemnation of social vices in relation to an implied virtue – each dominate some of the ten periods. All works of literature draw on one or other of these two aspects of the fundamental theme of world literature, or combine both. When the civilisation is growing, the quest for Reality predominates; and when it turns secular condemnation of follies and vices are to the fore. Literature that describes the quest is most found in the literature of the ancient world, the Middle Ages, and the Baroque, Romantic and Modernist periods, while literature that condemns social vices is most found in the classical world and in the Renaissance, Neoclassical, Victorian and 20th-century periods.

Also, *The New Philosophy of Universalism* has a section 'Literary Universalism', which reveals that all the genres of literature have a sense of order and approach Being, the timeless One: [342]

As all literature emerged from one point that preceded the Big Bang, from the first cell and from the first instance of human literature, literary Universalism sees all literature, the literature of all countries – as one, an interconnected unity, one supra-national literature. The poetry of each country is part of all nations' poetry, and all nations' poetry is one universal poetry that, at its deepest, reflects the infinite and timeless One.

In our time literature has been reduced to a materialist and social-humanist outlook with few aspirations or great insights. In poems, plays, short stories and novels, intuitional and rational Universalist *literati* express *experiences* of Being intermingled with rational interpretation. Revelations of Being are inevitably occasional. Much of Wordsworth's *Prelude* is about everyday life and social situations rather than about epiphanies of Being in mountain scenery, and there is a considerable amount of rational interpretation. Nevertheless, Universalist writers bring Being back into human life and focus on the existential contact between the individual and Being amid rational narrative, as distinct from rational philosophers' *thinking* about Being. Universalist literature focuses on experiences and revelations of Being in literary works, experiences which give added meaning to lives and suggest that human lives have purpose.

I have said that the intuition esemplastically approaches the One and perceives unity whereas the reason's analysis sifts and makes distinctions, disunites and separates into parts. Universalist literature, including poetry, appeals to the intuitive faculty and reflects the oneness of the universe, but it also appeals to the reason as it interprets experience rationally in a blend of the intuitional and rational, Romanticism and Classicism, and the opposites of sense and spirit in the Baroque Age. Universalist literature is the successor to Vitalism and Existentialism on the intuitional side, and to Empiricism on the rational side as it emphasises the empirical nature of the experience of Reality and of the scientific view of the universe.

The Universalist poet reshapes and restructures the apparently chaotic universe into a structure of order in the act of writing a poem. It could be said that he lets the universal order principle into literature. In ordering the universe, he approaches, or puts himself in readiness for, the One and receives back from it a glimpse or reassurance that the universe is an ordered one, with a purpose and a meaning. The Universalist poet restates the order in the universe in both intuitional (experiential) and rational (interpretative) terms.

Universalist poets who are imbued with this sense of order include Homer, Virgil, Dante, Shakespeare, Marlowe, Wordsworth, Shelley and Eliot. The worlds of Greek and Roman epic are presided over and ordered by the gods, and the world of Dante by God. Shakespeare has a profound sense of order, the disruption of which affects the "great chain of being". Marlowe sees order in terms of Heaven, Wordsworth in terms of the "Wisdom and Spirit of the universe". Shelley's sense of order can be found in 'Adonais' (which is about the death of Keats) – "the One remains, the many change and pass" – and Eliot's sense of order can be found in *The Four Quartets*, where (in 'East Coker') he wrote of "the still point of the turning world". Dostoevsky dealt with the nihilists who attempted to overthrow order, and their defeat reinstates the idea of order. Tolstoy relates war and peace to the fundamental order in society.

The forms or genres of a Universalist man of letters all reveal Being. Universalist poems investigate the universe with precise language and catch intimations of unity, of the presence of the One, Being. They offer sudden revelations of Being and capture Being in the moment. Epic relates the extremities of war and peace to notions of Heaven and Hell. Verse plays link order to the Being behind the divine right of kings, and to awareness of Being. Autobiography, diaries and short stories link the everyday to growing awareness of Being and contact with Being.

All these forms can be used to probe and investigate the universe. Just as miners find one drill is useful for one kind of coalface, and another for another, so different literary forms or genres are appropriate depending on the particular coalface of the universe the writer drills.

In *A New Philosophy of Literature* I set out the fundamental theme of world literature of a pair of opposites, one of which dominates in every generation: the quest for the One, and condemnation of social follies and vices. I anthologised my poems and stories on this theme in my *Selected Poems: Quest for the One* and *Selected Stories: Follies and Vices of the Modern Elizabethan Age*.

In literature, +A (the traditional metaphysical quest for the One, beginning c.2600BC) + –A (secular, social condemnation of follies and vices) = 0 (the oneness of world literature).

In philosophy, the traditional metaphysical approach involves the quest for metaphysical Reality, the One:[343]

For 2,700 years philosophy and the sciences have had two conflicting approaches.

In *The New Philosophy of Universalism: The Infinite and the Law of Order* I showed[344] that ever since the Presocratic Greek philosophers of the 6th and 5th century BC there has been a tradition describing the inner quest for metaphysical Reality. Anaximander of Miletus wrote of 'the eternal, infinite boundless' (*to apeiron*). Xenophanes and Parmenides wrote of space being a plenum, a fullness, and of the One, which Xenophanes called god (or God). Anaximenes wrote of *aither* (ether) and Heracleitus of a moving 'ever-living Fire'. To the Greeks, the *kosmos* was an 'ordered whole', and the universe was ordered and unified.

The early religions spoke of an ordered universe. The Indo-European Kurgan shamanism saw the Underworld, Earth and Sky united in the World Tree, and the early Indian *rta*, harmony in Nature, reflected metaphysical harmony. The *Upanishads* wrote of supreme good as being One. Zoroastrianism sought to 'gain the reign of the One'. The Chinese *Tao* was like an ocean, a sea of moving energy 'infinite and boundless'.

Plato carried on the metaphysical tradition of a hidden Reality. It was continued in Plotinus's One, Augustine's 'Intelligible Light', and Grosseteste's metaphysical 'Uncreated Light'. In the 15th century Christian philosophers such as Ficino turned to Neoplatonism. Rationalism took over the metaphysical tradition....

Alongside this quest for metaphysical Reality co-existed from early times a social, empirical view of reality, a scientific tradition which was begun by Aristotle. His *Metaphysics* was in three parts: ontology (the study of Being and existence), natural theology (the study of the prime mover, his equivalent of God, the existence of the divine) and universal science (the study of first principles and laws). He laid the groundwork for science, 'which studies Being'.[345] His early students spoke of metaphysics as '*ta meta ta phusika*', 'what comes after physics'. Originally metaphysics was on an equal footing with physics and came after it chronologically in Aristotle's works.... John Locke, George Berkeley and David Hume turned away from metaphysics (in its sense of 'the study of Being') and led to Utilitarianism and Pragmatism, logical positivism, linguistic analysis and the anti-metaphysical utterances of the Vienna Circle.

In our time the two conflicting approaches have expressed themselves in philosophy as: Intuitionism versus logical and linguistic analysis.

In philosophy, +A (the traditional intuitive vision of metaphysical Reality, the One) + –A (the rational, social, secular, scientific approach to the universe) = 0 (the oneness of world philosophy).

In history, as we saw in (14), the traditional metaphysical approach takes into account the vision of the Light which is central to all civilisations' religions. I *précis* pp.155–157 as a reminder of what I said:[346]

For 5,000 years history has had two conflicting approaches.

Attempts to describe and interpret the events of history by philosophical reflection belong to speculative philosophy of history. Speculative philosophy assumed that history is linear, that it has a direction, order or design and is not a random flux without a pattern. In my work[347] to prepare the ground for my study of civilisations I reviewed the traditional approach of describing world history in terms of metaphysical Reality (as expressed in religions) and the pattern of all civilisations.

In the religious texts of the ancient cultures there was always an attempt to explain the direction of human events and to interpret them. In the classical world the *Bible* interpreted human events in relation to Providence.... The involvement of Providence was not considered deterministic as leaders had free choice which either furthered the Providential purpose or caused a reaction against their own policies – which furthered the Providential purpose.

Alongside this speculative linear tradition was another tradition of history, that it is a study of past events connected with individuals, individual countries, specific empires and social movements....

I have continued both the metaphysical/Providential and secular/ speculative traditions in my own study of history, *The Fire and the Stones*, Part Two updated as *The Rise and Fall of Civilizations* (1991, 2008). I taught Gibbon, Spengler and Toynbee..., and saw a fourth way, of seeing world history as a 'whole history', a common historical process in different cultures, which I describe as Universalism. I see 25 civilisations going through 61 stages. Each civilisation then passes into a younger civilisation. The growth takes place through the importing of a metaphysical vision, as I have shown,[348] and when this vision begins to fail the civilisation turns secular and declines. I see history as both linear and cyclical:[349] cyclical because the Light recurs in different rising and falling civilisations, but linear because it spirals in a linear direction like a spiral staircase towards one world-wide civilisation, the coming World State. My rise-and-fall patterns are cyclical, but the direction of each civilisation's metaphysical vision is linear – progressing towards a World State.

In history, +A (the traditional metaphysical approach which gives importance to civilisations' religions and their vision of the Light) + –A (the secular, social approach to national events) = 0 (the oneness of world history).

In comparative religion, the traditional metaphysical approach involves the religious common experience of the Fire or Light:[350]

For 5,000 years all religions have had two conflicting approaches.

In the traditional ancient religious texts, from the *Egyptian Book of the Dead* to the *Upanishads* and the Neo-Taoist *The Secret of the Golden Flower*, there was an inner quest for metaphysical Reality, the Light. I have shown (in Part One of *The Fire and the Stones*, 'The Tradition of the Fire', and Part Two of *The Light of Civilization*, 'The Tradition of the Light', and also in *The Universe and the Light*, and *The Rise and Fall of Civilizations*)[351] that all the early religions – the religions of the Mesopotamian, Egyptian, Aegean-Greek, Roman, Anatolian, Syrian, Israelite, Celtic, Iranian, Germanic-Scandinavian, Indian, South-East Asian, Japanese, Chinese and Tibetan civilisations – had a common experience of the Light or Fire, as did (judging from their 'stones', their megalithic ruins and statues) the religions of the Indo-European Kurgan, Oceanian and Central-Asian civilisations. (The other seven civilisations I do not regard as 'ancient'.)

I have shown (in *The Universe and the Light*)[352] the influence of religions as the driving force of the history of civilisations. I have described how the Light or Fire enters and takes hold in new civilisations:

> A contemplative mystic has a vision of the Fire which migrates to a new area and forms a new religion. This becomes associated with the State, and increases the power of a priestly class who performs its rites. Peoples are attracted to the Fire and a political unification takes place around it. The Fire inspires the civilisation's expansion. Foreign invaders create a revival of a past culture, the religion turns worldly and undergoes a Reformation and a new people adopt a heresy as the new orthodoxy. There is another expansion. Eventually the religion declines. This decline is associated with the decolonization. The Fire is now absent, the civilisation enters a conglomerate and is increasingly secularized. Eventually after a period of federalism, the civilisation is occupied and it ends up passing under its successor's religion.[353]

This vision happened in all 25 civilisations, and the one Light or Fire was seen by a contemplative before it passed into each civilisation. The Light or Fire is therefore common to all rising-and-falling civilisations. For how the Fire spread from civilisation to civilisation, see the chart of 25 civilizations and cultures in Appendix 2.

I then described the experience of the Fire, the growth of the religion and the consequent growth and development of the civilisation as a result of its driving force, in 10 religions: the Egyptian, Islamic, Hindu, Chinese, Japanese, Orthodox, Christian, Buddhist, Tibetan Buddhist and Judaistic religions.

I have set out in *The Light of Civilization*[354] 21 civilisations with distinctive religions and four civilisations with related religions (that are early phases or later branches of the 21 distinctive civilisations)....
In their worship within these religions and sects contemplatives all see the common Light.

Alongside comparative religion's emphasis on the inner quest for metaphysical Reality is another approach which focuses on outer observance in the services in their churches, temples and mosques. This approach concentrates on rituals in places of worship and emphasizes social religion. I have shown evidence[355] that this social approach has secularised English hymns after 1889: the Light or Fire has more or less vanished from hymns and been replaced by social, humanist subjects.

I have also set out[356] the times in each civilisation when the Light weakens during religions' and civilisations' times of decline under foreign threat or rule. In North America the only time of decline was during the period before and after the Civil War, c.1854–1896. In Europe the most recent time of decline was c.1880 to c.1991 when the Germans subjected Europe to two world wars and the Russians occupied Eastern Europe until the fall of the Berlin Wall. I believe that the incorporation of Eastern-European nation-states into the EU and the ratification of the Lisbon Treaty in the first decade of the 21st century have ended this European time of decline – and weakening of the Light in Europe.

Also, *The New Philosophy of Universalism* has a section 'Religious Universalism', which shows that the Fire or Light is the common essence of all religions:[357]

As all religion emerged from one point that preceded the Big Bang, from the first cell and from the first appearance of organized religion, religious Universalism affirms that all humankind will eventually be saved from the finite universe of time into infinite timelessness, not just members of one particular religion. Religious Universalism sees the prospect of all religions becoming merged into a one-world religion based on what all religions have in common: the infinite, timeless Being that has been intuitionally glimpsed by mystics of all religions as the "Divine" or inner Light. All human beings are equal in relation to the metaphysical Light that can fill their souls and order Nature's ecosystem.

I have dealt fully with the tradition of this common denominator in all religions elsewhere.[358] The vision of metaphysical Light or Being is found in every generation and culture. We have seen that it is a universal experience received in the universal being or intellect.

This experience can bring together Catholic, Protestant and Orthodox Christianity, Islam, Judaism, Hinduism, Buddhism, Jainism, Sikhism and Taoism and many other religions such as Zoroastrianism, which all share the experience. There can be regional and local variety round this universal experience as in the Greek time when Zeus became identified with the local gods of religions in Asia Minor, and in the Roman time when Jupiter became identified with local gods such as Zeus and Sol who had taken over from Mithras in Asia Minor. In this one-world religion (if it happens) there will be one experience of the "ever-living Fire" or metaphysical Light which all religions have in common and which has been stated slightly differently within each religion, and there will be a merging of God, Allah, Yahweh, Brahman, the Enlightened Buddha, Nirvana, Om Kar, the *Tao* and Ahura Mazda. The words of their prophets will be regional and local variations of the universal God: Christ, Mohammed, Moses, Siva, the Buddha, Mahavira, Guru Nanak, Lao-Tze (or Lao-Tzu).

In Chinese Taoism, as we saw on pp.4–6 +A (*yin*) + –A (*yang*) = 0 (the One *Tao*, the Void).

In comparative religion +A (the traditional vision of metaphysical Reality as the Light which is central to all religions) + –A (secular, social rituals in churches, temples, synagogues and mosques) = 0 (the oneness of world religions).

In international politics and statecraft, the traditional metaphysical approach involves religions' vision of metaphysical Reality and State religious rites to unite the people under the Fire or Light (as in the US's "In God we trust"):[359]

For 2,600 years – and probably for 5,000 years – international politics and statecraft ('the art of conducting affairs of state', *Concise Oxford Dictionary*) have been conducted by two conflicting approaches.

In the early stages of the ancient civilisations when the Fire or Light within their religion was managed by a priestly class who performed State rites, the State was dominated by the Fire or Light, and the vision of metaphysical Reality attracted surrounding peoples and led to political unification and expansion. The growing, expanding civilisation's diplomacy maintained the metaphysical vision enshrined in its religion, and in ancient Egypt, Mesopotamia, Iran, India, China and the other early civilisations the Pharaoh or priest-King or Supreme Ruler derived his authority from the chief god of the new religion. International politics and statecraft, and diplomacy, maintained the *status quo* of the unifying vision of metaphysical Reality, which saw the oneness of surrounding humankind, and perpetuated its local forms....

Alongside the tradition of international politics and statecraft that were controlled by religion's vision of metaphysical Reality and affirmed unified peoples and acknowledged the religion of local groups, now co-existed a new approach to international politics and statecraft that emphasized the secular and social relationships of sovereign nation-states as they competed with each other in continental wars. From the 19th century on nationalism was a force, and swiftly expressed itself through rival imperialisms in Africa and elsewhere....

Nationalism was challenged by a plan for a New World Order. Conceived by self-interested élitist families I have described collectively as 'the Syndicate' – most notably the Rothschilds and Rockefellers – who created the Bilderberg Group, they worked to bring in a world government that would loot the earth's resources....

The Syndicate are motivated by their own business profits, increasing their trillions into quadrillions at the expense of the peoples of the earth and are not interested in perpetuating a vision of metaphysical Reality that embraces the oneness of humankind....

There has long been a yearning for a genuine World State as opposed to one that would enable families of an *élite* to become quadrillionaires....

In *The World Government* I set out a blueprint for a democratic World State,[360] and we shall return to this in due course. The point now is that there is an honourable tradition of respected thinkers calling for a World State that would recognize the oneness of humankind as perceived in the unifying vision of metaphysical Reality. The World State is an ideal that would abolish war, famine, disease and poverty and see off the greed and self-interest of *élites* who care more about their family fortunes than the whole of humankind.

In international politics, +A (the traditional metaphysical approach of uniting a nation's people under State religious rites based on the vision of the Light) + −A (the secular, social approach to governing a nation's people) = 0 (the oneness of world international politics).

In culture, the traditional metaphysical approach involves the vision of metaphysical Reality:[361]

For 5,000 years culture – 'the arts and other manifestations of human intellectual achievement regarded collectively; the customs, civilization, and achievements of a particular time or people' (*Concise Oxford Dictionary*) – has been dominated by two conflicting approaches.

We have seen that the vision of metaphysical Reality entered each civilisation and passed into its religion, that all cultures had the Fire or Light in common. I reproduce a chart from *The Light of Civilization*[362] which shows this process very clearly (see Appendix 2). The Fire originated in the Central Asian civilisation and passed to the Indo-European Kurgan civilisation and after that passed between civilisations as the arrows show, ending in a 'world-wide civilisation' which will be the coming World State. The chart is subtitled 'The Fundamental Unity of World Culture' and demonstrates the underlying cultural unity the Fire or Light makes possible.

In the early civilisations, as we have seen, each civilisation's culture was unified like a tree whose branches are nourished from its religious trunk. Within all the civilisations, including the European civilisation, the arts – the tree's branches – expressed the vision of metaphysical Reality that had passed into its religion. Each civilisation's culture was originally a unity and in the European civilisation until the Renaissance its works of art – in philosophy, painting, music and literature – all expressed the civilisation's central idea, its vision of metaphysical Reality round which the civilisation had grown. This vision can be found in European thought and in the European arts: in the philosophy of Christian Platonism and Aristotelianism which looked back to Heraclitus's 'ever-living' Fire ('Fragment 30'), Plato's Fire or 'universal Light' which causes shadows to flicker on the wall of the cave, and Plotinus's 'authentic Light' which 'comes from the One and is the One'; in paintings such as Jan van Eyck's *Adoration of the*

Lamb, Fra Angelico's angels in *Christ Glorified in the Court of Heaven* and Michelangelo's Sistine Chapel ceiling; in the sacred choral music of Palestrina, Tallis, Byrd and Monteverdi and in the Hallelujah Chorus of Handel's *Messiah*; and in the literature of Dante, and in particular in Dante's sempiternal rose in his *Paradiso*. All expressed the sublime vision of Paradise. During the Renaissance Plato-inspired Ficino and Botticelli shared Dante's vision. Unity of culture continued during the Elizabethan time in spite of the Reformation and during the time of the Metaphysical poets (the stage-30 secession from the Renaissance Humanists by Cromwell's Puritans, who wanted more rather than less metaphysical vision), the time of Milton's 'God is light,... Celestial light' in *Paradise Lost*, bk 3.

Alongside the vision of metaphysical Reality which dominated the rise of each civilisation co-existed a secular, social approach whose artists presented a secular, social perspective in works of art. In the European civilisation the 18th-century Augustans espoused Enlightenment reason and social virtue, and condemned social vices, and the vision of metaphysical Reality took second place. Since the 18th century the metaphysical idea has weakened. This weakening was captured in works of art within philosophy, painting, music and literature: in the Vienna Circle's verification progress; in the French impressionists and Constable and Turner; in program music that tells a social story and evokes images; and in the 18th- and 19th-century novel. Works of art by Metaphysical poets are now outnumbered by works of art with social themes. The rise of the film has encouraged this trend: as a medium, film is better able to describe social events than inner visions. We are living in a time when the inner has given way to the outer in many areas of our lives.

Also in *The New Philosophy of Universalism* there is a section 'Cultural Universalism' that associates the secular outlook in declining cultures with 50 'isms':[363]

The process of secularisation has taken place in European civilisation today. The metaphysical sap of Christianity began to fail with the Renaissance and was drying up at the end of the 17th century. Since that time 50 'isms' or doctrinal movements have arisen, representing secular, philosophical and political traditions that indicate fragmentation, loss of contact with the One and disunity within the declining European civilisation. They are:

humanism

scientific

   revolution/reductionism

mechanism

Rosicrucianism

Rationalism

Empiricism

scepticism

Atomism/Materialism

Enlightenment/deism

Idealism

Realism

liberalism

capitalism

individualism

egoism

atheism

radicalism

utilitarianism

determinism

historicism

nationalism

socialism

Marxism

anarchism/syndicalism

Darwinism

accidentalism

nihilism

communism

conservatism

imperialism

totalitarianism

Nazism

fascism

Stalinism

pragmatism

progressivism

Phenomenology/Existentialism

stoicism

vitalism

intuitionism

modernism/post-modernism

secularism

objectivism

positivism

analytic and linguistic

   philosophy/logical empiricism

ethical relativism

republicanism

hedonism/Epicureanism

structuralism/post-structuralism

holism

These 'isms' or doctrinal movements were unthinkable before the Renaissance because Christendom was unified as the medieval philosophy curriculum demonstrated. The diversity of the 'isms' or doctrines reveals the multiplicity within which humans now live.

The 50 'isms' give a bird's-eye view of the fragmentation and disintegration of the once unified European and (insofar as European civilisation impacted on the world as a major part of Western civilisation) world culture.

In culture, +A (the traditional metaphysical vision of Reality, the Light, which is central to all religions and influenced art in all civilisations) + −A (the secular, social approach to subjects in the arts) = 0 (the oneness of world culture).

All this follows a variation of +A + −A = 0: +A (an original metaphysical approach and vision of Reality, the One) + −A (a secular, social approach) = 0 (the One universe seen through seven disciplines, the oneness of the disciplines and of the universe).

(16) Humankind's opposites
Throughout the different ages between birth and death (themselves opposites) human beings live amid opposites, all of which are reconciled within the unity of the universe, the One. Every pair of opposites can be seen as being reconciled within an underlying unity. There are many examples of such pairs of opposites or antonyms. Below is a list of 488 pairs of common opposites or antonyms in alphabetical order:

abate – increase

able – unable

abortive – successful

above – below

abridge – expand

absent – present

abundant – scarce

accept – decline, refuse

accurate – inaccurate

achieve – fail

add – subtract

adjacent – distant

admire – detest

admit – deny, reject

adore – hate

advance – retreat

advantage – disadvantage

against – for

agree – disagree

alive – dead

all – none, nothing

allow – forbid

ally – enemy

alone – together

always – never

amateur – professional

amuse – bore

ancient – modern

annoy – soothe

answer – question

antonym – synonym

apart – together

apparent – obscure

appear – disappear, vanish

approve – disapprove

arrive – depart, leave

arrogant – humble

artificial – natural

ascend – descend

ask – answer, tell

attack – defend

attic – cellar

attractive – repulsive

awake – asleep

awkward – graceful

back – front

backward – forward

bad – good

bare – covered

beautiful – ugly

before – after

begin – end

beginning – conclusion

behind – in front of

below – above

bent – straight

best – worst

better – worse

big – little, small

birth – death

bitter – sweet

black – white

blame – praise

bless – curse

blunt – sharp

body – soul

bold – meek, timid

boring – interesting

borrow – lend

bottom – top

bound – unbound, free

boundless – limited

boy – girl

brave – cowardly

break – repair

brief – long

bright – dim, dull

brighten – fade

bring – remove

broad – narrow

brother – sister

build – destroy

busy – idle

buy – sell

calm – stormy, windy, troubled

capable – incapable

captive – free

capture – release

careful – careless

cautious – reckless

centre – edge

chalk – cheese

change – *status quo*, remain

cheap – expensive

cheerful – sad, discouraged

child – adult

chilly – warm, hot

clean – dirty

clever – foolish, stupid

clockwise – anti-clockwise

close – open

clumsy – graceful

cold – hot

combine – separate

come – go

comfort – discomfort

complete – incomplete

complex – simple

compliment – insult

compulsory – voluntary

conceal – reveal

continue – interrupt

contract – expand

cool – warm

copy – original

correct – incorrect, wrong

courage – cowardice

courteous – discourteous

crazy – sane

create – destroy

cruel – kind

customer – supplier

damage – improve

dangerous – safe

dark – bright, light

dawn – sunset

day – night

daytime – night-time

dead – alive

decline – accept, increase

decrease – increase

deep – shallow

definite – indefinite

demand – supply

despair – hope

difficult – easy

dim – bright

disappear – appear

discourage – encourage

diseased – healthy

doctor – patient

doubt – trust

down – up

downwards – upwards

dreary – cheerful

drunk – sober

dry – moist, wet

dull – bright, shiny

dusk – dawn

early – late

earth – sky

east – west

easy – hard, difficult

effective – ineffective

elementary – advanced

employer – employee

empty – full

encourage – discourage

end – begin, start

enter – exit

even – odd

evening – morning

evil – good

excited – bored

exhale – inhale

expand – contract, shrink

expensive – cheap

export – import

exterior – interior
external – internal

fact – fiction
fade – brighten
fail – succeed
fake – real
fall – rise
false – true
familiar – unfamiliar
famous – unknown
far – near
fashionable – old-fashioned
fast – slow
fat – thin, skinny
feeble – strong, powerful
few – many
find – lose
firm – flabby
first – last
float – sink
floor – ceiling
fold – unfold
foolish – wise
for – against
fore – aft
forget – remember
forgivable – unforgivable
forgive – blame
fortunate – unfortunate
forward – backward
found – lost
free – captive, restricted
freeze – boil
frequent – seldom
fresh – stale

friend – enemy
frown – smile
full – empty
funny – sad

generous – stingy
gentle – rough
get – give
giant – tiny, small, dwarf
girl – boy
give – receive, take
glad – sad, sorry
gloomy – cheerful
go – stop
good – bad, evil
grant – refuse
great – tiny, small
grow – shrink
guest – host
guilty – innocent

happy – sad
hard – easy, soft
harmful – harmless
harsh – mild
hate – love
haves – have-nots
he – she
healthy – diseased, ill, sick
heaven – hell
heavy – light
help – hinder
here – there
hero – coward
high – low
hill – valley

hinder – help

honest – dishonest

horizontal – vertical

hot – cold

humble – proud

hungry – full

husband – wife

identical – different

ignite – extinguish

ignorant – educated

ill – healthy, well

immense – tiny, small

important – unimportant

in – out

include – exclude

increase – decrease

inferior – superior

inhale – exhale

inner – outer

inside – outside

instructor – pupil

intelligent – unintelligent

intentional – accidental

interesting – uninteresting

interior – exterior

internal – external

join – separate

joy – grief

junior – senior

just – unjust

justice – injustice

knowledge – ignorance

known – unknown

landlord – tenant

large – small

last – first

laugh – cry

lawful – unlawful, illegal

lazy – diligent, industrious

leader – follower

leave – stay

left – right

legal – illegal

lend – borrow

lengthen – shorten

lenient – strict

less – more

light – dark, heavy

like – dislike, hate

likely – unlikely

limited – boundless

little – big

live – die

lock – unlock

long – short

loose – tight

lose – find

loss – win

loud – quiet

love – hate

low – high

loyal – disloyal

mad – happy, sane

major – minor

male – female

man – woman

many – few

marry – divorce

mature – immature

maximum – minimum

melt – freeze

merry – sad

messy – neat

minor – major

minority – majority

miser – spendthrift

misunderstand – understand

more – less

most – least

nadir – zenith

narrow – broad

nasty – nice, pleasant

natural – artificial

near – far, distant

neat – messy, untidy

nephew – niece

never – always

new – old

night – day

night-time – daytime

no – yes

noisy – quiet, silent

none – some

noon/midday – midnight

north – south

notice – ignore, overlook

now – then

obedient – disobedient

odd – even

offer – refuse

old – new, young

on – off

open – closed, shut

opposite – same, similar

optimist – pessimist

optimistic – pessimistic

out – in

outer – inner

over – under

parent – offspring

part – whole

pass – fail

past – present

patient – impatient

peace – war

permanent – temporary

plentiful – scarce

plural – singular

poetry – prose

polite – rude, impolite

possible – impossible

poverty – wealth, riches

powerful – weak

praise – criticism

pre – post

predator – prey

pretty – ugly

private – public

prudent – imprudent

pure – impure, contaminated

push – pull

put on – take off

qualified – unqualified

question – answer

quiet – loud, noisy

raise – lower

rapid – slow

rare – common

real – fake

regular – irregular

relevant – irrelevant

respect – disrespect

rich – poor

right – left, wrong

right-side-up – upside-down

risky – safe

rough – smooth

rude – courteous

sad – happy

safe – unsafe

same – opposite

satisfactory – unsatisfactory

satisfied – dissatisfied

scatter – collect

second-hand – new

secure – insecure

seller – buyer

separate – join, together

serious – trivial

servant – master

shallow – deep

shout – whisper

shrink – grow

sick – healthy, ill

simple – complex, hard

single – married

singular – plural

sink – float

sit – stand

slave – master

sleep – wake up

slim – fat, thick

slow – fast

small – big

smart – stupid

smooth – rough

sober – drunk

soft – hard

some – none

sorrow – joy

sour – sweet

sow – reap

speed up – slow down

spend – save

start – finish

stop – go

straight – crooked

strengthen – weaken

stress – relax

strict – lenient

strong – weak

success – failure

sunny – cloudy

sweet – sour

synonym – antonym

take – give

tall – short

tame – wild

teach – learn

terrible – wonderful

them – us

there – here

thick – thin

throw – catch

tie – untie

tight – loose, lack

tiny – big, huge

together – apart

tolerant – intolerant

top – bottom

tough – easy, tender

transparent – opaque

trap – release

true – false

truth – falsehood, lie, untruth

ugly – beautiful

under – over

understand – misunderstand

unfold – fold

unknown – known

unqualified – qualified

unsafe – safe

up – down

upside-down – right-side-up

upstairs – downstairs

us – them

useful – useless

vacant – occupied

vague – definite

vanish – appear

vast – tiny

vertical – horizontal

victory – defeat

virtue – vice

visible – invisible

voluntary – compulsory

war – peace

wax – wane

weak – strong

wet – dry

white – black

wide – narrow

win – lose

wisdom – folly, stupidity

within – outside

wrong – right

yes – no

yin – yang

young – old

zenith – nadir

zip – unzip

All this follows a variation of +A + −A = 0: +A and −A (pairs of opposites or antonyms) = 0 (oneness, reconciled within the underlying unity of the One universe).

(17) Humankind unified in a World State
In the history of politics and international relations there are many examples of +A + −A = 0:

- +A (the Greeks) + −A (the Persians) = 0 (the hegemony of the 5th century BC).

- +A (the Romans) + −A (the Carthaginians) = 0 (the hegemony of the 3rd–2nd century BC).

- +A (the Romans) + −A (the British) = 0 (the hegemony in Britannia in the 1st century AD).

- +A (the British) + −A (the French under Napoleon) = 0 (the Great Power in 1815).

- +A (the British and Allies) + −A (the Germans and Allies) = 0 (the superpowerdom of the First World War).

- +A (the Allies) + −A (the Axis) = 0 (the superpowerdom of the Second World War).

- +A (the West and NATO) + −A (the USSR) = 0 (the superpowerdom of the Cold War).

- +A (the West) + −A (China) = 0 (the superpowerdom of the Pacific Cold War).

At some point, all humankind will unite in a democratic World State. This has been a dream for 2,000 years, kept going by Dante and Kant and in our own time in calls for a world government since 1945 by Truman, Einstein, Churchill, Eisenhower, Gandhi, Russell, J.F. Kennedy and Gorbachev.

First of all in *The World Government* (2010), then in *World State* and *World Constitution* (both 2018) I have called for a democratic, partly-federal World State with sufficient authority to abolish war, enforce disarmament, combat famine, disease and poverty, and solve the world's financial, environmental, climatic and virological problems. I call my World State a United Federation of the World (echoing Tennyson's "federation of the world" in 'Locksley Hall') and see it as replacing the UN, after turning the UN General Assembly into an elected Lower House of a World State.

The seven goals of a World State are:[364]

1. bringing peace between nation-states, and disarmament;
2. sharing resources among, and solving energy supply to, all humankind;
3. solving environmental problems such as global warming;

4. ending disease;

5. ending famine;

6. solving the financial crisis;

7. redistributing wealth to eliminate poverty.

World State has the following to say on a World Parliamentary Assembly and on a World Commission and World Senate:[365]

The Inter-national Level
Inter-national affairs are at present conducted by the UN General Assembly and Security Council, and a number of inter-national agencies which are offshoots from the General Assembly.

World Parliamentary Assembly
I have said that the UN General Assembly should become a lower house, the directly-elected World Parliamentary Assembly which would be involved in passing supranational laws proposed by the World Authority. The Assembly should be chosen on the basis of an initial allocation of 816 seats, along criteria stated by Clark and Sohn in *World Peace Through World Law* and modified to take account of, and adjust for, nuclear influence and Permanent Membership of the Security Council.

I can now set out the structure of the representation on the World Parliamentary Assembly. The mathematics looks like this:

Countries	Multiplied by	Members	Equals	Seats
4	x	30	=	120
3	x	15	=	45
10	x	12	=	120
15	x	8	=	120
20	x	6	=	120
30	x	4	=	120
40	x	3	=	120
48	x	1	=	48
+2	x	1	=	2
			Initial total	815

In future

+26	x	1	=	26
+8	x	1	=	9
			Eventual	
			total	850

I can now flesh out these figures, using the UN-sourced world population percentages. These are based on 1 July 2017 estimates by the UN Department of Economic and Social Affairs – Population Division.[366] The alphabetical letter at the beginning of each line below corresponds to the alphabetical letters against the relevant countries in the Appendix [in *World State*, not shown in *The Algorithm of Creation*]. The "largest" and "next-largest" countries mean "most populated". The representation in the World Parliamentary Assembly (initially calculated on the 2009 data and criteria and now comprehensively updated to 2017) looks as follows:

Letter in Appendix	Countries	No. each	Seats
A.	The 4 largest countries: China, India, the EU (on the basis of 26 members all marked A, excluding the UK and France, see B), the US	30 each	120
B.	3 Permanent Members of the Security Council: The Russian Federation, France, the UK	15 each	45
C.	The 10 largest countries after the 4: Indonesia, Brazil, Pakistan, Nigeria, Bangladesh, Japan, Mexico, Ethiopia, Philippines, Egypt	12 each	120
D.	The 15 next-largest countries: Vietnam, Democratic Republic of Congo, Iran, Turkey, Thailand, Tanzania, South Africa, Myanmar (Burma), South Korea, Colombia, Kenya, Argentina, Ukraine, Sudan, Uganda	8 each	120
E.	The 20 next-largest countries: Algeria, Iraq, Canada, Morocco, Saudi Arabia, Uzbekistan,	6 each	120

	Malaysia, Peru, Venezuela, Nepal, Angola, Ghana, Yemen, Afghanistan, Mozambique, Australia, North Korea, Taiwan, Cameroon, Côte d'Ivoire		
F.	The 30 next-largest countries: Madagascar, Niger, Sri Lanka, Burkina Faso (formerly the Republic of Upper Volta), Syria, Mali, Malawi, Chile, Kazakhstan, Ecuador, Guatemala, Zambia, Cambodia, Senegal, Chad, Zimbabwe, Guinea, Republic of South Sudan (separated from Sudan in 2011), Rwanda, Tunisia, Cuba, Bolivia, Somalia, Haiti, Benin, Burundi, Dominican Republic, United Arab Emirates, Jordan, Azerbaijan	4 each	120
G.	The 40 next-largest countries: Belarus, Honduras, Tajikistan, Israel, Switzerland, Papua New Guinea, Togo, Serbia, Sierra Leone, Paraguay, El Salvador, Laos, Libya, Nicaragua, Kyrgyzstan, Lebanon, Singapore, Eritrea, Norway, Central African Republic, Costa Rica, New Zealand, Turkmenistan, Republic of the Congo, Oman, Kuwait, Liberia, Panama, Mauritania, Georgia, Moldova, Bosnia and Herzegovina, Uruguay, Mongolia, Armenia, Albania, Jamaica, Qatar, Namibia, Botswana	3 each	120
H.	The 48 next-largest countries: Republic of Macedonia, Lesotho, Gambia, Gabon, Guinea-Bissau, Bahrain, Trinidad and Tobago, Mauritius, Equatorial Guinea, East Timor, Swaziland, Djibouti, Fiji, Comoros (an island nation in the Indian Ocean), Bhutan, Guyana, Solomon Islands, Montenegro, Suriname, Cape Verde, Brunei, Belize, Bahamas, Maldives, Iceland, Northern Cyprus, Barbados, Vanuatu (an island nation in the South Pacific), Samoa, São Tomé and Principe (an island nation in the	1 each	48

	Gulf of Guinea), Saint Lucia, Kiribati (formerly known as the Gilbert Islands, in the Pacific Ocean), Saint Vincent and the Grenadines, Grenada, Tonga, Federated States of Micronesia, Seychelles, Antigua and Barbuda, Andorra, Dominica, Marshall Islands, Saint Kitts and Nevis, Liechtenstein, Monaco, San Marino, Palau (an island nation in the Pacific Ocean), Tuvalu (formerly known as the Ellice Islands, in the Pacific Ocean), Nauru (formerly known as Pleasant Island, in the South Pacific)		
I.	2 non-UN-members with largest populations: Palestine, Puerto Rico	1 each	2
	Initial total		815
J.	26 dependent territories outside the UN and not yet independent with associate status until they become independent, seats held in reserve for when they become independent: Macau, Western Sahara, French Polynesia, New Caledonia, Guam, Curaçao (an autonomous region of the Netherlands in the Caribbean), Aruba (an autonomous region of the Netherlands in the Caribbean), US Virgin Islands, Jersey (UK), Isle of Man (UK), Guernsey (UK), Bermuda (UK), Cayman Islands (UK), American Samoa, Northern Mariana Islands (a commonwealth in political union with the US, in the Western Pacific), Greenland (Denmark), Faroe Islands (Denmark), St Maarten (Netherlands), Saint-Martin (France), Gibraltar (UK), Turks and Caicos Islands (UK), British Virgin Islands (UK), Bonaire (an autonomous region of the Netherlands in the Caribbean), Cook Islands (New Zealand), Anguilla (UK), Vatican City (in view of its diplomatic importance)	1 each	26
K.	9 disputed territories not on the UN-based list:	1 each	9

	Indian Kashmir (Jammu and Kashmir), Tibet, Pakistani Kashmir (Azad Kashmir), Kosovo, Chechnya, Transnistria, Abkhazia, Nagorno-Karabakh, South Ossetia		
L.	13 island groups with populations under 12,000 to be represented by colonial power: Wallis and Futuna (France), St Barthélemy (France), Saint Pierre and Miquelon (France), Saint Helena, Ascension and Tristan da Cunha (UK), Montserrat (UK), Sint Eustatius (Netherlands), Falkland Islands (UK), Norfolk Island (Australia), Christmas Island (Australia), Saba (Netherlands), Niue (New Zealand), Tokelau (New Zealand), Cocos (Keeling) Islands (Australia), Pitcairn Islands (UK)		0
	Eventual total		850

It must be stressed that this is a flexible list. There would be movement within it. For example, as more nation-states enter the EU, those that enter would lose their individual representation, leaving a vacancy in their group for a country in the group below it to be promoted. Associate dependent territories in J. and K. obtaining independence will have one member each and those passing within another nation (as Hong Kong did when it became a part of China) would lose their independent representation. The 850 seats are a maximum that would not be reached immediately. There is scope for future developments within the designation of an eventual tally of 850 seats, and there is no reason why there should not in due course be 850 seats or more if changes to national arrangements in the coming years require such an increase. The above list is therefore a transitional list to cope immediately with an ever-changing scenario.

Offshoots

All the offshoots from the UN General Assembly would continue to operate: the Economic and Social Council, the International Criminal Court, the International Court of Justice and the UN organs (UNDP,

UNHCR, UNICEF and UNEP) and specialised agencies (FAO, UNESCO, WHO and WTO).

Of these, the International Court of Justice, based in the Peace Palace at The Hague, Netherlands, the primary judicial organ of the UN, would continue to settle legal disputes submitted to it by nation-states and give advisory opinions on legal questions placed before it by the replacement for the UN General Assembly, the World Parliamentary Assembly. Like the tripartite Assembly/Executive Council/Secretary-General, the International Court of Justice would continue to resolve legal problems arising from conflicts in the first instance.

World political parties

All members of the World Parliamentary Assembly would belong to one of the world political parties. There would be a World Centre/Right Party, a World Socialist Party and a Liberal Centrist Party. In addition, no doubt, there would be a World Green Party, a Far-Left Party, a Far-Right Party and a Party for World Sceptics. Different world parties would have different views on the management of world resources and energy supply, and on the level of world taxation and redistribution.

UN Executive Council

The UN Security Council would continue to operate as a veto-less Executive Council. It would continue to have 5 Permanent Members and 12 Non-Permanent Members elected for two-year terms. The World Parliamentary Assembly would make recommendations as does the UN General Assembly, and the Executive Council would decide as does the Security Council. The UN Secretary-General and his Secretariat would continue to implement.

This tripartite UN structure would continue to field all conflicts requiring resolution in the first instance. Only the intractable problems on which the Executive Council could not reach agreement would be passed up to the supranational level. The UN Peacekeeping Forces would continue to operate under the Executive Council as under the Security Council in containable conflicts that do not require the involvement of the supranational level.

The UN Executive Council's (ex-Security Council's) 17

representatives would be elected by the World Parliamentary Assembly. Five representatives would represent each of the Permanent Members, 5 of the next 10 largest nation-states would have rotational representation (so that all 10 are represented, but only 5 at any one time), and the other 8 would be chosen by the World Parliamentary Assembly from representatives of other nation-states. This veto-less Executive Council would decide whether each stage of the disarmament process was satisfactorily completed.

It is important to stress that, from the point of view of the nation-states, life under the UN, and their experience of the UN and the international level, would go on pretty much as before except that the UN General Assembly would operate as a directly-elected Parliament. Life within each nation-state or regional grouping, whether loose, or tight (like the EU following the ratification of the Lisbon Treaty), would go on just the same, and nation-states would continue to pursue their self-interest within their implementing of the supranational authority's seven goals. Civilizations would carry on progressing through their various stages. At the inter-national level, the life of civilizations would continue.

There would be just one fundamental change, that from time to time the supranational tier would come into play. When a nation-state or civilization had a conflict that the UN could not deal with and might lead to war, as has happened in the case of conflicts involving 124 countries since 1945 according to one selection, and 162 conflicts since 1945[367] according to another, the conflict would be referred to the supranational level whose higher tier would enforce peace and impose a settlement just as if two American states had had a disagreement and had gone to court and been bound by the court's legal ruling.

The Supranational Level

The supranational level would provide the context within which the nation-states and their civilizations operate.

The executive, the World Commission – the supranational authority's equivalent of the European Commission – would propose laws, monitor the implementation of all directives and world policies, and represent the World Authority in the international institutions of global governance (the G8, OECD, the Bank for International

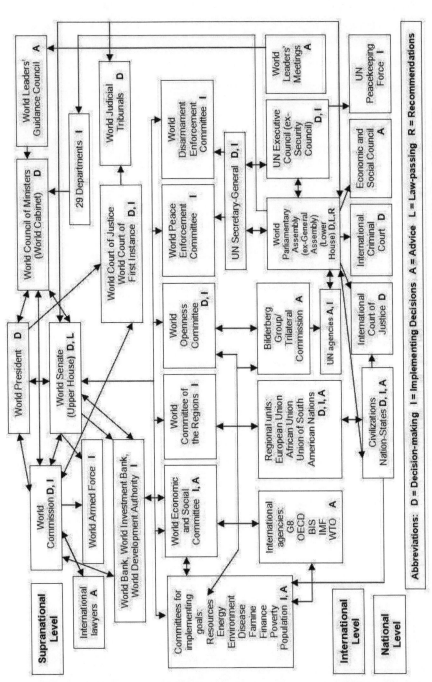

Diagram/Flow Chart of the Supranational Authority: The Structure of the World State.[368]

Settlements, the World Bank, the International Monetary Fund, the World Trade Organization and all the agencies of the UN). The World Commission's proposals would all relate to the achievement of the goals listed, including the enforcement of peace and disarmament, while guaranteeing the security of nation-states.

World Commission

The World Commission should be composed of 27 members, about the maximum number that can comfortably sit round a table and take part in world-authority meetings to enforce peace and conduct a world government. The 27 World-Commission representatives should be appointed on the basis of taking into account: regional representation; population percentages; topographical extent; Permanent Membership of the UN; and possession of nuclear warheads to be disarmed.

The data for these criteria can be found in the Appendix [not shown in *The Algorithm of Creation*]. The world population percentages for each nation-state are in Appendix E3, and historical movements of percentages are in Appendix E2. A fundamental approach to regions is in Appendix E1: Asia has 59.6 per cent of world population; Europe 9.8 per cent; North America 4.8 per cent; Latin America and the Caribbean 8.6 per cent; Africa 16.6 per cent; the Middle East and North Africa 3 per cent; and Oceania/Australia 0.5 per cent. As we shall shortly see, I have broken these regions up into smaller representational denominations. (I have done likewise in arriving at different representational regional zones or constituencies for the World Senate below.)

There are different schools of thought regarding regions. Sometimes Central America is linked to the Caribbean to form a region, and this region is distinguished from South America. The Rockefellerite Club of Rome, one of the undemocratic international agencies, put forward 10 regions in 1973.[369] These 10 zones, identified during the Cold War, included "centrally-planned Asia", meaning Communist Asia (Mongolia, North Korea, North Vietnam and China).

China's emergence as the most populated nation with industrial and nuclear power has required a rethinking of regional zones. It must not be forgotten that the main object of the World Commission is to deliver world peace through international law, and so it must include representation from all the nation-states that have been party to recent wars and conflicts. A balance has to be struck between representing the entire population of the world (as in the World Parliamentary Assembly) and representing the most powerful nation-states that need to sit round a table in order to deliver world peace and disarmament with the greatest degree of effectiveness.

The principle should be that members of the World Commission should be nominated by the governments or organisations of the countries/regions on the basis of no more than one member per region (as distinct from nation-state). I am basing my regions on the regions in the list in Appendix E4, but have varied some regions to achieve a scheme that works in relation to delivery of the seven goals.

I propose that the 27 members of the World Commission should be drawn up on the basis below. I have given reasons for my proposal in brackets. The abbreviation R means that there are regional considerations for the nomination; PM means Permanent Member of the UN; and N means that there are nuclear/disarmament considerations behind the nomination; wpp means world population percentage. The number 1 means one member to be nominated by a particular nation-state. Thus 1 China means one member nominated by China. My proposed members are as follows:

1   China (PM, N, wpp 18.5%)
1   Communist Asia (Mongolia, North Korea, N, Laos, Vietnam)
1   Russian Federation (PM, N, largest country in world)
1   Central Asia (ex-USSR)
1   Canada (topographical size) and Greenland
1   USA (PM, N)
1   UK (PM, N)
1   France (PM, N)
1   Western Europe including Germany (R)
1   Eastern Europe (R)
1   Central America and Caribbean (R)
1   North-East South America (including Brazil, N, R)
1   South-West South America (including Argentina, R)
1   Northern Africa (R)
1   Southern Africa (R)

1   Arab Middle East (R)

1   Israel (N)

1   Iran (N?)

1   Pakistan (N)/Afghanistan (R)

1   India (N, wpp 17.9%)

1   West-Central and South-East Asia (Bangladesh to Cambodia, R)

1   East Asia (Japan, South/North Korea, Philippines, R)

1   Indonesia (wpp 3.5%)/Malaysia

1   Australia/Oceania (R, wwp 0.5%)

1   UN (secondment of executive expert)

1   World Bank (secondment of expert)

<u>1</u>   Agencies of global government (secondment of expert)

27   Total

The allocation of governments'/agencies' nominations is a balance between regional representation, population percentages and world influence. It can be criticised from a number of angles.

Have I been too generous to the Russian Federation? I would say that the territory covered by the Russian Federation and the ex-USSR states east of the Caspian Sea and in Eastern Europe is perhaps four times as large as India.

Have I been too ungenerous to the US, the sole superpower? I would say that the principle has been one member per region, and there are certainly two regions (East/West) within the US. However, Canada only has one member and has larger territory. Australia is nearly as large as the US and has to share one member with Oceania (including New Zealand).

Have I given too much weight to nation-states with nuclear weapons? I would say that a World Commission committed to complete disarmament cannot have under-representation from the powers that produced 69,401 nuclear warheads by 1985, and reduced them to nearly 24,000 in 2009 and 15,375 by autumn 2016, and according to the Federation of American Scientists to 15,350 in the course of 2016 and to 14,900 in 2017,[370] which are still in storage throughout the world, whose co-operation it needs.

Is it right to include Israel, the Middle East and Iran and thus allocate three members to the Middle-Eastern region, the same number as Central and South America? I would say that the greatest threat to world peace is in the Middle East, where the World Commission needs to influence Islamists in Iraq, Palestine, Israel, Iran, and of course in Central Asia (Afghanistan and Pakistan).

Is it right to subordinate North Korea to one of three Communist-Asian nation-states? I would say that North Korea needs steadying, and that it would be wrong to allocate one seat to it for breaking rules but equally wrong to ignore it and isolate it.

Similar "cons" can be raised about each nation-state listed as nominating a member to the World Commission and I would say that for each "con" there is a "pro" that gives each listed nation-state a better right to nomination than the right of any other nation-state or region I have not included on the list.

There may be arguments about the detail of my list, which can easily be corrected, but the important thing is: the principle behind the choice of nominating countries/regions/agencies is right. In principle, I have put forward a workable list. In the interests of administrative efficiency, the representation on the World Commission should not be increased beyond 27.

World Senate

The World Senate would be a directly-elected upper house, like the American Senate. The relationship between the World Commission and the World Senate would be like that of the European Commission and the European Parliament – only there would be a bicameral system, one chamber (the World Parliamentary Assembly) being at inter-national level and the higher chamber (the World Senate) being at supranational level and dedicated to implementing the seven goals of the supranational authority by passing the appropriate laws to bring in and enforce a peaceful world.

The World Senate would be elected on a basis of regional zones, with weight being given to more scrupulously defined regions than the ones that would nominate the World Commission. The criteria would include regional representation, population percentages, topographical area, possession of nuclear weapons to be disarmed, political clout and environmental considerations.

I propose that there should be in the first instance 46 zonal constituencies, each of which would elect 2 Senators, making a total of 92 seats. I have based these regional zones on the lists of regions in Appendices E4 and E5, which include countries in each region. The numbers in what follows refer to the number of zonal regions or constituencies, each of which would have 2 Senators.

4   China: East, West, North (including Mongolia), South (PM, N, wwp 18.5%)

3   Russian Federation: East, Central, West (PM, N)

2   Central Asia: North-Central (including Kazakhstan), South-Central (including Pakistan)

3   Canada: East, West, North (including Greenland and Arctic)

4   USA: East, West, North, South, including dependent territories (PM, N)

1   UK (PM, N)

1   France (PM, N)

4   Europe: 2 North-Western (including Scandinavia, Germany), 1 Eastern, 1 Southern (i.e. Northern Mediterranean – Spain, Italy, Greece/Aegean)

1   Central America, including Mexico

1   Caribbean

3   South America: North-East (Brazil, Bolivia, Paraguay), South (Argentina, Chile, Uruguay and Antarctica), North-West (Peru, Ecuador, Colombia, Venezuela, North Coast)

4   Africa: North, East, West (including Atlantic Ocean), South (NB Middle Africa distributed among these 4)

2   Middle East: Northern (including Turkey, Iraq), Southern (including Egypt, Saudi Arabia)

1   Israel

1   Iran

3   India, North (including Nepal), Central, South (including Sri Lanka, Maldives and Indian Ocean)

2   East-Central Asia (Bangladesh, Myanmar/Burma, Thailand, Laos, Vietnam, Cambodia)

1   South-East Asia (Malaysia, Indonesia, wpp 3.9%)

2   East Asia (Japan, North/South Korea, Philippines)

2   Australia: East, West
__1__   Oceania (New Zealand, islands and dependent territories)
46   Total

Again, there may be disagreements regarding the detail and the number of seats allocated to a particular region, bearing in mind its population or topographical size. But I have done enough to demonstrate that 92 Senators can fairly represent all the nation-states and regions in the world. If this scheme were implemented there would be scope for 4 more zones to even up unfair allocations where regions feel hard-done-by, which would eventually give a Senate of 50 zones and 100 Senators.

The Senators would work closely with the World Commission, and with the World Parliamentary Assembly. The Assembly would send up to the World Commission matters beyond its competence, conflicts which would require enforcement, and the Commission would pass its proposals to the World Senate for them to become law. It is envisaged that some Senators would be experts in international law.

As in the case of the American system, each party would select a Senate leader who would appoint members of his or her party to Senate committees. These would include a World Committee of the Regions, a World Economic and Social Committee, a World Peace Enforcement Committee and a World Disarmament Enforcement Committee which would liaise with the World Commission, the World Court of Justice and the Assembly.

There would also be World Committees to implement the remaining six of the seven goals: a World Resources Committee, a World Energy Committee, a World Environmental Committee, a World Disease Committee, a World Famine Committee, a World Finance Committee, a World Poverty Committee and a World Population Committee.

World Openness Committee

The Senate Committees would include the very important World Openness Committee, which would review the position of the agencies and scrutinise all world personnel for any relationship with Rockefellers and Rothschilds dynasties and their self-interested world agendas. In view of the enmeshing of the New World Order, the Syndicate of the *élites* and Freemasonry, this Openness Committee would be alert to undue Freemasonic influence and would particularly look for any hidden links to élitist agencies which might suggest that a candidate for world officialdom would be linked to their known wish to control the Earth's resources and energy supply, and effect population reduction by violent means, which would be a crime under World Authority international law.

Working in conjunction with the World Resources Committee, this Openness Committee would co-control the return of all privately-owned energy to the people of the Earth, including all Russia's oil and gas which was acquired by the Rockefellers dynasty in the early 1990s for about one-tenth of its market value.[371] Multinational corporations would be permitted to continue conveying energy, but would have to surrender private ownership to the World Authority.

Under this scrutinising regime the *élites* would continue to operate and the Rothschilds dynasty would be free to continue to control 187 world central banks (if they do), but any manipulation of candidates, officials or policies would be identified and exposed as a breach of international law for which there would be severe penalties. The World Openness Committee would also receive advance copies of all agendas of meetings of agencies such as the Bilderberg Group and the Trilateral Commission, and all minutes made of their meetings. It would be a legal requirement that two representatives of the Senatorial World Openness Committee should attend all meetings of these groups and report back to the World Commission, to ensure that the work of all agencies was transparent.

The Syndicate of the *élites* would be allowed to go on functioning but its activities would be controlled and members would be excluded from secret decision-making and would be subject to investigation by the civil police and law enforcement.

The Ministers in a new World State's government would be:[372]

World Council of Ministers
Vetting by the World Openness Committee would also apply to the World Council of Ministers, the world Cabinet, which would be drawn

from the directly-elected World Parliamentary Assembly.

There would be 29 Departments, and therefore 29 World Ministers, the absolute maximum number that can comfortably sit round a Cabinet table and take part in high-level World-Authority meetings.

Unlike the EU, where the Council of Ministers is an assembly of national Ministers, the World Council of Ministers would comprise ministers of World Departments:

- World Finance;
- World Treasury;
- World Peace;
- World Disarmament;
- World Resources;
- World Environment;
- World Climate Change;
- World Health (ending disease);
- World Food (ending famine, implementing crop-growing programs);
- World Regions, Communities and Families;
- World Work and Pensions;
- World Housing;
- World Economic Development;
- World Regional Aid and International Development (ending financial crises);
- World Poverty (eliminating poverty);
- World Population Containment (not reduction);
- World Energy Regulation;
- World Transport (world aviation, roads and rail);
- World Law;
- World Oceans;
- World Space;
- World Education;
- World Citizenship (law and order);
- World Culture and History;
- World Sport;
- World Unity in Diversity;
- World Dependent Territories;

- World Foreign Policies (liaising with nation-states' Foreign Ministers); and
- World Human Rights and Freedom (guaranteeing individual freedoms, including freedom from population reduction under the new system).

All these Departments would work closely with Senatorial Committees operating within their field.

World President

The World Council of Ministers would meet as the World Cabinet under the leadership of the World President, who would be elected every four years like the President of the United States.

The World President would be nominated by one of the world parties. There would be several candidates, one for each party, and each would be vetted by the World Commission and the Senatorial World Openness Committee for any links to élitist agencies that might suggest that a candidate would be likely to pursue élitists' agenda regarding world resources, energy supply and population reduction rather than a philanthropic agenda.

There would be a world election campaign involving visits to the main regions and television coverage of speeches that would go out to the whole world. There would be a world ballot involving polling stations in key cities and rural areas, postal voting and online voting.

The successful World President would be sworn in by the World Commission and a judge from the World Court of Justice. The President would lead the world Cabinet to achieve the seven goals, paying attention to the democratic sway of the electorate. He could stand for a second term and serve if elected.

World Guidance Council

A World Guidance Council of elder statesmen and distinguished world figures would assemble every three months, or more frequently if a crisis demanded, to give advice to relevant members of the World Commission, World Senators and World Council of Ministers. These wise elders would be chosen on merit and expertise, with consideration being paid to regional representation, population percentages and political clout, on the recommendation of the World Commission, World Senate and World Council of Ministers.

World Leaders' meetings

In addition to these guidance meetings, sometimes there would be a business meeting involving World Leaders of nation-states in key regions or their foreign ministers.

There would in particular be a Regional Leaders' meeting for the leaders of the 13 main regions: North America; Europe; Japan; Oceania; China; Tibet; the Russian Federation; South America; Islam; Africa; India; South-East Asia and Central Asia. There would be regular live link-ups between the World President and the Regional Leaders. There would be smaller groups of Regional Leaders, on the lines of the G4 and G8.

An appropriate member of the World Commission would be present at all meetings of World Leaders, along with the Minister of the appropriate department and a Senator. These meetings of World Leaders would be like the European-Council meetings.

World banking

The World Bank would be overhauled to operate at a supranational level. It would continue to make loans to poorer countries for capital programs to reduce poverty. The World Investment Bank would also be overhauled to streamline a new supranational role.

My *World Constitution* begins with a Preamble:[373]

We, the assembled peoples of the World – resolved to form a federal union of humankind; inaugurate a universal peace, freedom from war and menacing weapons among all nation-states; redistribute earth's natural resources and energy equitably; protect the environment from pollution and global warming; banish disease and famine; consolidate wealth, prosperity and financial security; eliminate poverty; advance universal justice and welfare, tolerance, human rights and social progress for the common good; preserve all civilizations' religious traditions; and found a new epoch of human unity – do ordain this Constitution to establish the United Federation of the World.

We can say,

- +A (the West) + −A (the East) = 0 (an authoritarian New World Order), or
- +A (the West) + −A (the East) = 0 (a democratic World State).

All this follows a variation of +A + −A = 0: +A (supranationalism) + −A (nation-states) = 0 (the political union of all humankind in the One universe).

(18) Humankind's spiritual survival after death

The human being who has evolved in consciousness as well as in body to live in his or her soul, know objective consciousness poets know and open to the Light leads a spiritual life in his or her transpersonal consciousness, the eleventh level of consciousness:[374]

> transpersonal consciousness – psychic and paranormal consciousness (telepathy, precognition, prophecy, healing), awareness of memories from the womb, past lives or another world seen in the near-death experience, mediumistic contact with departed spirits, rare glimpses seen through the centre known as the spirit which traditions claim has lived before (a spiritual state).

Rationally the self knows the possibility of survival after death. Despite +A + −A, the self knows it is in a non-duality of matter and mind, and sees the after-life in terms of the quantum vacuum:[375]

> A "system of general ideas" must be dominated by one of four metaphysical perspectives: Materialistic monism (that matter gives rise to mind, or scientific materialism); dualism (that matter and mind are separate substances, the Cartesian position); transcendentalist monism (that mind or consciousness gives rise to matter so that the ultimate stuff of the universe is consciousness, an Idealistic view);[376] and metaphysical non-duality (that a metaphysical Reality gives rise to both matter and mind, that a unity gives rise to a diversity). Whereas monism implies no diversity, non-duality implies both a unity at the transcendent metaphysical level of the infinite and diversity at the immanent physical level.

My "system of general ideas" is a non-duality. There is unity at the metaphysical, infinite level and diversity at the finite level. My system of coherent general ideas explains how unity became diversity by manifestation into matter, light, consciousness and the vegetative world of plants, how the One manifested into many....

The rational Universalist philosopher has to consider all questions which may affect his view of man's role in Nature and the universe. That means laying out all that is known – "every known concept" – whether it has been verified or not, and then arriving at an explanation in terms of a "system of general ideas" that satisfies every conceivable eventuality, including possible survival after death. There have been scientific experiments to monitor and test the phenomena of parapsychology. These tests have provided some scientific support for a multi-levelled view of consciousness and in some cases have confirmed the experiences of the transpersonal or universal/cosmic consciousness (levels 11 and 12) reported by those who have had experiences of telepathy, precognition, communication from the dead, near-death experiences and immersion in the mysterious Light.

Mediums report a world in which the departed are still alive but in a different state of "existence". They speak of their hair colour, of living in bungalows, of being able to smoke, share precise memories and have detailed up-to-the-minute awareness of what is happening in our lives. They are present in our rooms, in this universe. If what they convey is true, the departed must be within this universe to be present with us and not in an inaccessible multiverse. So where is "the other side"? It can only be within the level of Being.

It is rationally and theoretically not impossible that on death spirits return from Existence to a "land", or an "island", within the ocean of Being, where they can be contacted by the living and that they can communicate by a kind of radio wave. It may be that for the initial time after their departure many are present, unseen, in our midst, inches away from us in the invisible air, within Being rather than Existence but able to witness what we do and say. Many of the features of the paranormal may be found in this ocean of Being, for example, telepathy (or the power of minds to communicate at a distance). Also precognition, for past, present and future belong to time and are surrounded by a timelessness to which the future has already happened.

The quantum vacuum is not another dimension. It is present all round us now, though invisible. We have seen that the same is true of the infinite. Nothing has to "unfold" from the infinite for Universalism sees the universe as being governed by three dimensions of space and one of time – and by the infinite, which is hidden and all about us like the quantum vacuum. The infinite we encounter may *be* the quantum vacuum in which Being becomes Existence on the border of the infinite and the finite, and which has manifested from the infinite the surfer breasts on the tip of our expanding, accelerating universe, perched on the crest of a shuttlecock-shaped wave.

Without invoking other dimensions, I should point to the possibility that, just as all atoms are interconnected in the quantum vacuum (as was pointed out by Bohm, who held that consciousness is inherent in matter – I would say that the unitary infinite manifested into a diversity of finite consciousness and matter), so may the consciousnesses of humans act like atoms and be similarly interconnected. Waves between them may account for telepathy, where one atom-like human connects with another atom-like human on an identical frequency and a thought is transmitted from one brain via a radio-like wave and is received in another brain that is tuned to receive it far away, non-locally over a great distance, and simultaneously.

Just as particles such as bosons surface from the quantum vacuum, live a short time and then disappear back into the quantum vacuum where they continue their "virtual" existence in a potential rather than a material state, so the consciousnesses of humans may emerge from the quantum vacuum on conception, live and then return to it on death – and continue a virtual, potential existence there. And perhaps the disembodied, disembrained consciousnesses of the physically, materially dead can send telepathic thoughts like radio waves which are transmitted and received in a living medium's mind when it is tuned to the transpersonal level of consciousness (level 11).

The assumption of reductionist, Materialist science is that all consciousness and mind is brain function and that when the brain dies there is no consciousness to survive death and transmit messages in the quantum vacuum. To the rational Universalist there is a universal being and a spirit, as all the religious traditions have taught, which return to the infinite and timelessness – eternity – below or behind space-time on death and perhaps live in the quantum vacuum of Being within the infinite timelessness. Perhaps we do not "pass away" from the earth on leaving our body but remain here unseen, tied or magnetically drawn to the place which dominated our interest in life and witnessing succeeding events until we cultivate wider interests which allow us to move on elsewhere. Being may be the invisible side of Existence, and the infinite the invisible side of the finite. The spirit (*pneuma*) St Paul wrote of[377] was supposed to have only one life. Hindus and some other traditions believe it reincarnates, though not immediately, and therefore lives many times. And that we have far memories, sometimes in the form of traumas, of our past lives. According to the teaching of traditions that affirm reincarnation, when the body it occupies becomes old and, its skin tattered, wears out, the spirit transmigrates and is reborn into a fresh body of a baby to make a fresh start. Perhaps eventually our core is reborn to learn new things on a planet which is a "classroom of the soul". It may be that the spirit re-lodges in a new human consciousness and anchors itself in the brain, somewhere below the claustrum.

Shamanism, the religion of Central Asia at the dawn of civilisation c.50,000 years ago and still to be found in certain societies today such as the Eskimos, has always held that animals have spirits.[378] If so, the same principle may work in biology. Considering "every known concept", the rational Universalist holds that it is not inconceivable that the mammals, reptiles, birds, insects and plants we have considered from a Darwinist point of view may similarly die back into the quantum vacuum when their bodies are old and worn out, to transmigrate to fresh young bodies on conception and make a fresh start. To the rational Universalist the true reality of evolution may be found in the quantum vacuum rather than in the vegetation and forests of continents emerging from Ice Ages.

To the rational Universalist it is not impossible that a part of our consciousness survives death and that we have spirits that are immortal. There is a consensus in religious literature that the spirit is the enduring, timeless part of our consciousness which may return to timelessness, an ocean of Being, on death to live in the quantum vacuum like a fish. Tradition, on the other hand, sees the soul (levels 7, 8 and 9) as a human's individual centre that connects feeling and thought. According to the Christian tradition it survives death, but according to the tradition of the Jewish Kabbalah, which is as old as Abraham, at least c.1700BC and probably older,[379] the soul perishes with the body whereas the spirit survives death. It is not impossible that the *genus Homo*, who evolved from the apes, developed consciousness that has spiritual awareness and has an enduring spirit (level 11) that returns to the infinite and timelessness – the quantum void, or Being, or collective unconscious to which all consciousness returns from space-time on death.

Because metaphysics includes "every known concept" the rational metaphysical philosopher admits such possibilities even if he has no direct evidence of them from his own experience. To the reason, these are valid possibilities and at the rational, philosophical level may have validity as postulates even though they cannot be verified.

Intuitively the self *experiences* the "metaphysical non-duality", sees consciousness as transmitted to the brain in light photons so the brain is a receiver of consciousness, and sees the after-life in terms of an ocean of Being and Light:[380]

The intuitional self, residing outside the rational, social ego, is able to experience the workings of the Whole in a way that can elude the rational metaphysical philosopher. I have said that the reason analyses and fragments the All or One into bits whereas intuition perceives the Whole, reunifies the fragments. Intuition's movement is from Existence to Being. The intuition perceives what the reason cannot perceive, that as it operates within a unity of Being the infinite order principle does not merely shape growth but can also shape our actions and fortunes if we put ourselves in harmony with its ordering force. The order principle works for good conditions (such as growth and development) to prevail in all life, the intuition perceives, and to the intuition order, which created the bio-friendly conditions in which we live, can make good things happen to us by detecting and advancing the yearnings in our cells. Our wishes and conduct influence the Whole and are influenced by it. That is the intuition's great hunch....

The intuitional system contains a sea of energy or latent Light that has manifested into subtle, end-of-spectrum physical light, matter and consciousness. The same manifestational process, once out of Being and in Existence and directed by the universal order principle, shows itself to be a Light-inspired drive of organisms through matter to higher and higher levels of self-organising and ascending hierarchical wholes.

The intuitional universe is a perception of the intuitional self or consciousness. The received view of consciousness is that it is brain-dependent and ends when its brain stops functioning. However, according to the intuitional universe it is not impossible that each consciousness may be transmitted as light or photons through the brain rather than be produced by the brain. It may control and regulate the brain's electrical rhythms, regarding the finite brain as Shelley's "dome of many-coloured glass" which "stains the white radiance of eternity".[381] It may be that the white radiance of the universal being, spirit and soul come from outside and anchor and lodge within the many-coloured dome of the operating brain and appear to be dependent on the brain as brain function.

If consciousness is transmitted into the brain and the brain acts like a radio receiver, it is probably received by photons above the brain, in a "Bose-Einstein condensate".[382] This condensate, or substance produced by condensation, was proposed in a theory developed in 1924–1925 by Einstein and the Indian physicist Satyendra Nath Bose on the statistical behaviour of a collection of photons, and it accounts from the streaming of laser light. Photons satisfying Bose-Einstein statistics are called bosons. Electrons satisfying Fermi-Dirac statistics are called fermions. These photons or bosons, including the Higgs field of bosons, have been undetected until now and were sought by CERN in its July 2008 experiment. If consciousness *is* received in the brain, the transmitted current of universal energy perhaps arrives on carrier photons or bosons. The energy stimulates and drives the neurons and pushes their electrical impulses into interconnectedness and binding with each other. The activity of the neurons may thus be a consequence of the process of consciousness bosons, not its generator. And consciousness may be a tenant squatting in the home of the brain, which it deserts to go elsewhere, leaving the brain empty and lifeless.

To be completely clear, what I am suggesting is that consciousness may be composed of hitherto undetectable bosons (photons that obey Bose-Einstein statistics) and that the discovery of bosons by CERN would make this proposal testable as a scientific fact.

I also want to make clear that metaphysical Light begins in the infinite, beyond physics, and, I submit, manifests into the finite, the universe after the Big Bang. At some stage, I submit, metaphysical Light manifests into finite bosons. CERN has sought this finite manifestation. Its particle chamber replicates the finite conditions *after* the Big Bang, by which time the infinite has manifested into the finite. It has tried to capture the moment of manifestation. I submit that the bosons of consciousness had an infinite origin, and have the infinite contained within their finite manifestation in accordance with the Law of Unity and the Law of Survival. According to metaphysics, in view of their infinite origin in Being, all finite phenomena contain within them that which is infinite. This was the point I argued with Bohm....

The intuition brings its own intuitional slant to non-duality for it is deeply involved in the workings of consciousness. It will be helpful to list the metaphysical perspectives I identified alongside the various views of consciousness to make it absolutely clear what I am proposing and not proposing as regards the intuitional universe as a system of light:

- Materialistic monism (that matter gave rise to mind or consciousness, scientific materialism) – mind or consciousness is Materialistic brain-function within electrical neurons dependent on the brain, and dies when the brain dies;
- dualism (that matter and mind are separate substances, Descartes' position) – mind or consciousness is composed of a substance separate from matter, and soul/spirit survives death whereas matter dies when the body dies and is recycled;
- transcendental monism (that mind or consciousness gave rise to matter, that consciousness is the stuff of the

universe, Idealism) – mind or consciousness is transmitted from manifesting metaphysical Reality and is perhaps composed of photon-like particles tinier than neutrinos such as photinos, but probably bosons, which form a halo-like system of light above the brain that uses brain-body, and soul, spirit and universal being survive death whereas matter is recycled;

- metaphysical non-duality (that the infinite metaphysical Reality, a unity, gave rise to both matter and mind/consciousness, a diversity although from a common origin) – mind or consciousness is transmitted from manifesting metaphysical Reality and is perhaps composed of photon-like particles tinier than neutrinos such as photinos, but probably bosons, which form a halo-like system of light above the brain, a Bose-Einstein condensate, that uses brain-body, and soul, spirit and universal being survive death whereas matter (atoms, cells and neurons) are recycled into a new form.

Universalism is a metaphysical non-duality, and this last perspective is the one I favour. In intuitional metaphysical non-duality, consciousness appears ambivalently, depending on its level....

The general system of ideas sees consciousness as split like the body, which is partly autonomic brain and partly bodily function. Consciousness is part finite, autonomic brain, and part infinite, a ring of light. It is a Bose-Einstein condensate, a form of "condensation" of light that is similar to the condensation of matter but it differs as it has an independent existence and survives death. A network of photons, also known as a field of bosons, *aither* or ether, links all living beings, who can absorb information from the infinite across its information-bearing particles. So where do we go when we "pass over"? Through a connecting "tunnel" into a level of Being that is otherwise separated from Existence. Looking into Existence from Being is like looking through a one-way mirror. We in Existence cannot see into Being, but departed spirits can see us in Existence. Spirits may continue in Being and their consciousness may survive.

If part of our consciousness is autonomic and another part can survive death, then the system we live in which includes the infinite, Nature's system and our everyday experience are all understandable and life did not begin as a pointless accident. Rather, the bio-friendly universe and evolution have been shaped by order. If part of consciousness survives death life has a meaning and purpose for it is related to an end. The metaphysical system leaves open the possibility, in a philosophical category, that part of consciousness *does* survive death and rejoin the infinite. This allows individuals to see that their lives have meaning and purpose as they advance through their projects and by self-transcendence towards the goal of life: knowledge of the metaphysical Light of Reality and of the possibility of an after-life – which they can embrace at the level of theological belief but which is nevertheless meaningful in philosophical terms. Metaphysics restores meaning to life and to the developing metaphysics of higher consciousness.

As Reality can be experienced intuitively, known in the intellect, as Light, our consciousness may be a high-frequency system of boson light and its development can be seen as an aspect of an orderly drive that is linked to the Light. The Universalist philosopher holds that the meaning of life among the species is the drive to high (or deep) consciousness, and that the meaning of an individual life is to see it in relation to the full context of life in the universe, as purposeful in relation to its projects. As Light conveys information to consciousness, it may be that we live in an *aither* of tides like sea-sponges in the sea, as filled with Light as sponges are filled with tides. Non-duality sees our bodies and consciousnesses as emerging from the same source, an ocean of Being and Light.

Metaphysical living opens to the possibility of a fifth dimension after death, an after-life in infinite Being for which, motivated by the order principle, each soul, cleansing itself, prepares and to which it returns:[383]

Bohm saw our three-dimensional world as an explicate order of objects, space and time. He believed we live in a multi-dimensional world[384] – an aspect of the multiverse – and that the implicate order gives rise to our physical, psychological and spiritual experience and is in a fifth dimension of which we are largely unaware. (Time is the fourth dimension.)...

We have seen that there may be "radio transmissions" between spirits of the dead that occupy a part of the infinite and living consciousnesses on earth. There is a view that such an "after-life" is in a fifth dimension. In philosophical terms it can be located in Being rather than Existence, and if Being is a fifth dimension then the Universalist philosopher will not disagree with this view.

Life on earth, we have seen, involves a mixture of order and randomness. The religious view depends on the quasi-empirical experience of Being manifesting in metaphysical Light and order. Atheists exaggerate the prevalence of randomness at the expense of order. If there were no order, there would be chaos and nothing would exist, live or develop. Religion tends to attribute providential importance to order, and religion's problem is to cope with randomness....

In his consciousness, moving between 12 levels or seats of his self, he [the surfer] knows that the aim of life is to arrive at the truth of why we are here in the universe, to replace himself by leaving behind children and to live in harmony with Reality. He knows that the purpose of life is the *genus Homo*'s thrust, by self-improvement, self-betterment and self-transcendence, from ape to superconsciousness which can experience Reality, the metaphysical Light, and to cleanse and heighten his own consciousness to prepare for the survival of part of his self after death. He knows that the meaning of life is to know with certainty and profound joy that life is not a chance, random accident but is the careful work of a caring order principle that is bio-friendly and imbues human activity with a profound meaning as humans journey towards the survival of part of their consciousness beyond this life, to a new form of existence which is only a breath away. In his multi-levelled consciousness he knows that the aim, purpose and meaning of *his* life mirror the aim, purpose and meaning of life in his projects and self-appointed tasks, his domestic harmony and his existential – or rather, universal and Universalist – oneness with the All or One, with Being, within the scheme of manifestation. Now within his universal being, his intellect, and now within his soul breaks the metaphysical

Light, filling his consciousness and his limbs with a profound meaning that brings with it "a peace that passeth understanding". He feels wonderfully serene and accepts the conditions of existence. He loves life and *his* life, and though solitary and aware that he can fall off the crest of his wave before his next breath, he is confident that when he departs this life he will live on unseen in infinite Being (in accordance with the Law of Survival which states that all beings do not cease to be contained within the unity of Being) and return to it on death, and he counts himself happy.

In non-duality, +A (mind and 12 levels of consciousness) + −A (materialistic matter) = 0 (metaphysical non-duality, a unity that gave rise to both mind/consciousness and matter via evolution through cells and DNA).

All this follows a variation of +A + −A = 0: +A (metaphysical non-duality of matter and mind in a unity that transcends materialism) + −A (contact with the ocean of Being, the Light) = 0 (the One universe that includes a fifth dimension and the spirit's survival after death).

(19) The ultimate fate of the universe

The surfer, surfing at the edge of an expanding universe and breasting the infinite while his feet are in space-time (see (8)), from where we are now after 13.82 billion years can look ahead and speculate on what will happen during the next 5 billion years while our sun "goes out" and becomes a black dwarf, and beyond.[385] From what we have seen the ultimate fate of the universe depends on what happens from now, and there are three possible outcomes:

1. If the universe is closed, i.e. spherical with curved space, then gravity will eventually stop the universe's expansion, and the universe will collapse back to a point and end in a 'Big Crunch';
2. If the universe is open, with no curved space, then it will expand for ever, driven by dark energy until the acceleration becomes so speeded-up it overwhelms the gravitational, electromagnetic and strong forces and rips apart in a 'Big Rip' in a universal Heat Death and then a Big Freeze;

3. If the universe is flat,[386] balanced between the two with flat space in all directions and with zero curvature, then it will expand for ever at an ever slower rate, leaving behind the earth's present stars, but will then speed up and face the same fate as an open universe: Heat Death,[387] and it will then eventually slowly come to a stop at zero velocity, in a cold dark end, in a 'Big Freeze'.[388] A flat universe expands for ever if there is no dark energy, but at a continually decelerating rate and eventually approaching zero. With dark energy the expansion rate of the universe slows down as a result of gravity but although there is a possibility it may eventually increase and meet the same fate as an open universe and will rip apart, the likelihood is that it will end in a hot 'Big Crunch'[389] (the heat caused by collapsing and collisions) and gradually collapse back to a point.

Of these three possible outcomes, it looks at present as if our shuttlecock-shaped flat universe[390] will expand because of its dark energy until the braking effect of dark matter and gravity slow it down. The expansion rate of the universe suggests it is open but other measurements by the Wilkinson Microwave Anisotropy Probe suggest the universe is very close – within a margin of 0.4 per cent of error – to being flat.

Although a school of thought maintains that the universe could end between 2.8 and 22 billion years from now,[391] and another school of thought sees the universe doubling every 12.2 billion years[392] and sees hypothetical dark energy dominating the universe's expansion in 850 billion years' time and turning the universe cold,[393] there is a consensus that stars will continue to form for 10^{12} to 10^{14} (1–100 trillion) years, but eventually there will be no more gas to form stars and stars will run out of fuel and cease to shine.[394] This process will have begun and be fairly advanced in 10 trillion years' time. It has been calculated that as soon as 1 trillion years from now the only stars will be red dwarfs, which can live 1–20 trillion years. It has also been calculated that the last star will cease to shine in 10–100 trillion years' time, when the universe will start to go dark. If the flat universe is propelled by dark energy without dark matter intervening, the universe will continue to expand for ever but will be lifeless, cold and dark. If the flat universe's dark energy is counteracted by dark matter, which is the more likely, the universe's expansion rate will slow down and then increase and it will suffer the fate of an open universe: Heat Death, then a Big Freeze and, assuming a negative cosmological constant, a negative density and positive pressure, would recollapse to a Big Crunch, a dimensionless singularity.

In the final darkness there will be a final burst of light (as collapsing causes collisions) and then darkness for all eternity while the black holes evaporate away. Large black holes and collapsing superclusters of galaxies would take 10^{106} years to evaporate,[395] assuming protons decay. If protons do not decay, this evaporation would take 10^{139} years.[396] There are speculative possibilities that the end might take beyond 10^{2500} years if protons decay, and $10^{10^{76}}$ years if protons do not decay, but these are unevidential.

In black holes the laws of physics break down and space and time do not exist. The final darkness will be akin to the One, the *Tao*, the physical Reality in which space and time do not exist and are not properties of Nature, and the ingredients of all information are stored, including, perhaps, ingredients of remnants of all human consciousness. The universe would collapse back into how it started, to the Void with a singularity, a reverse contraction of its early expansion.

To put it another way, the universe will slowly grow darker and be dominated by black holes and as they emit Hawking radiation black holes will eventually disappear. The universe will approach zero temperature (7×10^{-31}K, 0.7 million-trillion-trillionths of a degree above absolute zero),[397] and rather than expand for ever will end in a Big Freeze and eventually (like the Big Rip) collapse back into a point.

We have ignored the later models than the Big Bang – Steady State, the multiverse, string theory, and the multiverse Big-Wave models. The end for our part of the flat universe can come in 5 billion years' time when the sun burns up in a Heat Death before the universe enters a long slow Big Freeze.

We can say of the universe, +A (all order) + –A (randomness) = 0 (the unity of the One universe). We can also say of the end of the universe, +A (the universe's dark energy) + –A (the universe's dark matter) = 0 (the Void with a singularity, the Oneness of the ex-universe).

All this follows a variation of +A + −A = 0: +A (expansion and acceleration of the universe through dark energy) + −A˙ (gravity and dark matter) = 0 (balanced expansion and eventual Heat Death and then a Big Freeze, followed by a return of the universe to the point from which it came).

(20) The end of the universe

The future is that humans will be extinct in 7.4 million years' time, according to Richard Gott's formulation of the probabilistic Doomsday argument,[398] but will last at least another 5,000 years. In 1 billion years' time the solar luminosity will be 10 per cent higher than at present, and lack of oxygen, which will be back at the level it was before the Great Oxidation Event, will cause most life to die. In 4 billion years from now the increase in the earth's surface temperature will cause a greenhouse effect, heating the surface enough to melt it. All life on earth will then be extinct as a result of Heat Death.

There is a general consensus that in 5 billion years' time the sun will go out and be a black dwarf. There is a general consensus that in 10–100 trillion years' time, after a final burst of light all stars will fade and die and black holes will evaporate within 10^{106} years' time (assuming protons decay), and space and time will cease.

Our universe seems to be a flat universe and will expand for ever unless dark matter counteracts its expansion with gravity. Trillions of years in the future, long after the earth has been destroyed, the universe will drift apart until the formation of galaxies and stars ceases. In 100 trillion years from now, every star will be a black dwarf, and there will be a completely dark universe with no stars. There will be a Big Freeze and, assuming there is gravity, the universe will peter out and Creation will cease in a reverse Big Bang and implode back to a singularity.

As stars use up their hydrogen fuel reserves, gravity will cause their cores to contract and become hotter and become red giants. They will then burn to carbon, and shrink to white dwarfs.[399] About 98 per cent of stars will become white dwarfs, including our sun. In about 5 billion years' time, according to NASA our sun will exhaust the hydrogen fuel in its core and swell to a red giant. It will engulf the inner planets and scorch the earth, then shed its outer layers and its exposed core will remain and appear a white dwarf. Eventually the sun will go out and become a black dwarf, and life on earth will by then have come to an end before the universe collapses back via a Heat Death and then a Big Freeze into a point.

There are now considered to be 2 trillion galaxies in the observable universe, and 400 billion stars in our galaxy, the Milky Way. It has been calculated that the last star in the universe will cease to shine in 10–100 trillion years' time, and that the universe will then go dark and end in 10^{106} years' time (assuming protons decay).

If the flat universe is propelled by dark energy without dark matter, the universe will continue to expand for ever, but it will be lifeless, cold and dark.

If the flat universe's dark energy is counteracted by dark matter, which evidence shows as being more probable, its expansion rate will slow down and then increase and it will suffer the fate of an open universe: Heat Death, a Big Freeze and, assuming a negative cosmological constant, a negative density and positive pressure, it would re-collapse to a Big Crunch, a dimensionless singularity.

There is a general consensus that in the final darkness there would be a final burst of light and then darkness for all eternity while the black holes evaporate away. In black holes the laws of physics break down, and space and time do not exist. The final darkness will be akin to the One, the *Tao*, metaphysical Reality in which space and time do not exist and are not properties of Nature, and the ingredients of all information known during the life of the universe will be stored invisibly, perhaps including the ingredients of all remnants of all human consciousness.

The universe would collapse back into how the universe started, to a Void with a singularity, a reverse contraction of its early expansion: $+A + -A = 0$.

The Creation and end of the universe: the Theory of Everything as
$0 = +A + -A = 0$

The Creation of the universe before the Big Bang was: 0 (the singularity within the Void of Nothingness) = +A (+p, one of the first pair of particles) + −A (−p, one of the first pair of particles). In the same way the end of the created universe will be, in reverse: +A (one of the last pair of particles) + −A (one of the last pair of particles) = 0 (the One universe which has become a point in the infinite Void).

At the end of the universe, the process of Creation will be reversed. The algorithm for before the Big Bang and the end of the created process can be put together: the whole process from Nothingness/ the One Void before the Big Bang back to Nothingness/the One Void after the Big Crunch can be expressed in terms of the full algorithm of Creation as: $0 = +A + −A = 0$.

There is a view that a universe preceded our universe, whose Big Bang came out of the last two particles of the previous universe, and that another universe will succeed ours, coming out of the last two particles of our universe, suggesting the algorithm should be $0 = +A + −A = 0 \equiv 0 = +A + −A = 0 \equiv 0 = +A + −A = 0$ (the equals sign with three lines, $\equiv$, meaning that universes are equivalent to our universe but not necessarily equal).

There is also a view that more than one or an infinite number of universes preceded ours and more than one or an infinite number of universes will succeed ours, suggesting the algorithm should be expressed in infinitely-occurring or -repeating language as:... $0 = +A + −A = 0$ ... (the three dots signifying that our universe has occurred infinitely and will continue to occur infinitely with different rather than identical things happening within each universe). Although it can have three dots before it and after it, the algorithm of Creation remains the same: $0 = +A + −A = 0$.

There is also a view that there is a parallel universe or more than one or an infinite number of parallel universes in our one time or in more than one or an infinite number of fourth dimensions (times). For consideration of this highly-speculative and complicated view, see pp.301–304.

However, Universalism focuses on the unity of our universe and humankind, and regards a universe or a succession of more than one or an infinite number of universes before and after ours as a speculation which may have a place in the mathematics of possibilities but is unevidential. It must be noted that the algorithm of Creation, the universal algorithm $0 = +A + -A = 0$, remains the same for all other hypothetical universes, though when applied to an infinite number of hypothetical universes it will have three dots before it and after it.

The main thing is, the life cycle of our universe should still be expressed as the universal algorithm, $0 = +A + -A = 0$, and the algorithm of Creation always remains the same.

3

The Algorithm of Creation and a Theory of Everything

A Theory of Everything drawing on 100 key conditions for 4 variables

The algorithm of Creation's problem-solving instructions to the universe as a Theory of Everything

We have seen how the algorithm of Creation has given instructions and rules to the universe to calculate an answer to the problem: how to create and maintain, and in due course end, a universe and life, including its evolving to humankind; and whether the universe is an accident or has a purpose. At their simplest, the trillions of instructions to every aspect of the universe include:

- +A (day) + –A (night) = 0 (a 24-hour revolution of the spinning earth within the underlying unity of the One universe).
- +A (a life) + –A (its death) = 0 (a life span on the earth within the underlying unity of the One universe).
- +A (time) + –A (timelessness, eternity) = 0 (the One universe with its events from which time is derived and its underlying eternal Void).
- +A (the finite expanding universe) + –A (the surrounding, permeating infinite) = 0 (the One universe with its finite expanding space and the surrounding infinite the surfer encounters).
- +A (male) + –A (female) = 0 (the One universe with its means of procreation to renew species).

The instructions for the Creation and the development and end of the universe I have set out amount to a consistent Theory of Everything. The algorithm of Creation has provided a set of instructions and rules that will help calculate an answer to a problem: how did the universe rise and develop, and what will be its fate? The origin and creation of the universe and the entire life of Nature, animals and humankind can be seen as permutations of +A + −A = 0 (the algorithm of Creation applied to the stages after the Big Bang rather than to the full origin, development and end of the universe, 0 = +A + −A = 0).

Putting together the +A + −A = 0s of the 20 stages to give one algorithm of Creation

Before setting out a Theory of Everything I need to put together the 20 variations of +A + −A = 0, the instructions to particles, atoms and cells to produce in order to advance the universe in its 20 stages, to see how similar they are:

1. The structure of what is:
   - 0 (the One universe) = +A (infinite Non-Being and Being) + −A (finite Existence).

2. The origin and creation of the universe: Form from Movement:
   We need to invert the 0, and the +A and −A, to contrast virtual ripples before Existence with the real ripples after Existence began.
   - +A (the vacuum or Void or 'ocean of energy' in Nothingness, a Plenitude, a Fullness) + −A (virtual ripples or wrinkles, infinite movement within Nothingness, a virtual or potential structural plan for the coming universe) = 0 (the universe whose aftershock after the Big Bang materialised the virtual structural plan into real ripples or wrinkles, the structure of stars, galaxies and clusters of galaxies).

   We can now revert to beginning with 0 as we focus on what happened before Existence began.
   - 0 (the One point- or dot-sized universe) = +A (+p, one of the first pair of particles) + −A (−p, one of the first pair of particles);
   - 0 (the infinite universe) = +A (+p, the surviving one of the first pair of particles) + −A (−S, the spiral in, M−S, infinite movement limited by irregular movement); and
   - 0 (the One universe) = +A (+p in B, the surviving particle of the first pair of particles in Being) + −A (E, Existence).

3. The concept of the Big Bang:

We can now invert the 0, and the +A and –A, as Existence has begun.

- +A (the infinite singularity in Being's quantum vacuum) + –A (finite Existence's real particles) = 0 (the One universe).

4. The quantum vacuum and infinite order:
   - +A (the self-organising, self-regulating order principle in Being's quantum vacuum) + –A (Existence) = 0 (the One universe).

5. The immediate aftermath of the Big Bang: the shaping of the universe through expansionist light and contracting gravity:
   - +A (acceleration through hypothetical dark energy, with a critical density of no more than 5 or 6 hydrogen atoms per cubic metre) + –A (gravity with hypothetical braking of dark matter) = 0 (the One universe).

6. The formation of galaxies of stars and the observable universe:
   - +A (expanding universe of hydrogen atoms after inflation) + –A (gravitational cooling and condensing into gaseous clouds, galaxies and stars, their cores converted to helium) = 0 (the One universe).

7. The formation of the solar system and earth:
   - +A (expanding universe of gaseous matter) + –A (gravitational cooling and condensation into matter and planets revolving round sun, and wispy planets of hypothetical dark matter) = 0 (the One solar system).

8. Acceleration of the expanding universe, the surfer and the infinite:
   - +A (acceleration due to thrust from hypothetical dark energy of the expanding universe with a critical density within a factor of ten from $\Omega = 1$ and of no more than 5 or 6 hydrogen atoms per cubic metre, and a density of matter that is *less* than critical density (0.3 in relation to 1) + –A (gravity with hypothetical braking of dark matter) = 0 (the One universe).

9. The bio-friendly universe and order:
   - +A (40 orderly bio-friendly conditions, 326 constants and 75 ratios) + –A (random accidents and disorder) = 0 (the One universe that is creating the conditions for life).

10. The origin of life: the first cell:
    - +A (self-organising ordering of self-replication of cells and DNA into organisms) + −A (molecules from water heated by lightning and forming a cell) = 0 (the One universe).

11. Evolution and environment: expansionism and Darwinism:
    - +A (expansionist, self-organising ordering in self-replicating cells of organisms with purposive coded messages in DNA in bio-friendly environment) + −A (random, accidental, reductionist natural selection in harsh environment of at first rocks, Ice Ages and limited oxygen) = 0 (the oneness of all living things that evolved from the first cell in the One universe, life in relation to the One).

12. Ecology: Nature's self-running, self-organising system:
    - +A (Nature's self-organising system that advances and organises through self-improvement and self-betterment) + −A (random, accidental natural selection in harsh environment) = 0 (the oneness of all Nature that evolved from the first cell in the One universe).

13. Physiology's order within self-regulating body, brain and consciousness:
    - +A (body's, brain's and consciousness's self-organising system and homeostasis that advances and organises through self-improvement and self-betterment) + −A (random, accidental body, brain and consciousness reacting to harsh environment) = 0 (the oneness of all body, brain and consciousness that evolved from the first cell in the One universe).

14. Civilisations of history based on a vision of metaphysical Reality:
    - +A (orderly stages of history's civilisations based on a vision of metaphysical Reality, the Light) + −A (random, accidental history based on historical events) = 0 (the One universe with one world history).

15. Metaphysical and secular approaches in seven disciplines:
    - +A (a traditional metaphysical approach and vision of Reality, the One) + −A (a secular, social approach) = 0 (the One universe seen through seven disciplines).

16. Humankind's opposites:
    - +A + −A (pairs of opposites or antonyms) = 0 (oneness, reconciled within the One universe).
17. Humankind unified in a World State:
    - +A (supranationalism) + −A (nation-states) = 0 (the political union of all humankind in the One universe).
18. Humankind's spiritual survival after death:
    - +A (metaphysical non-duality of matter and mind that transcends materialism) + −A (contact with the ocean of Being, the Light) = 0 (the One universe that includes a fifth dimension and the spirit's survival after death).
19. The ultimate fate of the universe:
    - +A (expansion and acceleration of the universe through hypothetical dark energy) + −A (gravity and braking of hypothetical dark matter) = 0 (Heat Death followed by Big Freeze at the end of the One universe that is returning to one point or dot).
20. The end of the universe:
    - 0 (the One universe which has become a point or dot in the quantum vacuum's infinite Void) = +A + −A (the last pair of particles, the process of Creation in reverse).

What the 20 variations of 0 = +A + −A = 0 have in common

What the above 20 variations of 0 = +A + −A = 0 have in common can be listed as follows:

	+A	+ −A	= 0
1	infinite, Non-Being, Being	finite Existence	the One universe
2	'ocean of energy' +p +p − B +p in B	virtual ripples/ structural plan −p −S −E	the One universe with real ripples/structure of stars, galaxies and clusters of galaxies
3	infinite singularity	finite particles	the One universe

4	self-organising, self-regulating order principle	Existence	"
5	acceleration	gravity	"
6–8	expanding universe and gas	gravitational condensation into matter (galaxies, stars and planets)	"
9	bio-friendly conditions, constants, ratios	random accidents	the One universe creating the conditions of life
10	self-organising, ordering self-replication of cells and DNA	molecules	"
11	expansionist self-organising ordering in self-replicating cells of organisms and DNA	random natural selection	oneness of living things
12	Nature's self-organising	"	oneness of Nature
13	body, brain and consciousness self-organising	"	oneness of body, brain and consciousness
14	civilisations based on vision of the Light	random historical events	the One universe with one world history
15	metaphysical approach	secular, social approach	the One universe seen through seven disciplines
16	pairs of antonyms $\rightarrow$		oneness reconciled within the One universe
17	supranationalism	nation-states	political union of all humankind in the One universe
18	metaphysical non-duality of matter and mind	contact with Being, survival after death	the One universe that includes a fifth dimension and the spirit's survival after death
19	expansion and acceleration through dark energy	gravity and braking of dark matter	Heat Death followed by Big Freeze at the end of the universe

20	the last pair of particles, the process of Creation in reverse	$\longrightarrow$	the One universe as a point or dot in the quantum vacuum's infinite Void

The algorithm of Creation that works for all

All this boils down to: +A (infinite singularity, expanding, self-organising, ordering, expansionist, supranational, metaphysical acceleration) + −A (finite Existence, particles, gaseous matter, gravity, random accidents, natural selection, secular/social approach, nation-states, braking) = 0 (the One universe with oneness in Nature, body, brain, consciousness, history, humankind, fifth dimension for survival after death, finally a point).

This further boils down to: +A (an infinite singularity expands through hypothetical dark energy, self-organises, orders and accelerates with a metaphysical non-duality of matter and mind and a supranational approach to humankind) + −A (finite existence's hypothetical gravitational braking from dark matter, random accidents, natural selection, secular/social approach and a nationalist approach to humankind) = 0 (the One universe, at one with Nature, body, brain, consciousness, which can be spiritual and survive death, and at one with humankind and all disciplines and can create a World State before it returns to being a point).

The above pair of opposites (+A + −A) which are reconciled in a unification (=0) represents the +A + −A = 0 of the algorithm of Creation. What does the algorithm of Creation do? The above +A + −A = 0 sends instructions to the universe in a pair of opposites whose many variations are unified and reconciled in – and by – the harmony of the One universe.

We have seen that all the variations of +A + −A = 0 are in essence one variation, an algorithm or set of instructions to the universe, Nature and all living creatures including humans and their bodies, brains and consciousnesses essentially to solve the problem of how they should conduct their physical, mental and spiritual lives.

In short, the entire process of Creation from the atoms of matter to the self-replicating cells of Nature and the systems of homeostasis in all bodies, brains and consciousnesses is driven by a self-organising order principle which works in atomic co-ordination and the DNA of living creatures to make Nature work as one system. And +p + −p, the first two virtual/real particles, were self-organising and self-regulating. And so the algorithm of Creation is confirmed as being fundamentally +A (expansion of order, self-regulation and Light) + −A (contraction of gravity and braking) = 0 (the balanced matter and organisms of the One universe). And this points to its being the algorithm of the Theory of Everything that unites the four forces, including gravity, and quantum theory, which were all united within the singularity that underwent the Big Bang and (in the case of gravity) separated 10^{-43} seconds after the Big Bang, and can therefore reunite in the final singularity at the end of the universe.

A Theory of Everything

A Theory of Everything includes a unification of the four forces of quantum physics and relativity, which were united until 10^{-43} seconds after the Big Bang, and to replicate the temperature in the minute time they were united would require a particle chamber temperature that can never be reached:[1]

> There needs to be a complete theory of gravity that includes quantum mechanics. This has not been found. Calculations on quantum gravity lead to mathematical infinities. Much of recent physics has involved a largely unsuccessful attempt to bring together classical relativity and the new quantum theory, to integrate the very large and very small. Yet enough work has been done to show that the effects of quantum mechanics were crucial during the first 10^{-43} seconds after the Big Bang, when the universe had a density of 10^{93} grams per cubic centimetre. The era until 10^{-43} seconds after the Big Bang, the Planck era, is often also called the "quantum era", when the entire tiny universe would have been full of uncertainties and fluctuations, with matter and energy appearing and disappearing out of a quantum vacuum in the split second at the very beginning....
>
> Physicists are attempting a Grand Unified Theory by working towards the reunification of three of the four forces (excluding gravity), but the temperature at which they diverged, and might therefore

presumably reunite (somewhere between 10^{14} and 10^{27}K, and perhaps 10^{28}K, the temperature for the first 10^{-35} seconds after the Big Bang), is too high to replicate. The particle accelerator chambers would have to extend far into space to achieve such a temperature, and this looks a forlorn effort.

In an article in *The Daily Telegraph*,[2] Hawking claimed: "By colliding particles at the kinds of energies that would have been around at the Big Bang, we have unified three of the four forces: electromagnetics, and the weak and strong nuclear forces." The article was a transcript of a two-part television programme shown later, and Hawking repeated the claim word for word in the second programme. He should have said "theoretically and potentially unified" or "in low-energy consequences". It is understandable that, being disabled and only able to communicate at three words a minute, he should use words sparingly, but the unification of the three forces has only been solved in Hawking's mind. In *A Brief History of Time* he writes: "The strong nuclear force gets weaker at high energies. On the other hand, the electromagnetic and weak forces, which are not asymptotically free, get stronger at high energies. At some very high energy, called the grand unification energy, these three forces would all have the same strength and so could just be different aspects of a single force....A machine that was powerful enough to accelerate particles to the grand unification energy would have to be as big as the solar system....Thus it is impossible to test grand unified theories directly in the laboratory."

CERN have confirmed to me that the maximum temperature its particle chamber can reach is 10^{17}K. On 5 November 2007 a CERN expert wrote of the Large Hadron Collider (LHC): "The average collision energy at the LHC (with beams of protons) will be around 10TeV which correspond to about 10^{17}K. We don't know if and where the unification of forces will happen but the LHC is certainly the first-ever-instrument to provide such a high energy." The temperature at which grand unification takes place may be 10^{28}K. Although CERN have preferred to remain silent when I have twice written to challenge Hawking's claim, we can take it that there has been no practical grand unification and that Hawking was either talking in principle or overclaiming in the article and television programme, which many will have read or seen.

From September 2008 the LHC conducted the world's largest scientific experiment in the 17-mile-long circular tunnel 300 feet under the Alpine foothills near Geneva. Particles in proton beams raced round the route of pipes which contained a vacuum 11,245 times a second at 670 million miles per hour to smash headlong into each other at temperatures 100,000 times hotter than the centre of the sun (which is 13,600,000K)[3], i.e. 10^{12}, to reveal the first particles that existed in the split second after the birth of the universe in the Big Bang. The largest of four detectors, Atlas, which is bigger than Canterbury Cathedral, looked for the Higgs boson. It is hypothesised that this only exists at very high energies. Immediately after the Big Bang all particles are thought to have had no mass, and as the temperature cooled, it is suggested a field of theoretical bosons consisting of mass and little else came into existence which stuck to them, making them heavy. Theoretical bosons were proposed by Professor Peter Higgs, a particle physicist at Edinburgh University, in 1964. If they found the Higgs boson, it would become the standard model of particle physics. [It has since been found.] It would establish that there is a field like *aither*, which may be the ether that was not discovered in the 19th century, which may explain variability as well as mass. If they did not find it, then Nature had another way of giving particles mass – which would turn science on its head. It may be that there is an *aither* or metaphysical Light which is an ordering principle that gives particles their mass, operating like Higgs bosons, and that it has varying intensities round different particles and in different localities, embodying hidden variability. The huge experiment was expected to create 15 million gigabytes of data a year (equivalent to 21.4 million CDs), and scientists have had to create a new form of the internet to cope with it. Many of the findings might therefore not be known for some while.

Physicists hope that all four forces can be reunified to include gravity and create a Theory of Everything, but the temperature to be replicated, 10^{32}K, would be even higher than the temperature to reunify three forces, and the particle accelerators even larger. Nevertheless, if the technology could be devised, all four forces *can* be reunified as they were unified at the beginning of their life shortly before 10^{-43} seconds after the Big Bang.

A Theory of Everything must go beyond both quantum physics, which successfully explains much of the particle, and the general theory of relativity, which deals with the universe and gravity. Both these theories work well, and so a Theory of Everything must include both. All particle theories before string theory lead to infinities when gravity is included. These infinities cannot be renormalised and cause the equations to break down. The five main candidates for a Theory of Everything emerged from hypothetical, unevidenced string theory in the 1980s and from an underlying hypothetical, unevidenced "M" theory (M for Mystery) that has not yet been found. The main candidate predicts that the universe has 10 hypothetical dimensions instead of the three actual spatial dimensions, and the other seven have not been found.[4] Universalists leave multi-dimensional theories to one side as mathematical speculations that cannot be proved.

The search for a Theory of Everything, putting together the very large (relativity) and the very small (quantum mechanics) to explain quantum gravity, has progressed in my lifetime. As we have seen, Oppenheimer and Snyder introduced black holes in 1939. Penrose established that a universe could collapse into a singularity, a black hole of infinite density, and reversing this, running the film backward, Penrose and Hawking established that a universe could come out of a singularity (now a white hole). Later Hawking established that a black hole (if it exists) must have a glow of radiation round it as pairs of particles drawn towards it split. One, the positive particle, escapes, giving off radiation, while the other, the negative particle, disappears into the black hole. Black holes never lose their heat and so have the potentiality to explode. We can see how a Big Bang could come out of a singularity, but how did gravity behave in the singularity before the universe inflated? It must have held back the expanding force of light or radiation within the singularity until it could hold it back no longer. A Theory of Everything requires an expanding force of light or radiation within the singularity, against which quantum gravity acts as a contracting force. These two opposites must both have been present in the first singularity before the Big Bang. A Theory of Everything cannot be established until it is shown what was behind these two opposites, light and gravity, and what lay outside the singularity before the Big Bang: the infinite.

Three of the four forces are of roughly the same strength, but gravity is much weaker. (A magnet can pick up a chunk of iron on a mountain in defiance of the entire earth's gravitational pull.) The original symmetry when there was one superforce within the dot-sized universe shattered soon after the Big Bang, and whereas the fragments of the electromagnetic, weak and strong nuclear forces can be put together theoretically by weakening the strong nuclear force at high energies and strengthening the electromagnetic and weak forces at the same high energies,[5] gravity seems irreconcilable as it is so weak. I am convinced that the most likely path to a Theory of Everything is not by speculating regarding superstrings, M theory and ten or eleven dimensions and seeking to put together the very large and the very small in black holes – pinning hopes on a multiverse – but by revisiting the idea Newton worked on: that there is an expanding force in light which counteracts the contracting force of gravity. In other words, gravity is weak because of the push of light against it. If physicists move away from string and M theory and analyse the pushing photons of light in relation to the pulling of gravity, I believe they will find a Theory of Everything – but it must include the infinite from which the pre-Big-Bang point emerged. To assert with Hawking[6] that the earth had no beginning or singularity or boundary but has not existed forever, having somehow emerged like a bubble, is a colossal logical and scientific fudge.

For decades cosmologists have accepted the notion that the universe of space-time began as a singularity, whose temperature, density and everything else about it were infinite. If a Theory of Everything marries quantum and gravity, the marriage will have to permit actual infinity which fills the point or singularity from which the universe began.[7]

At a conceptual level, our survey of quantum theory has made evident zero-energy's links with the infinite. The quantum vacuum contains the latent potentialities of existence: virtual particles. It is finite within the inflated balloon of the universe, but it emerged from the infinite singularity in the first second after the Big Bang and has its origin in the infinite. In this sense it is at one with the limitless, boundless infinity outside the balloon of the expanded singularity. Having emerged from the Void that preceded the Big Bang, its emptier, pre-space-time form, the quantum vacuum now contains more virtual particles than when there was a Void or empty nothingness.

Quantum theory plays down the significance of when time along with space began at the moment of the Big Bang, and of when time will end if the universe ends in a Big Crunch (as is thought not to be the case [since found to be likely because of the gravitational braking of dark matter]). However, before time began, underlying time and continuing after time ends was, is and will be the infinite timelessness that pre-existed the universe and will survive it: the Void from which the hot beginning and inflation happened and which has yielded, since the Big Bang, a seething quantum vacuum.

Overlaid on this underlying timelessness are spatial events as we saw. Time is a succession of 10^{22} (10 thousand million million million) spatial events every second. That colossal number of events has happened every second since the earth cooled 4.55 billion years ago. In this Leibnizian sense, time has no reality of its own but is an artificial measurement of spatial events which succeed each other. Einstein, discussing time with his friend Besso on a May evening in 1905, saw time as relative, like H.A. Lorentz's "local time", and he linked time with space and matter, the curvature of space distorting time. Space to Einstein did not have a primary independent existence as it did for Newton, and time was also secondary rather than primary.[8] Within Einstein's mathematical theory, clock time is an artificial measurement of spatial events. I would say that the underlying timelessness is primary and absolute, and the measuring of spatial events by time is as relative as the finite is within the infinite. Time is not moving towards disorder (a point accepted by most modern cosmologists and theoretical physicists including Einstein, Hubble and Bohm) as the universe appears to be expanding forever and as it is overlaid on timelessness and infinity, from which it emerged....

We have been holding order at the back of our minds. The evidence for the universal principle of order, or order principle, in physics, then, is the order within the quantum vacuum which may have produced our universe from zero energy and may counterbalance gravity with an expanding force of light which may control chemical composition and growth. This order is reflected in Einstein's constant and in

Bohm's order. This hidden order may be found to control the apparent randomness of the uncertainty principle by determining which nuclei decay through hidden variability. It may explain how the fall-out from the Big Bang created such perfect conditions for life to form.

All this follows a variation of +A + −A = 0: +A (an expanding force of light and quantum physics of light with virtual photons of Light becoming real photons of light and a hypothetical to-be-discovered order principle, a fifth force that holds together 326 constants and 75 ratios, see pp.69–85) + −A (relativity theory and a contracting force of gravity with electromagnetic, weak and strong nuclear forces) = 0 (the One universe).

Einstein attempted to reunite these four forces:[9]

Einstein spent the last thirty years of his life working unsuccessfully, trying to re-integrate gravity with the other three forces, attempting a unified field theory in which the strong and weak forces, electromagnetism and gravity would be united in an underlying field.[10]

But because Einstein never found a hidden principle of variability does not mean that it cannot be found. The Universalist philosopher detects within the hitherto undiscovered hidden variability workings of a universal principle of order that is behind the apparent randomness of all nuclei and the uncertainty principle. May this hidden principle also include the workings of Newton's expanding force? And may it be associated with metaphysical Light?

Einstein sought order and a Theory of Everything in a unified field theory – a hidden principle of variability. This work has presented the algorithm of Creation as instructing this.

It follows a variation of +A + −A = 0: +A (order in hidden variability which unifies strong and weak forces, electromagnetism and gravity) + −A (apparent randomness of nuclei and the uncertainty principle) = 0 (the One universe with a unified Theory of Everything).

The idea is right; it is the detail and contextual perspective that need to be provided, which eluded Einstein and Bohm and which (having spoken with Bohm) I am now continuing.

An informed observer's input into a Theory of Everything: rational and intuitional approaches to Reality and the universe

Observation of the universe comes into the statement of a Theory of Everything, and an informed observer can have both rational and intuitional approaches to Reality and the system of the universe. It is important to include the informed observer's contextual perspective and input into a Theory of Everything.

(1) The rational approach to Reality, metaphysical philosophy and the system of the universe as light: the Law of Order and the Law of Randomness

Drawing out the threads of what has gone before, we find a clear narrative involving a Law of Order:[11]

1. Before the Big Bang there was an eternal, ever-moving boundless or timeless infinite Reality, the Void or One, an "ever-living Fire".

2. The universe manifested from this infinite Reality as a point or germ, the first singularity, and, soon after the Big Bang, expanded by inflation, and the expansion has accelerated. A constant may save the universe from slowing down and perishing in a Big Crunch. The universe is finite and shaped like a shuttlecock or cone, open at the top. Space-time is within it.

3. Outside the universe, all round it and within it is the boundless, infinite Reality, a sea of energy which the surfer breasts. It consists of Light (Uncreated Light), and may be *"aither"* (ether) or dark energy, a moving process or flux. This contains the latent possibilities of Being and Existence from which our universe was thrown up. It manifests into a quantum vacuum, an emptiness filled with order.

4. There is some evidence that there was a predisposition to order in the very early universe. This includes land mass and climatic conditions which favour life. There are 40 bio-friendly laws and many ratios and constants that were exactly right for life and allowed human life to happen. The order principle can be detected at all stages in the universe's life and exists alongside the apparent randomness and uncertainty.

5. The order principle, or Law of Order, has been associated with

and may be contained within the photons of physical light, which may travel within metaphysical Light as Newton thought (his "spirit" and "ether"). This multi-layered Light may manifest into an expanding force of light that counteracts gravity, stimulates chemical composition and stirs plants and the bodies of organisms to growth. It also controls the four forces, particles and all forms of matter.

6. It is not clear how life began but the drive of evolution, an order and life principle, from one cell via apes and Ice Ages to modern *Homo sapiens'* body, brain and consciousness has perfected many self-regulating, self-organising physiological systems. It has developed higher (or deeper) consciousness, part of which can know Reality. The ordering thrust of evolution may be borne to cells by the information in photons, a tenet of Expansionism.

7. All living things share oneness through DNA, and take part in a self-running, self-organising system in which all creatures know their place and what to do. The order principle can be seen to operate through competition and co-operation: in the food webs, niches, symbiosis and healing plants of Nature's vast ecosystem. The focus of science is on the whole planet. The earth is now in an interglacial within a continuing Ice Age.

8. The constant of homeostasis is self-regulated in the self-organising systems of both body and brain. Photons carrying information enter the brain through photoreceptor cells and are converted into chemical energy along neural circuits. There are 12 levels of consciousness. Mind may be partly an effect of brain and partly a system of light.

The scientific view of "what is" can be expressed more succinctly:

The finite universe manifested from an infinite, timeless, boundless Reality or Being which surrounds and permeates it and is both transcendent and immanent. Within the infinite the photons of physical light stimulate all growth, creating bio-friendly conditions for life through a Law of Order which works in Nature's local ecosystems, keeping all creatures in their place and driving to higher consciousness.

Philosophy, then, hears a message from the scientific view of the universe and reshapes and restructures itself accordingly. If there is a universal order principle or Law of Order that develops brains through evolution and self-organisation into higher and higher (or deeper and deeper) consciousness, then philosophy needs to reflect this principle. We have seen that the previous models of philosophy are at variance with this model of the universe. Rationalism, Idealism and Intuitionism, and Empiricism, Realism and Logical Analysis by themselves do not reflect the findings of the scientific evidence and my focus on the universal order principle. The reason, sense impressions, intuition and logic are not sufficient by themselves for the task of describing the universe of the 21st century. The operation of the universal order principle or Law of Order means that only manifesting metaphysical philosophy reflects this model of the universe.

In the rational approach to Reality light is fundamental to the order in the universe:[12]

We have seen that Being is a cosmic sea of energy, the potentialities of Existence. The study of Being in philosophy is ontology, which is not Materialistic or phenomenal (i.e. the study of visible existence). Many philosophers have seen Being as fundamental to the universe. In the 13th century, Robert Grosseteste wrote of Being as the Uncreated Light[13] which is the potentiality of form, physical light, and which, he claimed, is God. In this he was following a tradition that since Plato has seen Being as metaphysical Light. Heraclitus wrote of the world order as an "ever-living Fire", an idea close to the Uncreated Light. Plato wrote of "Universal Light",[14] which he imaged as the fire in the cave, meaning the Light behind the phenomenal world which appears as shadows. Augustine wrote of the "Light Unchangeable"[15] ("God is the Intelligible Light"). Mechthild wrote of the "Flowing Fire and Light of God".[16] Dante wrote of "Eternal Light",[17] "pure intellectual Light"[18] and the "sempiternal rose".[19] The Light of Being is known in enlightenment in the world's religions. We have seen that Newton thought that metaphysical Light was associated with physical light, and sought an expanding force of light that would counteract gravity.

Physical light was extremely important to the early universe. For the first 10^{-37} seconds there was no matter or radiation, just a force that was perhaps dark energy hurtling the universe into expansion. A short time later this force or dark energy in the vacuum transformed itself into radiation and empty space was filled with physical light and matter, mostly physical light.[20] During inflation the universe suddenly increased in size by a factor of ten trillion quadrillion (10^{28}), a factor a hundred thousand times larger than the number of galaxies in our observable universe (10^{23}).[21] The initial radiation cooled into protons, neutrinos and photons, which coalesced from the early physical light that filled the vacuum. Protons decay into photons, positrons and neutrinos. Before a second had elapsed, protons were generated and when nuclear activity processed a quarter of these into helium there were a billion times more photons than baryons. The microwave background radiation is composed of photons. Physical light, radiation and photons were fundamental while matter formed. When matter coalesced it comprised a tiny amount of what filled the empty space caused by inflation. Today, as we have seen, matter comprises 4 per cent of the universe, with dark energy (an anti-gravity force) and dark matter (matter that remains undetected because it does not emit electromagnetic radiation) thought to fill the rest.

In today's universe, light has wave-particle duality, sometimes behaving in waves and other times as particles. Matter also has wave-particle duality. De Broglie first saw the wave properties of matter in terms of photons in view of the wave-particle nature of electromagnetic radiation.[22] Both light and matter emerged from the radiation of the early universe, and both have wave-particle duality. Atoms have electrons and exist among charges and magnetic fields, and it is still not fully understood how the atoms of matter and the cells of organisms emerged alongside the photons of light.

The Universalist philosopher proposes that the essential potential substance of the finite universe – the dark matter, if it exists, the glue which holds galaxies together and which exerted the force that may have been dark energy of the first 10^{-37} seconds – is *aither*-like metaphysical Light, from which, he proposes, manifests physical light. Newton saw light as an expanding force which counteracts gravity, and since dark energy (if it exists), thought to comprise 73 per cent of the matter of the universe, is just such an anti-gravity force it is reasonable to associate dark matter/energy and light and to see dark matter/energy as perhaps being a latent potential of light. Dark matter (if it exists), thought to comprise 23 per cent of the matter in the universe, does not emit electromagnetic radiation and may consist of neutrinos, weakly interacting massive particles, rather than electrons and protons. Neutrinos, detected experientially in a nuclear reactor in 1953, have no charge and have a mass close to zero (perhaps about a millionth of the mass of an electron), always travel at the speed of light and can pass through the earth and our bodies.[23] Photinos also have a mass close to zero.

Both photons and neutrinos are created from the decay of protons. Neutrinos may be associated with photons in being emitted by the radiation following the Big Bang and there are the same number of neutrinos as photons (10^{87}).[24] It looks as if as both travel at the speed of light and exist in equal numbers, and as both are created in decay there is a symmetry between the equal amounts of neutrinos and photons, neutrinos representing the dark and counterbalancing photons, light. Dark matter and dark energy may be the manifestation of the metaphysical Light, the potential of photons, and may be composed of particles smaller than neutrinos such as photinos if they exist. (According to supersymmetry, photinos are ½ spin particles that partner photons.)

In our algebraic, dialectical thinking, just as [in this variation] +A + −A = 0, so photons + neutrinos = a balance between light and dark, a physical *yang* (light, −A) + *yin* (dark, +A).

It will be helpful to list the four kinds of light that operate during the manifestation process:

- Nothingness/Fullness – infinite metaphysical Light;
- Non-Being – the infinite Void outside the universe: the potentialities of infinite/finite Being, infinite metaphysical Light, virtual tiny photons;
- Being – the potentialities of finite Existence, perhaps dark matter/energy, virtual neutrinos or photinos manifesting into actual neutrinos or photinos beyond gamma rays on the

electromagnetic spectrum which convey the order principle into consciousness, seen as metaphysical Light, Uncreated Light or Intelligible Light; and

- Existence – the finite manifestation of Being: photons which photosynthesise and convey the order principle into bodies and brains via physical eyes/retinas.

In terms of $+A + -A = 0$, virtual neutrinos/photinos of Being + actual photons of Existence = all manifested metaphysical-physical light.

It will also be helpful to list the parts of the self that receive the four kinds of Light:

- Nothingness/Fullness – purest infinite Light of finest density, too pure to be seen clearly in universal being, the level of consciousness that has an infinite component;
- Non-Being – a denser metaphysical Light but still too fine to be seen in the universal being;
- Being – neutrinos or photinos or tiny photons smaller than neutrinos received in the universal being (level 12) as illumination in the Western tradition and in the base-of-spine *chakra* within the subtle body in the Eastern tradition; and
- Existence – photons received by retinas and transmitted to brains and bodies via objective consciousness/the soul (levels 7, 8 and 9).

On this view, the metaphysical Light manifests from the very fine and subtle to tiny, denser "consciousness" neutrinos or photinos and to much denser physical photons.

In fact, the rational universe is a system of light as *The New Philosophy of Universalism*'s section 'The Rational Universe as a System of Light' describes:[25]

From the rational viewpoint, the Universalist philosopher holds that the universe is a system in which invisible substance has manifested into visible Nature and matter. Energy from the invisible Reality of Being manifests into Existence, whose phenomenal forms and systemic workings bear witness to an orderly origin. The intricacy of Nature's ecosystem, whose photosynthesising growth is spurred on by photons, suggests organising behind it as shadows suggest a sun. The Universalist philosopher holds that the metaphysical Reality is hidden behind the world of appearances and phenomena which conceals it just as the Chinese characters *Chin Hua*, "Golden Flower", conceal a central character *Kuang*, "Light". (In the 9th century AD it was heretical to speak of metaphysical Light in China, and so the character for Light was hidden between two other characters.)[26]

Universalist philosophers hold that the manifesting energy comes into the universe along with coded instructions for ordering phenomena, which may be borne by light. These instructions may include a drive for self-improvement or self-betterment in relation to the environment, so that each organism realises the next stage in its development. The electromagnetic radiation of different frequencies interacts with matter in different ways, and has done so since the very early universe when both electromagnetic radiation and matter emerged from the infinite via the Big Bang. The electromagnetic spectrum reveals an ordered system. The nearest rays are long radio waves (with a frequency of $10\text{-}10^5$ cycles per second). Normal radio waves have a frequency of 10^6 cycles per second. Short radio waves have a frequency of $10^7\text{-}10^{12}$ cycles per second. The cosmic microwave background radiation has a wavelength of about 1.1 millimetres (about 10^{-3}) and a frequency of 217 million cycles per second – over 10^8 cycles per second, which places it in the radio region.[27] Radar and TV come next, and then infra-red (with a frequency of $10^{12}\text{-}10^{14}$ cycles per second). Beyond is visible light with a frequency of 10^{14} cycles per second. Each colour within the spectrum of light has a different wavelength. Beyond that is ultraviolet (with a frequency of $10^{14}\text{-}10^{16}$ cycles per second) and beyond that X-rays (with a frequency of $10^{15}\text{-}10^{20}$ cycles per second) and gamma rays (with a frequency of $10^{18}\text{-}10^{23}$ and beyond cycles per second), all of which are absorbed by molecular oxygen, ozone and molecular nitrogen. Universalists hold that the invisible substance that manifests from the infinite and is present throughout the universe is beyond gamma rays on the spectrum.

Within this spectrum, and working alongside the four forces (the strong and weak forces, the electromagnetic force and gravity), Universalists hold, is an order principle, possibly in light as Newton thought, which via sunlight is responsible for all growth in our solar system. Not enough is known about the transmission of light over the vast distances of intergalactic space, and the theory of light overlaps with the science of cosmology. Today scientists are searching for one logical theory that includes all known terrestrial phenomena of light – and the intergalactic sources of light and their origin.

But it is known that light is converted to chemical energy. We have seen that in plants light is absorbed by chlorophyll, which is in the elongated chloroplasts of cells, and produces energy. In harvesting sunlight energy through the process of photosynthesis, plants also harvest order.[28] (In plants there is one exception to the rule that all photosynthesising cells contain chlorophyll: the bacterium *Halobacterium halobium*, which photosynthesises by using a coloured protein resembling the visual pigment rhodopsin).[29] Nature's vast plant ecosystem is controlled by light, whose information-bearing photons are perpetually converted into chemical energy.

It is also known that in animals and humans light is converted to chemical energy by being received in light-sensitive cells and organs, notably the retinas of the eyes, which are designed to receive light from the lenses.[30] We have seen that eyes are not just for looking out and for receiving images, they are crucially also for receiving light. The light-sensitive cells transform the light signals into a photochemical reaction which leads to electrical charges in cells and eventually nerve signals.[31] Sunlight is crucial to the maintenance of the feedback systems of the body and brain.

The reception of light in the light-sensitive cells of animals seems to be universal. Iguanas and lizards warm up by basking in the early morning sun. Sunlight penetrates through their eyes and radiates their internal circulatory system until their inner thermostat prevents the warming process from making them too hot. These animals are motionless until they have absorbed the light they require, and only then do they go about their daily business. For an iguana, the most important and essential event of their day is basking in the sun in complete stillness to absorb the light they need to function.[32]

Humans acquire the light they need without having to sunbathe. Even in blind humans, light penetrates to the light-sensitive or photoreceptor cells.[33] The light-sensitive cells in the vicinity of the retina still catch light even though the eyes of blind humans have lost their ability to see visual images. What happens if a man has his eyes gouged out like Gloucester in *King Lear*? Does light still penetrate to light-sensitive cells in the eye sockets? It apparently does if the removal of the eyeball has left sufficient photoreceptors near the back of the eyeball intact. What happens if a man has to live in the dark for a long period of time, in a windowless underground dungeon? Does he become like a plant wilting in a dark room? A plant needs light to survive, and is this also true of humans and animals? Humans and animals can be expected to live longer than plants under conditions of light deprivation. What happens to citizens of the Arctic region when there are not many hours of sunlight for parts of the year? All these questions need further investigation in relation to the light-sensitive photoreceptor cells.

This system of living things converting light into chemical energy operates universally throughout Nature. Light-sensitive photoreceptor cells and organs specialised to received light are found in every creature, even in unicellar forms. Primitive eyes are found at the edge of jellyfish, in the armpits of starfish and the gills of tube worms. Earthworms have light-sensitive cells within their skin that are connected to their nervous system, many near their front end, a few at their back and not many in the middle. From worms developed molluscs (for example, octopi), anthropods (spiders and insects) and vertebrates (birds and man). Molluscs have some sort of eye. The eye of the octopus and the human eye represent independent lines of evolution as do the wings of insects and of birds.[34]

The vertebrate eye in animals and humans has a lens which increases the eye's ability to collect light (besides making the image sharp). The iris is pigmented and opaque, the pupil is a transparent hole in its centre. The size of the pupil depends on light intensity (and also on psychological factors such as excitement or boredom). The eye is protected by the cornea and eyelids (or in birds, winking membranes). The lens focuses a sharp image of the world on the retina. Humans can focus on objects at different distances by changing the shape of the lens, which is achieved by changing the tension in the fibres supporting the lens. Octopi, fish and jumping spiders do likewise. The lens refracts light: blue light (short wavelengths) is deviated more than red (long wavelengths). Sensitivity to shortwave light and visual acuity are held in balance. An eagle's visual acuity is six times that of humans, a bee's one hundredth. The eagle sacrifices shortwave vision and the bee longwave vision.[35]

But while such adjustments are being made, light is turned into chemical energy via the eye. The absorption of photons in the eye initiates a photochemical reaction which leads to other chemical reactions and to an electrical change in cells. Photons cause chemical reactions in the eye's fumaric acid and turn it into maleic acid, and vice versa. Light-sensitive cells have to be combined with pigment-rich cells which extinguish light from one direction, assisting directional progress.[36]

Light also stimulates physical and chemical processes to have a profound effect on the direction of movement in animals and humans, and (in ways not yet known) stimulates an inner clock which enables organisms to behave predictably as regards the time of the day and the time of the year.[37] There is also a photobiology of the skin, and light can counteract rickets and nourish the skin with Vitamin D, which develops the skeleton. Light deprivation results in seasonal affective disorder ("SAD"). Sunlight can, however, have a deleterious effect as it can cause skin cancer and DNA can be damaged by exposure to ultraviolet light. In short, sunlight is central to the vast biological ecosystem which includes local ecosystems, for it controls growth and an orderly way of life.[38]

Looked at in philosophical terms from the perspective of the reason, evolution has taken place within bio-friendly, orderly conditions amid a continuous process of manifestation from the boundless infinite into Being and Existence of invisible metaphysical Light becoming physical light via the cosmological procedures which created suns, and our sun. The system provides for the development of bodies and brains with an order that, if not purposive, attends to every need and deficiency.

Just as there is a Law of Order, so there is a Law of Randomness:[39]

In philosophical terms, within a philosophical category, I have frequently mentioned the universal principle of order or the order principle, to which Universalism draws attention. A principle in physics, such as the uncertainty principle, is "a fundamental truth" or "a highly general or inclusive theorem or 'law', exemplified in a magnitude of cases" (*Shorter Oxford English Dictionary*). Part Two contains evidence for a magnitude of cases in which a principle of order is at work. A law, as in "the laws of Nature" or "the law of gravity", is "a theoretical principle deduced from particular facts, expressible by the statement that a particular phenomenon always occurs if certain conditions be present" and "the order and regularity in Nature expressed by laws" (*Shorter Oxford English Dictionary*). ("Regular" means "conforming to some accepted rule or standard".)

Part Two has shown that order is always present in an ecosystem's conditions, and I have already defined order as "the condition in which every part or unit is in its right place, tidiness" – which covers how fish, birds, reptiles and animals regularly share feeding and breeding sites to make food go round. The instances of order in Part Two confirm a regularity of natural occurrences of order in particular instances ranging from 40 bio-friendly conditions to symbiosis, medicinal plants and physiological systems, all of which conform to a rule. I can therefore now speak of "the *Law* of Order" in place of "the principle of order". I can say that Universalism draws attention to the Law of Order.

In philosophical terms, within a philosophical category, approaching the universe objectively through the reason, I can in fact identify two conflicting laws:

1. The Law of Order, which manifests from the infinite into the universe, perhaps as a fifth force, and probably operates through light via the eyes, is conveyed to all creatures. It controls the growth of all creatures, regulates their systems and pervades and controls their DNA. The Law states that order creates organised, methodical, highly structural, complex systems which require a lot of information to describe them and which are maintained by the active co-operation of the participants in the whole ecosystem which

includes all sub-systems (as outlined in Part Two).

2. Alongside it is the Law of Randomness which states that the universe is also permeated by an unmethodical tendency for the energy of organised complex systems to deviate from their self-organised directions. It states that the systems deviate either through adaptation or through faulty DNA replication into accidental mutations or into chaotic breakdown due to chance mishaps which may bring starvation and early sudden death. "Random" can be defined as "made, done without method or conscious choice".

The Law of Randomness is different from the Law of Entropy, another name for the second law of thermodynamics, which states that organised systems become less and less organised as time advances. The Law of Entropy[40] is about the running-down and decay of systems as a result of disorder at the molecular and atomic levels. It is not about their being subject to chance mishaps as when a chick is a random victim of a predator or a careless driver.

The Law of Order can be seen at work in a range of natural phenomena from bio-friendly conditions to contours of landscape, ecosystems, physiological systems, mammals, reptiles, insects, vegetation and plants. All living organisms show patterns of complex order that shape and ease their lives and their struggle against the conflicting Law of Randomness, which brings chaos, starvation and sudden accidental death.

As in my algebraic, dialectical thinking, just as $+A + -A = 0$ the two laws, though in apparent conflict, are two sides of a fundamental unity and they are held in balance like *yin* and *yang* by a constant. This can be expressed as +Order + −Randomness = unity. One law does not triumph over the other. Order contains within itself a propensity for the disorder of randomness – and, indeed, entropy – but due to the constant, neither randomness nor entropy succeed in destroying the universe at the end. Rather, randomness is a deviant from order's energy, which enables an organised system to adapt and mutate into a new form that is beneficial to the whole ecosystem, or hastens its end in the interest of maintaining a balance in the population of species, which is also beneficial to the whole ecosystem.

Thus the two apparently conflicting forces and laws, the Law of Order (which suggests an energy methodically organising all creatures from birth into their rightful place within a system) and the counter-balancing Law of Randomness (that chance mishaps affect the same range of creatures within the whole food-web without method) are complementary like *yin* and *yang*, and myths about Chaos and Order are found in many ancient cultures. The point is that both laws exist, and a view of the universe in terms of the Law of Randomness alone (the view of extremist, anti-Creationist Darwinists) does not truly and objectively reflect the universe.

Order may one day be established as a fifth force, an expanding force of initially dark, subtle light that is invisible to the outer eye like electricity but manifests from the infinite into the universe at the level of Being in so far undetected particles even smaller than neutrinos such as photinos. When explicit – manifested in Existence – it counteracts gravity, permeating all Nature, is passed on through generations by DNA and shapes evolution to an end that is self-organising and self-regulating. It may one day be as measurable and testable as the other four forces and be regarded as the most fundamental of all the forces as it balances gravity and is responsible for the growth of organisms via eyes and chloroplasts in chlorophyll. CERN has sought the field of Higgs bosons, which may in fact be the field of the order principle. It may have been discovered, measured and tested in 2008, although the results may not be known until a year or two later. [CERN confirmed on 4 July 2012 that they had independently observed a new particle, a boson consistent with the Higgs boson.]

Is there a teleological purpose behind the Law of Order in the universe? Aristotle in *Metaphysica* declared that a full explanation of anything must consider not only the material, formal and efficient causes, but also the final cause, the purpose for which the thing exists or was produced.[41] There was a move away from Aristotle's teleological explanations in the mechanistic 16th and 17th centuries, which did not echo Aristotle's argument that things developed towards ends internal to their own natures but held that biological organisms are machines devised by an intelligent being. In *Kritik der Urtheilskraft* (*Critique of Judgement*), 1790, Kant held that teleology can only guide inquiry rather than reveal the nature of reality.

Teleology is "the doctrine or study of ends or final causes, especially as related to the evidence's design or purpose in nature" and "such design as exhibited in natural objects or phenomena" (*Shorter Oxford English Dictionary*). To proceed rationally, I have set to one side the idea of intelligent design to focus on "what is" rather than speculate as to why it is as it is. However, while "order" can be defined and described in terms of regularity and the functioning of a system in the present ("every part or unit in its right place") it has other meanings which include "a state of peaceful harmony under a constitutional authority", and the question of the purpose of the present harmony cannot be far away. It is not known why Nature's system exists. We can only speculate as to its end and final cause, and human evolution is not yet complete. If we are not careful, thinking about the purpose of the universe can take us readily from scientific description, philosophical exegesis and metaphysical rationalism – whose language is neutral – into theology, religion and faith, different categories of thought.

I can however say that the many instances of order in Nature's system I have cited in Part Two rationally suggest that in the case of order rather than in the case of randomness there must be a teleological principle at work for if the ordering was not towards any end it would not have been successfully concluded in terms of the purpose for which the ordering was taking place. In the case of randomness, of course, there is no teleological principle at work for randomness just happens without necessarily leading to a goal. Furthermore, I can say that Nature manifested from the infinite and that humankind has evolved in a partly ordered, not totally random way, and that there may well be a purpose linked to the infinite substance, the "ever-living Fire" or metaphysical Light, the subjective experience of which we shall consider in the next chapter.

The rational answer to whether the universe has a purpose, in contrast to the intuitional answer, must be that in view of the Law of Order the infinite must have manifested into Nature for a reason, and that to the extent that evolution is controlled by order this reason must be associated with the evolution of humankind. It is therefore not impossible that each human being, linked by physical light to his or her source of growth and by metaphysical Light to the infinite, is fulfilling an infinite plan in his or her everyday finite living. In this chapter I am being rationally objective and the final answer must await a consideration of subjective intuition via the universal being in the next chapter, and is necessarily beyond the competence of the reason and the rational, social ego to fathom. Logically, however, if the universe is ordered it is purposive.

The rational approach to Reality – and to metaphysical philosophy, which includes both the infinite and the finite, and describes a non-duality (as we saw on p.206) – includes a rational approach to the four subdivisions of metaphysics (ontology, psychology, epistemology and cosmology) that identifies 14 basic laws:[42]

The leading 20th-century metaphysician Whitehead declared, as we saw, that "speculative philosophy" (i.e. metaphysics) "is the endeavour to frame a coherent, logical, necessary system of general ideas in terms of which every element of our experience can be interpreted". Every element of our experience must include all our experience of the finite world and all our experience (such as it is) of the infinite, all the elements of experience I listed under the 12 levels of consciousness. My system of ideas – manifestation from the infinite into Nature and bio-friendly conditions which helped life evolve – does include these.

A "system of general ideas" must be dominated by one of four metaphysical perspectives: Materialistic monism (that matter gives rise to mind, or scientific materialism); dualism (that matter and mind are separate substances, the Cartesian position); transcendentalist monism (that mind or consciousness gives rise to matter so that the ultimate stuff of the universe is consciousness, an Idealistic view);[43] and metaphysical non-duality (that a metaphysical Reality gives rise to both matter and mind, that a unity gives rise to a diversity). Whereas monism implies no diversity, non-duality implies both a unity at the transcendent metaphysical level of the infinite and diversity at the immanent physical level.

My "system of general ideas" is a non-duality. There is unity at the metaphysical, infinite level and diversity at the finite level. My system of coherent general ideas explains how unity became diversity by manifestation into matter, light, consciousness and the vegetative world of plants, how the One manifested into many.

Metaphysical philosophy includes both the infinite and the finite, as we have seen. (See the four-tiered process of manifestation.) Just as in algebraic, dialectical thinking +A + −A = 0, so the infinite (which is untestable and unmeasurable) + the finite (which is testable and measurable) = the All or One. Metaphysics is the science of the All; a "system of general ideas" that covers the universal Whole, which includes all known experiences and concepts. Universal science was sketched by Plato and followed by Aristotle,[44] but is now the science of the infinite and the finite – the science I have outlined in Part Two. "Every known concept" includes the paranormal in relation to a sea of Reality, and the after-life, whether or not there is proof. The very fact that it is a concept makes it worthy of consideration.

But here, as I am following the rational strand of metaphysical philosophy, a cautionary note. The reason concocts a system of general ideas which explains the reason's place in the rational universe. The reason's approach to the experiences of the 12 levels of consciousness does not embrace *all* the experiences encountered in those 12 levels. It does not include levels 9, 11 and 12 and may not include 10. We must wait until the next chapter to include the remainder of the experiences of all levels of consciousness. But the reason does see the individual in relation to a rational universe.

I said that today metaphysics has four subdivisions, which cover different approaches to Being. They are in descending hierarchical order:

- ontology (the study of Being), which includes traditional metaphysics;
- psychology (what can be experienced of Being) which includes some traditional moral philosophy;
- epistemology (what can be known about Being), which includes some ethics; and
- cosmology (the structure of the universe), which includes traditional natural philosophy.

The reason has its own approach to this metaphysical scheme, as we shall now see.

Rational metaphysical Universalism offers a universe which is permeated by the infinite but ruled by reason as Reality is a sea of Being which follows mathematical laws. The rational study of ontology is therefore a study of Being manifested into the universe whose structure is rational. To the rational philosopher, the study of cosmology is of its compliance with rational, mathematical laws. The rational metaphysical philosopher uses a rational psychology and sees the rational, social ego, which uses the reason or higher mind, as *thinking* about infinite Being. His scheme allows the inclusion of the soul but his medium is the reason. He is more concerned with Existence and, as our moral judgements are psychological, with practical ethics and conduct (the branch of philosophy known as moral philosophy). Ethical problems, what is right and what is wrong, have social contexts and therefore involve the rational, social ego. He sees language, and therefore linguistic analysis, in terms of communicating socially, the world of the rational, social ego. The rational metaphysical philosopher uses a rational epistemology by which to seek certainty and probability that the infinite exists. It is rational to consider the position of the spacesuited surfer who "knows" the infinite by crouching on the "wave" of the advancing universe. The rational philosopher, within the psychology of his reason, thinks that infinite Being of ontology must be in front of the surfer, and knowing it epistemologically he sees ontology (Being) as giving rise to Becoming (the universe of cosmology). His universe is up-to-date but still has a strong affinity with the rational Enlightenment.

It is important to grasp that the four-tier manifestation process, the four subdivisions of modern metaphysics, the 12 levels of consciousness and non-duality are all integrated in the reason's system. The last two stages of the manifestation process correspond to all the subdivisions of metaphysics and to non-duality of matter and consciousness – whose 12 levels correspond to the same last two stages. The process can be summed up as follows:

Manifestation process	Subdivisions of metaphysics	Experiences of 12 levels of consciousness	Non-duality	
Nothingness/ Fullness				
Non-Being				
Being	ontology	levels 11–12	unity	
	psychology	levels 5–10		
Existence	epistemology		diversity (consciousness/ matter)	
	cosmology	levels 1–4		

Rational Universalism looks back to Classicism or neo-Classicism, which emphasised order, clarity, balance, restraint and sense of beauty. Classicism trusted in the powers of the mind, especially in the reason. Rational metaphysics has Classical attitudes in emphasising the order in the universe, and with it a clarity about "what is" and the system of the universe. It balances the infinite and the finite (on which it focuses) on the one hand, and on the other hand the contradictions in Nature: $+A + -A = 0$. Classicism is restrained in being objective about Nature, and perceives the beauty of bio-friendliness and the symmetry I remarked on in Part Two. Rational metaphysics, trusting in the reason, sees philosophy as "thought-based", as did Classicism. Classicism has a social view of humans in relation to evolution.

The rational approach to metaphysics looks for laws within the system. The reason analyses and fragments the All or One into bits, whereas intuition perceives the Whole and reunifies the fragments. The All is a whole and has laws. In rational metaphysical psychology (what can be experienced of Being according to the knowledge of the rational, social ego) there are 14 basic laws which are part of the invisible fabric of the universe and applications of the metaphysical perspective:[45]

1. the Law of Unity – all within infinity and eternity co-exist;
2. the Law of Interpenetration – an infinite scale cannot be divided into finite sections, all of which must contain infinity;
3. the Law of Symmetry – the infinite is symmetrical in all dimensions;
4. the Law of Cause and Effect – if all things are One, every

cause has an effect on everything else and most things that happen to human beings have a cause within human thought;

5. the Law of Existence and Non-Existence – as everything that exists has a cause, if a cause is created then its desired effect will follow;

6. the Law of Equilibrium – if two systems are created out of the One, their energy will return to unity like a pendulum coming to rest;

7. the Law of Degrees – in an infinite series of levels, all "facts" or parts of the same dimensionality are on the same level;

8. the Law of Pyramidic Construction – the individual parts of the All or One are joined together in a pyramidic structure;

9. the Law of Parallelism – all know everything obeys the same laws;

10. the Law of Enclosure – the higher encloses the lower as the Absolute encloses the universe, but the lower cannot enclose the higher;

11. the Law of Convergency (or "narrowing") – progress proceeds from chaos to perfection;

12. the Law of Finite Perception – the infinite and eternal can only be perceived from a finite degree, in apparently isolated parts;

13. the Law of Reciprocal Action – harmony depends on both sides giving and receiving; and

14. the Law of Factual Area – the fewer facts, the higher the degree as in a pyramid.

All these 14 laws assume that the All or One – the infinite and the finite – is a unity in which everything corresponds and everything has its allotted place. Our perceptions of the infinite are blinkered as we are finite, but as we shall see in chapter 10 we can know the infinite through universal being.

All this follows a variation of +A + –A = 0: +A (the Law of Order and self-organising and self-regulating) + –A (the Law of Randomness and accident and harsh environment) = 0 (the One universe seen rationally).

This approach solves 20 problems of philosophy and advances a Theory of Everything:[46]

The rational approach to the scientific view of the universe and the universal order principle has cleared up a number of problems that philosophers have pondered since the time of the Presocratic Greeks, the topics that fall under the definition of "metaphysics". By way of summary I can now list 20 problems on the left and solutions given by the rational approach of metaphysics after the dash:

1. the "first principles of things", i.e. the source, first element – the "ever-living Fire" or metaphysical Light, the Void/ sea of Being which manifested into light, atoms and consciousness;

2. the "first cause" of the universe – the Big Bang out of a pre-particle in the infinite/particle, rapidly followed by inflation;

3. non-duality – unity in transcendence/the infinite, diversity in immanence/the finite without the difficulty of monism (that everything is either matter or mind, i.e. Materialism or Idealism);

4. dualism – false dichotomy of mind and body, light and darkness, false as all opposites are reconciled in non-duality, emerging by manifestation from the One into diversity;

5. the origin of the universe – in the infinite pre-manifestational "ever-living Fire" or Light;

6. knowing – epistemological knowing of the infinite through the concept of the surfer;

7. substance, "the essential material", the first material – the latent pre-manifestational, subtle "ever-living Fire" or metaphysical Light in the real unity of Nothingness, Non-Being (the Void, the vacuum field within S) and Being (B within S) which manifests into the diversity and multiplicity of physical light, matter and consciousness;

8. ultimate Reality – Nothingness which manifested into Non-Being and Being, which then manifested along with the universal order principle into Existence;

9. ultimate material – the "ever-living Fire" or metaphysical

Light which manifested into physical light, matter and consciousness;

10. essence (to Aristotle that which a thing really is when it exists and which if altered would destroy the unity of the thing) – the derivation of all phenomenal forms of Nature from the latent "ever-living Fire" or Light of pre-manifestational Being;

11. Being – the potentialities of Existence in the infinite "ever-living Fire" or Light (B within S);

12. Existence – manifestation from the infinite into the finite;

13. the reason for Existence – within Order rather than Randomness, the goal of the universal order principle within evolution and consciousness;

14. the purpose of the cosmos – within Order rather than Randomness, to create bio-friendly conditions in which self-regulating, self-organising organisms can develop and humans can develop higher consciousness;

15. personal identity – to the rational philosopher, the rational, social ego which is the centre of personality and identity, with regard given to the soul (but to the intuitional philosopher, the whole of the multi-levelled self whose deepest seat is universal being);

16. change – the flux of Becoming which has manifested from the moving sea of Being;

17. time – part of a unified space-time fabric curved by gravity, the flux or succession of spatial events which manifested from the timeless infinite;

18. space – relational to a unified space-time fabric curved by gravity, also co-existence of events or qualifications, not as primary as Newton held;

19. free will – the power to act in the finite world without the constraint of necessity or fate, the choice of independent consciousness; and

20. mind – the seat of the rational, social ego that thinks and of all the 12 levels of consciousness.

The Big Bang settles the debate in Plato (*Laws* X) as to whether the motion in the universe is imparted or self-orginated and the debate in Aristotle (in *Physics and Metaphysics*) as to whether the first cause was a Prime Mover or Unmoved Mover. The infinite was an Unmoved Mover, and the Big Bang was a Prime Mover which set the world of Creation in flux, and it was imparted from the infinite Unmoved Mover rather than self-originated accidentally. The Unmoved Mover is the infinite One in which exists Non-Being, from which Being emerged into Existence in an emanational manifestation. The universe could under different circumstances conceivably not exist, and so to that extent it is contingent. Its existence has a cause which can be associated with the 40 conditions, suggesting it needed many right conditions to apply before life could be created – and that they *did* apply. As the right conditions applied the universe does not appear contingent. The universe's causation was *in esse*, within Being, but also *in fieri*, within Becoming, for as soon as Being became Existence with the Big Bang it became a process, a Becoming. Universalism supports Aquinas in seeing the universe's causation as being both *in fieri* and *in esse* – and as I have just said, in contingency – rather than Aristotle who saw it as solely *in esse*. I leave aside Plato's argument (in *Timaeus*) that there was a demiurge or creator, for this requires us to leave the philosophical category and stray into theology or faith.

In a celebrated image, as we have seen [on pp.160–161] Descartes saw philosophy in terms of a tree. In one of his last writings, the Preface to the French edition of the *Principles*, Descartes wrote that "all philosophy is like a tree, whose roots are metaphysics, whose trunk is physics, and whose branches, which grow from this trunk, are all of the other sciences, which reduce to three principal sciences, namely medicine, mechanics and morals". Descartes was re-establishing the Aristotelian–Christian synthesis of metaphysics, physics and the sciences.

Universalism, to adapt Descartes' metaphor, is a tree whose roots are in the infinite and metaphysical, whose trunk is the finite universe and whose branches are the sciences which express the infinite and the finite. In this emphasis on the infinite context of the finite, Universalists differ from Aristotle and Whitehead.

(2) The intuitional approach to Reality, metaphysical philosophy and
the system of the universe as Light: the Law of Order and the Law
of Randomness

The intuitional approach to Reality and to metaphysical philosophy,
which includes the infinite and the finite and describes a non-duality
– includes an intuitional approach to ontology and cosmology
that *experiences* the unity of the universe, the One, with intuitional
knowledge; and to all four subdivisions of metaphysics (ontology,
psychology, epistemology and cosmology, see p.273), all of which are
experienced intuitively in terms of Reality as the metaphysical Fire or
Light:[47]

> Crucially, metaphysical Universalism is also intuitional and subjective.
> Besides looking at the universe through sense data and the reason,
> Universalism also relates to it intuitionally for we are part of the
> oneness of the universe, not separate from it. The proposals of the
> reason are confirmed by the direct experience of individual intuition.
> The intuitional Universalist accepts the rational view but introduces
> "add-on" or "overlay", intuitional experience which strengthens it. I
> shall now follow the "add-on" or "overlay" Intuitionist tradition of
> metaphysical philosophy.
>
> Whitehead spoke of speculative philosophy (i.e. metaphysics)
> as "the endeavour to frame a coherent, logical, necessary system of
> general ideas in terms of which every element of our *experience* can be
> interpreted". We are now looking to see how the system of the universe
> puts our *experience* in touch with Reality, and how our experience of
> Reality gives us a framework in which all our experience can be fitted....
>
> Intuitional Universalism holds that, quite simply, it is possible
> for humans to experience and *know* Reality, the "ever-living Fire" or
> metaphysical Light which I established rationally. This hidden "ever-
> living Fire" or Light can be experienced individually and existentially
> in a manifesting form. Ontology (see the four subdivisions of modern
> metaphysics), such an important aspect of the structure of the universe,
> can be known within our consciousness when our consciousness is at
> level 12: universal (or cosmic, or unity) consciousness.
>
> This is a most important consideration. Communion between

consciousness and the metaphysical "ever-living Fire" or Light cannot happen unless the individual enters the highest, or rather deepest, level of consciousness, which is below the rational, social ego. The rational metaphysical philosopher can only *think* about Reality because he remains in his rational, social ego in level 8. He describes Reality mathematically. The intuitional metaphysical philosopher goes deeper, or transcends the rational and reaches a transpersonal level, and can open existentially to the "ever-living Fire" or Light in inner depths....

Intuitional epistemology (the study of what can be known about Being) differs from rational epistemology, which is based on a concept of the rational, social ego's: the concept of the surfer breasting the infinite. Intuitional epistemology is based on direct experience of the "ever-living Fire" or Light from a different centre, the universal being in which it is possible to see with universal consciousness. It is therefore appropriate to deal with intuitional epistemology before intuitional psychology.

Sitting quietly with closed eyes, the philosopher waits patiently for the inner darkness to open, and, as he goes deeper in contemplation and his breathing becomes slower, sitting as still as a stone, Light breaks within his inner dark, first shafts from below the horizon and then the full white intensity of bright light like a sun. All the cultures and civilisations record the powers this mysterious energy pours into the universal being: powers of knowledge, healing and vision. In universal or cosmic consciousness (level 12) *Homo sapiens* is able to open to the Reality behind the universe in this life, as first shamans and later mystics have always taught. This Reality is stated in many traditions as a hidden Light. It seems that the universal being within consciousness – a deeper brain centre than the cortex – is able to admit an energy from the quantum vacuum which fills it with health, a sense of meaning and purpose as a harbour admits the sea.

A crucial point is that the universal being is in fact "the intellect". This originally had nothing to do with the reason. "*Intellectus*" in Latin means "perception" (and therefore "understanding", "comprehension"), and it is an intuitional and intuitive faculty which perceives universals and meaning, and therefore understands, whereas the reason is a logical faculty that analyses particulars and often sees meaninglessness. The "intellect" is a perceptive faculty that lies outside the five senses. The claim is that when it perceives the "ever-living Fire" or metaphysical Light (which is Shelley's "Intellectual Beauty") it receives wisdom in an area of the brain which is separate from the sensory combinations of neurons, outside the five senses, and this infused, revealed, intuitional knowledge then influences the reason. Today the word "intellect" has been corrupted. It has almost lost its original meaning, and is used as a synonym for "reason" to mean "the faculty of reasoning, knowing and thinking, as distinct from feeling". An "intellectual" is regarded as having a highly developed reason. The movements of logical positivism and linguistic analysis marked a further turning from the intellect to empirical reason....

In intuitional psychology (what can be experienced of Being and in what part of the self we experience it) the intuition of the intellect or universal being *experiences* Being, the "ever-living Fire" or Light, in contrast to rational psychology in which the rational, social ego *thinks* about Being.

The human multi-layered consciousness and personality are anchored in the brain. The human, rational, social ego is above ape-old instinct, while somewhere beneath the cortex are the hierarchical levels of consciousness 1–12. To see where the intellect or universal being is located within the structure of the self of identity and personality we need to consider the seats of consciousness, which correspond to differing permutations of brain activity and which I arrange in descending hierarchical order:

- universal being or intellect (*intellectus*) – in universal consciousness, into which illumination shines from the One (level 12);
- spirit (*pneuma*) – in transpersonal consciousness, possibly relating to Non-Being (level 11), more probably relating to Being (level 11);
- reason (*nous*) – higher mind (levels 7, 8);
- soul (*psyche*) – higher feeling (levels 7, 8, 9);
- ego – rational, social ego (levels 4, 5, 6, 7);
- sense – controlled sense-impressions (level 5);
- body (*soma*) – body consciousness of body-brain system,

  robotic automaton (level 4); and

- instinct – (levels 1–3).

This arrangement solves several philosophical problems regarding the mind which have traditionally been posed as questions. Is mind brain-function? Or does mind use the brain? Is it true that neuroscience cannot explain I-hood? Is the "I" or self a construct of memory? Is the self lacking in substance, is it a mass of impressions? How do we achieve mental and bodily activity, and a stable self? The answer to all these questions is surely that different levels of consciousness use different parts of the brain and that different permutations and combinations of the neurons operate from different centres which connect to differing layers or strata of Reality at different times. At different times, all these questions may be answered "Yes". Mind seems to be brain-function in levels 1–4. Mind seems to use the brain in levels 5–12. "I"-hood seems to defy neuroscience in levels 1–4. The "I" seems to be a construct of memory in level 5. The self is a mass of impressions in level 5. The self is stable in levels 6–9.

So it is time to ask again the question whether consciousness is an effect of the brain like homeostasis. The answer is "No", not in levels 5–12. Does consciousness control the brain with the aid of photons as homeostasis controls both body and lower brain? Possibly, but this has not been proved. There is a philosophical case for seeing this as happening, for consciousness should be within the system of the One which includes Nature's visible ecosystem and the more invisible system of air like *aither* (or the Higgs field of bosons) full of massless photons entering brains and controlling bodies, and entering plants and controlling photosynthesis. Does consciousness control homeostasis? Possibly, but again a case can be made out in philosophical terms as the science on this matter is in its infancy – even though photonology is being applied successfully to machines.

I submit that consciousness controls the brain and homeostasis in levels 5–12 within the philosophical category of the rational metaphysical system, and the system I offer includes this partial aspect of consciousness's behaviour. The autonomic parts of consciousness from levels 1 to 4 are an aspect of the autonomic functions of the brain. Lower or shallow consciousness is enmeshed in the body and physical brain's electrical activity, while higher or deep consciousness contains within itself the possibility of controlling the brain and homeostasis, and through them the body. The saying "mind over matter" captures the serene consciousness imposing the calm of homeostasis on the brain – rather than being a mere physiological effect of the physiological calm of homeostasis.

Having made clear the two-tier nature of consciousness (which is partly autonomic and partly controlling) I can list the seats of consciousness with the levels of manifestation to which they connect:

- Nothingness, the All or One – infinite metaphysical Light of finest density, too pure to be seen in universal being or intellect (level 12), mystical/"divine";
- Non-Being – infinite, denser metaphysical Light but still too fine to be seen in the universal being (level 12) or spirit (level 11), spiritual;
- Being – infinite/finite, sea of neutrino-like or photino-like particles of metaphysical Light received in universal being or intellect (level 12), and psychic/paranormal energies received in spirit (level 11) as discarnate entities can be expected to be in the sea of Being, spiritual; soul (levels 7, 8, 9), archetypes (including Ideas if they exist), psychological; and
- Existence – natural phenomena and natural light seen by the physical, rational, social ego (lower mind, levels 5, 6, 7) and bodily instinct known to levels 1–4, physical.

An intuitional approach to metaphysics identifies 14 laws that are practical applications of the intuitional metaphysical perspective and can be experienced, and that are counterparts to the 14 rational laws (see pp.254–255):[48]

The intuitional self, residing outside the rational, social ego, is able to experience the workings of the Whole in a way that can elude the rational metaphysical philosopher. I have said that the reason analyses and fragments the All or One into bits whereas intuition perceives the Whole, reunifies the fragments. Intuition's movement is from Existence to Being. The intuition perceives what the reason cannot perceive, that as it operates within a unity of Being the infinite order principle does not merely shape growth but can also shape our actions and fortunes if we put ourselves in harmony with its ordering force. The order principle works for good conditions (such as growth and development) to prevail in all life, the intuition perceives, and to the intuition order, which created the bio-friendly conditions in which we live, can make good things happen to us by detecting and advancing the yearnings in our cells. Our wishes and conduct influence the Whole and are influenced by it. That is the intuition's great hunch. To the All's 14 rational laws [see pp.254–255], intuition adds a complementary 14, which are all practical applications of the metaphysical perspective:[49]

1. the Law of Order – the infinite orders the finite but the finite cannot order the infinite;

2. the Law of Randomness – everything is random that has not been ordered and organised by the Law of Order;

3. the Law of Harmony – when the self is in harmony with Being it attracts happiness, as in Taoism;

4. the Law of Contradictions – all contradictions are ultimately reconciled within the overriding unity, just as $+A + -A = 0$;

5. the Law of Unity – as Non-Being and Being have unity, all manifestation and multiplicity are contained within the unity, are interconnected and are one with the timeless infinite;

6. the Law of Multiplicity – multiplicity is contained within primordial unity and does not cease to be so contained when manifested into many;

7. the Law of Survival – as multiplicity does not cease to be contained within the unity of Being when it manifests, all beings who manifest do not cease to be contained within the unity of Being and return to it on death;

8. the Law of Illusion – those who do not know they are contained within the unity of Being may be under the illusion that no part of their being or consciousness will survive death;

9. the Law of Correspondence – as Being has unity, all phenomena of Existence manifest from the unity of Being

in accordance with the Law of Unity, all phenomena share Being and are linked together and correspond to each other within the unity of Being in such a way that they all contribute to the universal harmony, and our outer world therefore reflects the thoughts, beliefs and attitudes of our inner world;

10. the Law of Symbolism – as all phenomena manifest from the One in accordance with the Law of Unity and the Law of Correspondence, the lower domain can always be taken to symbolise the reality of the higher order in which can be found its profoundest cause, or to put it more simply "as above, so below", and all phenomena can be intuitively perceived as symbols of aspects of the One and reveal Being;

11. the Law of Attraction or Thinking – as Being is a unity and the order principle shapes phenomena and therefore events, by working in harmony with the order principle the self can attract what it thinks (or prays for) and expects provided it takes initial concrete action to implement the new condition, while most things that happen to human beings have a cause within human thought;

12. the Law of Prosperity or Abundant Supply – as Being is a unity and the order principle has dominion over events as well as phenomena, by working in harmony with the infinite order which means opening to the infinite, the self receives plenty to supply all needs provided it allows new energies to flow in which means cutting out dead wood so new growth can come through and giving readily to make room for what plenty will bring;

13. the Law of Sowing and Reaping – as the universe of Being is filled with both order and randomness, our actions either work with order or against it and have appropriate consequences, and what our being sows we will reap, either for ordered good or random ill; and

14. the Law of the Good of the Whole – as the universe is a unity of infinite Being imbued with an order principle that operates for the good of the Whole, if we work with it and our being

acts virtuously and selflessly for the good of the Whole, filled with the infinite our being maintains an equilibrium through feedback and receives good back for ourselves that we did not seek.

These practical laws are all more everyday applications of the intuition's perception of the infinite pervading the finite. Using their intuition in this way, the religious can claim that prayer, giving and rectitude all work in a practical sense, while those outside religion can claim equivalent benefits as their thought attracts desired circumstances to their lives.

The intuitional self achieves its perception of the infinite by the painstaking way it assembles its vision. Whereas the reason analyses the whole of Nature and fragments it into bits the intuition works in the opposite direction. It examines the bits, perceives the whole and reunifies the fragments as an archaeologist pieces together potsherds of a broken urn. We have seen that Coleridge referred to this faculty of the mind as "the esemplastic power of the imagination", the word "esemplastic" coming from the Greek *"eis en plattein"*, "shape into one". The intuitional imagination takes the pieces and puts them back together again. The intuitional philosopher (of whom Coleridge himself in his philosophical reflections was an example) pieces science's fragmentation of Nature back into One Whole, where all the paragraphs form part of a whole structure like a stuck-together urn.

The intuitional philosopher does not rationalise – or extrapolate by reasoning – the Law of Order, but by esemplastically piecing the analysed bits back into a whole, intuitionally perceives the workings of order in the All or One, which includes the infinite and the finite in accordance with algebraic thinking (+A + −A = 0). Receiving the metaphysical Light in his universal being, the intuitional philosopher intuitionally knows that it brings order into his *pneuma* and *psyche* which is reflected in his behaviour – a cleansing of the appetites, aversion to unnecessary clouding of the senses with alcohol or polluting them with irrelevant loud music or television programmes – and intuitionally locates the origin or order in the infinite which, his intuition tells him, manifested into metaphysical Light. He arrives at the same conclusion as the rationalist metaphysical philosopher, the difference being that whereas the rationalist metaphysical philosopher has arrived at his conclusions by using the concepts of his reason (the concept of the surfer), the intuitional metaphysical philosopher reports on his direct experience.

Both the reason and intuition alike affirm a Law or Order and its opposite, a Law of Randomness; a Law of Harmony and its opposite, the Law of Contradictions; and a Law of Attraction or Thinking which, if used correctly, can lead to the Law of Abundance or Prosperity. Reason may acknowledge the other laws, including the Law of Unity, but as a more theoretical, analytical concept. The rational and intuitional philosophers use different faculties within the whole self to gain such insights. One of the difficulties with modern Analytic philosophy is that it operates theoretically and analytically and has lost touch with intuition.

The intuitional universe is a system of Light:[50]

Whereas in the rational universe, infinite Reality manifests and penetrates physical eyes and chloroplasts as an objective event, in the intuitional universe infinite Reality, the metaphysical Light, manifests and is received within the universal being or intellect, in the inner eye of contemplation, as a subjective experience.

In the Yogic practices of some forms of Hinduism and Buddhism the Light travels through the *chakras* or *cakras*, psychic centres or centres of the invisible subtle body which, according to Hindu tradition, operates in conjunction with the body's plexuses near key parts of the spine. The top *chakra* is the crown *chakra*, which to Hindus is the seat of illumination. The seven Hindu *chakras* are from bottom up: root (base of spine), spleen, navel, heart, throat, brow and crown (which is associated with the pineal gland). In Buddhism there are four *chakras*. To Hindus (and Theosophists, who adopted the *chakra* system), the Light enters at the base of the spine and flows up the central nervous system from *chakra* to *chakra*, a process known as the rising of the serpentine fluid of Kundalini, and explodes into illumination in the crown *chakra*, which in Western thought is the intellect.[51]

In Western thought, the metaphysical Light does not travel *up* the spine. That is a Yogic phenomenon. By going into contemplation, the self shuts down many of the parts or centres within the self and brain: instincts, sense, ego. Detached from sense impressions and desires, it moves deeper from soul to spirit to universal being or intellect, which opens as a mirror to reflect the incoming rays of "every-living Fire". What the contemplator sees is a reflection of a dazzling metaphysical Light that has manifested from the infinite, bringing energies from deep within the cosmos. These energies transform the self, turning it away from low desires and burning impurities which religions categorise as the accretions of "sin" from the pure soul beneath so that the recipient frequently loses the desire to smoke, drink alcohol or in some cases be promiscuous. The energies cause virtues to flourish and eliminate psychological problems and fear of death for they bring assurance that humans have an immortal spirit. The energies bring serenity and "the peace that passeth understanding", a sense of the unity of the universe and of human beings' oneness with all humankind and all Creation.

The intuitional system contains a sea of energy or latent Light that has manifested into subtle, end-of-spectrum physical light, matter and consciousness. The same manifestational process, once out of Being and in Existence and directed by the universal order principle, shows itself to be a Light-inspired drive of organisms through matter to higher and higher levels of self-organising and ascending hierarchical wholes.

The intuitional universe is a perception of the intuitional self or consciousness. The received view of consciousness is that it is brain-dependent and ends when its brain stops functioning. However, according to the intuitional universe it is not impossible that each consciousness may be transmitted as light or photons through the brain rather than be produced by the brain. It may control and regulate the brain's electrical rhythms, regarding the finite brain as Shelley's "dome of many-coloured glass" which "stains the white radiance of eternity".[52] It may be that the white radiance of the universal being, spirit and soul come from outside and anchor and lodge within the many-coloured dome of the operating brain and appear to be dependent on the brain as brain function.

If consciousness is transmitted into the brain and the brain acts like a radio receiver, it is probably received by photons above the brain, in a "Bose-Einstein condensate".[53] This condensate, or substance produced by condensation, was proposed in a theory developed in 1924–1925 by Einstein and the Indian physicist Satyendra Nath Bose on the statistical behaviour of a collection of photons, and it accounts from the streaming of laser light. Photons satisfying Bose-Einstein statistics are called bosons. Electrons satisfying Fermi-Dirac statistics are called fermions. These photons or bosons, including the Higgs field of bosons [since found], have been undetected until now and were sought by CERN in its September 2008 experiment. If consciousness *is* received in the brain, the transmitted current of universal energy perhaps arrives on carrier photons or bosons. The energy stimulates and drives the neurons and pushes their electrical impulses into interconnectedness and binding with each other. The activity of the neurons may thus be a consequence of the process of consciousness bosons, not its generator. And consciousness may be a tenant squatting in the home of the brain, which it deserts to go elsewhere, leaving the brain empty and lifeless.

To be completely clear, what I am suggesting is that consciousness may be composed of hitherto undetectable bosons (photons that obey Bose-Einstein statistics) and that the discovery of bosons by CERN would make this proposal testable as a scientific fact.

I also want to make clear that metaphysical Light begins in the infinite, beyond physics, and, I submit, manifests into the finite, the universe after the Big Bang. At some stage, I submit, metaphysical Light manifests into finite bosons. CERN has sought this finite manifestation. Its particle chamber replicates the finite conditions *after* the Big Bang, by which time the infinite has manifested into the finite. It has tried to capture the moment of manifestation. I submit that the bosons of consciousness had an infinite origin, and have the infinite contained within their finite manifestation in accordance with the Law of Unity and the Law of Survival. According to metaphysics, in view of their infinite origin in Being, all finite phenomena contain within them that which is infinite. This was the point I argued with Bohm.

Evolution and brain function may have been spurred on and be consequences of the wave transmissions of the "ever-living Fire" or metaphysical Light which may manifest into physical light. As we have seen this may contain coded messages of order impelling all creatures to the next stage of their growth. Bohm was thinking along similar lines.

He said that "Light can carry information about the entire universe"[54] and he held that information can move at superluminal speeds, speeds faster than the speed of light (which is just over 186,000 miles per second). When this happens, as G.N. Lewis, a physical chemist, argued in the 1920s, there can be immediate contact at great distance as time slows down, distance is shortened and the two ends of a light-ray may have no time and no distance between them. This quantum non-locality is to physics what synchronicity is to psychology.

Thus, the invisible, latent, infinite "ever-living Fire" or metaphysical Light may travel faster than the speed of natural light, pouring order, truth and wisdom into consciousness and information into matter at speeds greater than the speed of light. Matter Bohm described as "condensed or frozen light". Matter forms when two photons collide. Matter is formed by light and there is a good scientific case for regarding matter as frozen light.

This is such an important idea that we should dwell on it. A moment's thought will reveal the connection between matter and light in Einstein's famous equation, E (energy) = m (mass) c^2 (the square of the speed of light in a vacuum). This equation tells us that mass is related to energy via the speed of light, that the speed of light fixes the "conversion factor" as to how much each kilogram of matter is worth in terms of energy.[55] Bohm, Einstein's protégé, took up this idea in conversation: "All matter is a condensation of light into patterns moving back and forth at average speeds which are less than the speed of light. Even Einstein had some hint of that idea. You could say that when we come to light we are coming to the fundamental activity in which existence has its ground, or at least coming close to it."[56] The transfer of order from light to living processes or organisms takes place via the absorption of sunlight into eyes and its transformation into disordered heat radiation.[57]

I would add that natural light is condensed metaphysical Light. And that the process of manifestation is a kind of condensation. This hidden network of the manifested Light may therefore be a network of "air waves". The Light is around humans and permeates them all the time. Just as neutrinos pass through human bodies as they have so little mass, so massless photon-like boson particles of manifested Light that may be tinier than neutrinos must also be able to pass through human bodies. Tinier-than-neutrino particles such as photinos/bosons, supersymmetric counterparts to photons just as gravitinos are counterparts to gravitons, may emerge from the hidden Reality beyond the quantum mechanic waves and space particles. All creatures are like sea-sponges living in a sea of Light. It is in fact a network of transcendent and immanent energy, manifesting from the infinite and flowing finitely into humans' intellects and into all chlorophyll.

The intuition brings its own intuitional slant to non-duality for it is deeply involved in the workings of consciousness. It will be helpful to list the metaphysical perspectives I identified alongside the various views of consciousness to make it absolutely clear what I am proposing and not proposing as regards the intuitional universe as a system of light:

- Materialistic monism (that matter gave rise to mind or consciousness, scientific materialism) – mind or consciousness is Materialistic brain-function within electrical neurons dependent on the brain, and dies when the brain dies;

- dualism (that matter and mind are separate substances, Descartes' position) – mind or consciousness is composed of a substance separate from matter, and soul/spirit survives death whereas matter dies when the body dies and is recycled;

- transcendental monism (that mind or consciousness gave rise to matter, that consciousness is the stuff of the universe, Idealism) – mind or consciousness is transmitted from manifesting metaphysical Reality and is perhaps composed of photon-like particles tinier than neutrinos such as photinos, but probably bosons, which form a halo-like system of light above the brain that uses brain-body, and soul, spirit and universal being survive death whereas matter is recycled; and

- metaphysical non-duality (that the infinite metaphysical Reality, a unity, gave rise to both matter and mind/consciousness, a diversity although from a common origin) – mind or consciousness is transmitted from manifesting metaphysical Reality and is perhaps composed of photon-like particles tinier than neutrinos such as photinos, but probably bosons, which form a halo-like system of light above the brain, a Bose-Einstein condensate, that uses brain-body, and soul, spirit and universal being survive death whereas matter (atoms, cells and neurons) is recycled into a new form.

Universalism is a metaphysical non-duality, and this last perspective is the one I favour.

Having reviewed the Law of Order and seen the universe as a system of Light, we are now ready to consider the intuitional approach to Reality within the four subdivisions of metaphysics:[58]

The intuitional approach to metaphysics differs from the rational approach. The rational approach to metaphysics was based on a system of general ideas that interpreted every known concept and all experience from the perspective of the reason and the rational social ego. My rational system is based on non-duality, an infinite unity which manifested into becoming in the Form from Movement process and into a finite diversity of matter, manifested light and consciousness.

The intuitional approach to metaphysics, by contrast, is based on a system of general ideas which interprets all experience from the perspective of the universal being or intellect. My intuitional system is based on an *experience* of the infinite unity which manifests into becoming and into the finite as subtle, manifested, boson-like Light which enters intellects and chloroplasts and energises all creatures.

As in the case of the rational approach to metaphysics, a word of caution is necessary. The system of general ideas the intuition concocts explains intuition's place in the intuitive universe. Intuition's approach to metaphysics does not embrace *all* experiences encountered within the 12 levels of consciousness. It bypasses much rational experience. The intuition's approach to experience must be combined with the reason's approach, and will then encompass *all* the levels of activity in all the 12 levels of consciousness, each of which has a centre that seems like a self. The self moves between levels of consciousness and centres, each of which relate to the universe in different ways. The intuition and reason, making separate approaches to aspects of the universe at different times and, at different times, engaging in all the experiences I listed under the 12 levels of consciousness, cumulatively have a very full knowledge of the universe and Nature's system.

Today modern metaphysics can be said to have four subdivisions, which in descending hierarchical order are:

- ontology (the study of Being);
- psychology (what can be experienced of Being);
- epistemology (what can be known about Being); and
- cosmology (the structure of the universe).

I summed up the rational approach to these four subdivisions, and their integration with the manifestional process. Taking into account the 12 levels of consciousness, intuitional, Expansionist, Universalist metaphysical philosophy restates the four subdivisions of metaphysics in descending hierarchical order:

- ontology – studying the infinite or timeless Being through experiences of Reality, " the ever-living Fire" or metaphysical Light, in the intellect of level 12 (source, All-One/Non-Being/ Being);
- transpersonal (or spiritual) psychology – how the spirit of level 11 can know the potentialities of Existence in Being, particularly how it can know psychic or paranormal phenomena and what seem to be far memories of distant lives (source, Being);
- epistemology – how the mind knows Reality as an experience by transcending the reason and rational, social ego so that the top levels 7–12 of consciousness can free themselves from Existence – detach themselves from the world of the senses – and know Being (source, Being/Existence); and
- cosmology – the study and description of the structure, science and theory of the universe (source, Existence).

The intuitional multi-levelled self occupies different levels and operates within different seats of consciousness. The intuitional Universalist studies cosmology but experiences the infinite and knows what he studies from within, relating to it as an organic process rather than as a fixed system as perceived by the reason. Sensing the One intuitively from a seat of consciousness other than the reason, he feels at one with the One as did the Romantic poet Wordsworth (whereas Hegel, his inspiration, thought the One).

Intuitional metaphysical Universalism sees the purpose of language as not only to communicate socially within Existence and focus on linguistic analysis (rational metaphysical Universalism's view), but also to explore and reveal Being as well as Existence. To the intuitional philosopher, language is not an end in itself but a means to an end. It focuses on the universe and Nature, and serves the purpose of explaining their system of general ideas. To the intuitional philosopher, language is therefore outward-looking, not inward-looking.

The intuitional approach to metaphysics is thus one of direct experience from a deeper level of the self than the reason or rational, social ego. In fact, intuitional Universalism, as distinct from rational Universalism, has an existential dimension.

There are 14 universal experiences that can be known intuitionally, and which are available to all humankind:[59]

1. the universal experience of the unity and order of the infinite and the universe via intellect;
2. the universal experience of "ever-living Fire" or metaphysical Light, being-in-the-One;
3. the universal drive to self-improvement and self-betterment, which realises possibilities, leads to self-transcendence and helps bring in the next stage in one's development;
4. the universal order principle's advancement towards higher (or deeper) consciousness;
5. the universal freedom to break habit and choose one's future through self-transcendence, as the deep subconscious drive to self-improvement is reflected by willpower and acts out coded instructions;
6. the universal consciousness of purpose and awareness of meaning;
7. the universal self-transcendence into new levels of consciousness via projects;
8. the universal being's inner calm and serenity;

9. the universal inner awareness of one's own purpose;
10. the universal sense of the unity of all humankind;
11. the universal order principle's self-organising reflected in one's own life;
12. the universal order principle's readiness for the One (i.e. being in a state of readiness to experience Being);
13. the universal being-out-of-time (i.e. the mystical sense of the timeless unity of the universe); and
14. the universal harmony within all humankind (a political idea of benevolent – not malevolent – world government).

To the Universalist, man is an unfinished being. Man is ever in process towards his possibilities.

All this follows a variation of +A + −A = 0: +A (the Law of Unity – Non-Being and Being have unity, all manifestation and multiplicity are interconnected with unity and are one with the timeless infinite) + −A (the Law of Survival – as multiplicity does not cease to be contained within the unity of Being when it manifests, so all beings who manifest do not cease to be contained within the unity of Being, and return to it on death) = 0 (the One universe seen intuitionally).

(3) The reunification of the rational and intuitional approaches
A blend or unification of rational and intuitional Universalism reconciles 12 opposite pairs with apparent dualism:[60]

Universalism combines both rational and intuitional wings, thereby uniting the metaphysical and the scientific:
- trust in the reason, emphasis on the finite universe, a view of Nature as order, clarity, balance, restraint, sense of beauty, a social view of humans via the rational, social ego, objectivity (Classicism/rational metaphysics); and
- trust in the intuition and shaping imagination, awareness of the infinite behind the finite, revolt against reason, philosophical tradition and the fragmentation and reductionism of science, a view of Nature as organic dynamic process, a view of humans as individuals, attempting to live in all levels.

In balancing the two, Universalism brings together conflicting traditions: reason and intuition; the metaphysical and the empirical/scientific; Classicism and Romanticism. In blending the rational and intuitional and in superimposing the intuitional on the rational, Universalism sees that they complement each other as a description of the whole from different parts of the self, one emphasising the finite universe, the other the manifesting infinite, in accordance with algebraic thinking ($+A + -A = 0$). Universalist philosophy thus implements the combination of rational and intuitional metaphysics.

The blend or reunification of rational and intuitional Universalism works through 12 opposite pairs, each contradiction representing $+A + -A = 0$:

1. trust in the reason – analysed fragments of science, and
   trust in intuition and shaping imagination – making the universe/Nature whole;
2. emphasis on the finite universe, and
   emphasis on the infinite behind the finite;
3. a view of Nature as an ordered ecosystem, and
   a view of Nature as an ordered system of organic dynamic process;
4. emphasis on reason and mathematics, and
   Expansionist revolt against reason, philosophical logic and language, and the fragmentation and reductionism of science;
5. rational clarity about what is, and
   intuitional experience of what is;
6. sense of rational balance between the infinite and finite and all disciplines, and
   revolt against past balance to return to the Whole;
7. restraint in being objective about Nature, and
   emphasis on being subjective about Nature;
8. sense of the beauty of bio-friendliness and symmetry, and
   sense of wonder at the phenomena of Nature;
9. view of humans as social in the concept of evolution, and
   view of humans as individuals experiencing Reality;
10. rational, social, ego in social activities, and

universal being or intellect in solitary situations but at other times using all levels of self;

11. focus on objectivity, and
    focus on subjectivity;

12. language used as an end, to communicate socially, and language used as a means to an end, to reveal Being.

The reunification of rational and intuitional Universalism makes possible a manifesto of Universalism with 15 tenets:[61]

1. focus on the universe rather than logic and language;

2. focus on the universal order principle in the universe, a law which may act as a fifth force;

3. the universe/Nature manifested from the infinite/timelessness;

4. the universe/Nature and time began from a point and so everything is connected and one;

5. the infinite/timelessness can be known through universal being below the rational, social ego;

6. reunification of man and the universe/Nature and the infinite/timelessness;

7. reunification of fragmented thought and disciplines;

8. reunification of philosophy, science and religion;

9. focus on the bio-friendly universe, not a multiverse;

10. affirming order as being more influential than random accident;

11. affirming the structure of the universe as unique, its cause being the universal order principle from the infinite/timelessness/Void/Being/"sea" of energy;

12. affirming the eventual reunification of humankind;

13. affirming humankind as shaped by a self-organising principle so it is ordered and purposive;

14. affirming all history and culture as being connected, and one-world government and religion; and

15. affirming that life has a meaning.

The reunification of rational and intuitional Universalism can restate metaphysics as a blend and unification of two wings:[62]

- ontology – studying a static concept of Reality in the rational, social ego, and experiencing Reality as a dynamic, moving process through the universal being or intellect;
- psychology – seeing the self in terms of the rational, social ego and seeing a multi-levelled self in a dynamic process whose variations can open to transpersonal spiritual experience;
- epistemology – how the mind knows Reality as a concept through the reason and rational, social ego, *and* how the intellect experiences and knows Reality as an experience of metaphysical Light; and
- cosmology – the structure of the physical universe, and esemplastically shaping the fragments of scientific knowledge to return the fragments to a whole.

The reunification of rational and intuitional metaphysics allows for the reunification of science and metaphysics:[63]

Bergson in his *Introduction to Metaphysics*, 1903, called for "a much-desired union of science and metaphysics". Metaphysical science is science that connects to every possible concept, not just to every existing concept, and which therefore relates to the infinite. In the Middle Ages science and metaphysics were united.

The foundations of most modern objectivist, positivist and reductionist science have a Materialist ontology, for most science sees the phenomenal world as the only reality. Its epistemology is therefore sense data and, more recently, mathematics. The foundations of holistic science have a wholeness ontology, seeing the wholeness of the finite universe and its interconnectedness as the only reality. Its epistemology is intuitional inner knowing of interconnectedness within the finite universe. The foundations of a new metaphysical science, such as the one for which Bergson yearned, have a metaphysical ontology for it sees the infinite as the source of Reality, from which all else has manifested. Its epistemology is inner knowing of the metaphysical Light which manifests from the infinite. The foundations of a new metaphysical science, its ontology and epistemology are quite different from those of reductionist and holistic science.

A metaphysical science will question and challenge 150 years of reductionism and Materialism in the sciences and philosophy, and will carry metaphysics into many areas of science and philosophy. Metaphysical science should research, test, assemble evidence and do mathematics on the following hypotheses:

1. The origin of the universe: seeing the hot beginning (Big Bang) in relation to the "ever-living Fire" or metaphysical Light.

2. Reality in the microworld: seeing Newton's expanding force, Einstein's cosmological constant, dark energy (if it exists), the principle of hidden variability and the origin of mass as effects of the manifesting *aither*-like metaphysical Light's varying intensities in different localities.

3. Electromagnetic spectrum: seeing the high-frequency, gamma end of the spectrum in terms of the neutrino-like or photino-like bosons of manifested metaphysical Light.

4. Brain physiology: seeing consciousness as a system of high-frequency boson light which connects to all rays on the electromagnetic spectrum in the region of 4 cycles per second.

5. Mind-body problem: seeing the mind and consciousness as transmitted photon-like bosons bearing coded information that are not dependent upon physiological processes but which use the brain transmissively for their own metaphysical purposes, i.e. seeing the mind as independent of the body.

6. Synchronicity: seeing coincidences as evidence of a unified interconnectedness of matter and consciousness through the *aither*-like manifesting Light.

7. Mysticism: seeing religious experiences and mystical states of consciousness as contact with the metaphysical Light, which manifests into and pervades the universe, pouring universal energy into the universal being or intellect of humans.

8. Order principle: seeing the infinite and Being (the metaphysical Light which can be known intuitively or existentially through contemplation) as an ordering principle which sends photons into retinas and plants' chloroplasts, whence they pervade DNA.

9. Evolution: within order as opposed to randomness, seeing all evolution in terms of the order principle and teleological evolutionary power of the neutrino-like, photino-like boson particles of manifested metaphysical Light, and its influence over a low-level 'mechanism' that controls DNA exchange and chromosome division and evolution's drive for self-improvement and self-betterment.

10. Philosophy: seeing philosophy as subject to metaphysical science, as framing a system of general ideas that include all possible concepts, the whole universe at all hierarchical levels and life as a whole, in terms of the manifesting infinite's metaphysical Light.

Through this agenda metaphysics returns science to its metaphysical foundations in the infinite. Metaphysical science will move science away from granular Materialism where visible matter occupies only 4 per cent of the matter in the universe, and away from reductionism. The discovery by CERN of a new ordering invisible Reality in the microworld would overthrow Heisenberg's uncertainty principle and explain how particles have their mass. It is an idea waiting to be discovered and verified, just as the philosophical idea of the atom waited 2,500 years after it was first proposed to be discovered and verified as a scientific entity in the 19th and 20th centuries.[64]

The reunification of rational and intuitional metaphysics reconciles 5 sets of contradictions:[65]

As regards the union of metaphysical and scientific traditions within philosophy, I have endeavoured to blend the metaphysical tradition with the scientific view of the universe to achieve that union.

It can now be seen that the scientific universe is a manifestation of the metaphysical "boundless", and the coming-together and union of the two divergent strands of the philosophical tradition can be seen as the metaphysical tradition's subsuming of the scientific tradition. It can also now be seen that the appropriate term for this union is "metaphysical" as metaphysical Reality manifests into scientific reality. It is therefore appropriate to describe this union in terms of "the metaphysical view of the universe".

Within this unification I have endeavoured to unite the rational and intuitional metaphysical philosophies.

As regards the union within modern philosophy, I have endeavoured to bring together Linguistic Analysis and the Phenomenological view of Being. Language will focus on a precise description of Being's ordering of the universe and Nature, the phenomenal world of finite Existence. Language will therefore become outward-looking and will be a means to an end, to reveal infinite Being as well as finite Existence. Meaning and symbolism have layers that correspond to the infinite and finite levels and the 12 levels of consciousness. Literature, and in particular poetry, should contain levels of Being *and* Existence, as does Wordsworth's *Prelude*. Linguistic analysts will of course use language with their customary precision, but in an outward-looking way.

Universalism's aim is not to reject or refute but to bring out the one-sidedness of different approaches by integrating them into a more encompassing frame. It may appear to reject one-sidedness but its aim is to reconcile and unify contradictions and opposing views within a systematic Grand Unified Theory of Everything.

To sum up, Universalism has reconciled five sets of contradictions:

1. science and metaphysics – through metaphysical science;
2. science and philosophy;
3. metaphysical and scientific traditions (Plato's and Aristotle's traditions) – by reconciling infinite Reality and the scientific view of the universe;
4. rational and intuitional philosophies within metaphysics – by stating rational metaphysics and overlaying intuitional metaphysics; and
5. linguistic and Phenomenological philosophies – by reconciling language and Being so language creates neologisms to define and describe the infinite perspective with precision and accuracy.

This process amounts to a Metaphysical Revolution. I formulated a Metaphysical Revolution to return philosophy to metaphysics as long ago as 1980 and declared it in 1991. The Metaphysical Revolution will continue to work for the reconciliation of the five sets of contradictions.

Of movements, Whitehead wrote in *Introduction to Mathematics,* 1911:[66] "Operations of thought are like cavalry charges in a battle – they are strictly limited in number, they require fresh horses, and must only be made at decisive moments." The decisive moment is long overdue to effect a sudden and fundamental change in science and philosophy. The Universalist movement has begun at a decisive moment, and its Metaphysical Revolution is its crusading banner.

All this follows a variation of +A + −A = 0: −A (the *yang*-rational approach to the universe as a system of light) + +A (the *yin*-intuitional approach to the universe as a system of Light) = 0 (the One universe that unites rational thinking and intuitional understanding).

An informed observer must be aware of the need to balance, and reunite, the rational and intuitional approaches to the universe before proposing a Theory of Everything.

Assembling a Theory of Everything

We can now focus on a Theory of Everything that is objective, and capable of being approached both rationally and intuitionally.

Einstein's suggestion that gravitons exist gives a way forward to a Theory of Everything. Let us remind ourselves of what we read on p.45:[67]

After rejecting quantum theory at the fifth international conference on electrons and photons at Solvay, Brussels, in October 1927 (which marked the triumph of quantum mechanics), in 1930 Einstein had a bruising meeting with Bohr which ended in his grudgingly accepting quantum mechanics.[68] Einstein worked on quantum gravity and proposed hypothetical, unevidenced "gravitons", gravity particles which may also act as gravity waves. His general theory of relativity had predicted black holes from whose intense gravitational fields light cannot escape, and also massless gravitons travelling like photons at the speed of light.[69] If gravitons are proved to exist, the gravitational force could be described in terms of an exchange mechanism. It would operate like the electromagnetic force of repulsion between two electrons due to the exchange of a virtual photon between them, like the weak force in which the decay of a neutron is caused by the exchange of a virtual W-particle, and like the strong force between quarks, due to the exchange of a gluon. There would then be a unified theory of all force, a huge step to a Theory of Everything.

However, the gravitational 'force' – strictly speaking, gravity is not a force between masses but an effect of the warping of space and time in the presence of mass – has a very long range and appears to propagate at the speed of light, and so if gravitons exist they can be expected to be massless, which creates a mathematical problem. Gravitons may be particles that also act as waves.

The fact that light and the gravitational 'force' appear to be at the same speed ties them together: $+A$ (light) $+ -A$ (gravity) $= 0$ (the underlying unity of forces in the universe – the gravitational force (waves travelling at the speed of light) and its exchanging hypothetical gravitons between spin-2 particles, bosons; the electromagnetic force and its exchanging virtual photons between charged electrons; the weak force and its exchanging virtual W-particles and Z bosons between neutrons; and the strong force with its exchanging gluons between quarks, see the previous paragraph and p.45.

The requisites for a Theory of Everything include: $+A$ (quantum physics and the particle) $+ -A$ (general theory of relativity dealing with the universe and gravity) $= 0$ (the One universe).

The New Philosophy of Universalism has things to say about finally assembling a Theory of Everything:[70]

Physicists hope that all four forces can be reunified to include gravity and create a Theory of Everything, but the temperature to be replicated, 10^{32}K, would be even higher than the temperature to reunify three forces, and the particle accelerators even larger. Nevertheless, if the technology could be devised, all four forces *can* be reunified as they were unified at the beginning of their life shortly before 10^{-43} seconds after the Big Bang.

A Theory of Everything must go beyond both quantum physics, which successfully explains much of the particle, and the general theory of relativity, which deals with the universe and gravity. Both these theories work well, and so a Theory of Everything must include both. All particle theories before string theory lead to infinities when gravity is included. These infinities cannot be renormalised and cause the equations to break down. The five main candidates for a Theory of Everything emerged from hypothetical, unevidenced string theory in the 1980s and from an underlying hypothetical, unevidenced "M" theory (M for Mystery) that has not yet been found. The main candidate predicts that the universe has 10 hypothetical dimensions instead of the three actual spatial dimensions, and the other seven have not been found.[71] Universalists leave multi-dimensional theories to one side as mathematical speculations that cannot be proved.

The search for a Theory of Everything, putting together the very large (relativity) and the very small (quantum mechanics) to explain quantum gravity, has progressed in my lifetime. As we have seen, Oppenheimer and Snyder introduced black holes in 1939. Penrose established that a universe could collapse into a singularity, a black hole of infinite density, and reversing this, running the film backward, Penrose and Hawking established that a universe could come out of a singularity (now a white hole). Later Hawking established that a black hole (if it exists) must have a glow of radiation round it as pairs of particles drawn towards it split. One, the positive particle, escapes, giving off radiation, while the other, the negative particle, disappears into the black hole. Black holes never lose their heat and so have the potentiality to explode. We can see how a Big Bang could come out of a singularity, but how did gravity behave in the singularity before the universe inflated? It must have held back the expanding force of light or radiation within the singularity until it could hold it back no longer. A Theory of Everything requires an expanding force of light or radiation within the singularity, against which quantum gravity acts as a contracting force. These two opposites must both have been present in the first singularity before the Big Bang. A Theory of Everything cannot be established until it is shown what was behind these two opposites, light and gravity, and what lay outside the singularity before the Big Bang: the infinite.

Three of the four forces are of roughly the same strength, but gravity is much weaker. (A magnet can pick up a chunk of iron on a mountain in defiance of the entire earth's gravitational pull.) The original symmetry when there was one superforce within the dot-sized universe shattered soon after the Big Bang, and whereas the fragments of the electromagnetic, weak and strong nuclear forces can be put together theoretically by weakening the strong nuclear force at high energies and strengthening the electromagnetic and weak forces at the same high energies,[72] gravity seems irreconcilable as it is so weak. I am convinced that the most likely path to a Theory of Everything is not by speculating regarding superstrings, M theory and ten or eleven dimensions and seeking to put together the very large and the very small in black holes – pinning hopes on a multiverse – but by revisiting the idea Newton worked on: that there is an expanding force in light which counteracts the contracting force of gravity. In other words, gravity is weak because of the push of light against it. If physicists move away from string and M theory and analyse the pushing photons of light in relation to the pulling of gravity, I believe they will find a Theory of Everything – but it must include the infinite from which the pre-Big-Bang point emerged. To assert with Hawking[73] that the earth had no beginning or singularity or boundary but has not existed forever, having somehow emerged like a bubble, is a colossal logical and scientific fudge.

For decades cosmologists have accepted the notion that the universe of space-time began as a singularity, whose temperature, density and everything else about it were infinite. If a Theory of Everything marries quantum and gravity, the marriage will have to permit actual infinity which fills the point or singularity from which the universe began.[74]

At a conceptual level, our survey of quantum theory has made evident zero-energy's links with the infinite. The quantum vacuum contains the latent potentialities of existence: virtual particles. It is finite within the inflated balloon of the universe, but it emerged from the infinite singularity in the first second after the Big Bang and has its origin in the infinite. In this sense it is at one with the limitless, boundless infinity outside the balloon of the expanded singularity. Having emerged from the Void that preceded the Big Bang, its emptier, pre-space-time form, the quantum vacuum now contains more virtual particles than when there was a Void or empty nothingness.

Quantum theory plays down the significance of when time along with space began at the moment of the Big Bang, and of when time will end if the universe ends in a Big Crunch (as is thought not to be the case). However, before time began, underlying time and continuing after time ends was, is and will be the infinite timelessness that pre-existed the universe and will survive it: the Void from which the hot beginning and inflation happened and which has yielded, since the Big Bang, a seething quantum vacuum....

Gravitons. Einstein's theory of curved-space time required the existence of gravity waves, "ripples" in space-time's fabric which, he predicted, would be made of – it has to be said, so far undetected – gravitons (particles with zero mass and spin 2). According to hypothetical, unevidenced string theory, which details strings of particles and is an aspect of the multiverse we have left to one side, there could be lines of strings of gravitons. String theory says that every particle associated with a force (such as a graviton) must have a partner (a gravitino). Physicists have conjectured that supersymmetry plus string theory gives superstrings, from which a Theory of Everything may come. The reunion of gravity with the other three forces must in theory be possible as all four were unified at the beginning of the universe, gravity separating at 10^{-43} seconds, the shortest time interval that can be measured, when the temperature was 10^{32}K. If 10^{28}K can only be reached at the present level of technology by using an accelerator bigger than the solar system, 10^{32}K will require one proportionately larger, and at the present level of technology this can never be replicated on earth and therefore can never be proved outside mathematical theory. The unification, though the minutest part of a split second in duration, lasted long enough to determine the conditions in which the galaxies, stars and life evolved. If gravitons exist, uniting gravity and quantum gravity, they have provided just the right conditions for the evolution of galaxies, stars and life.[75]...

We have seen that a Theory of Everything has eluded theoreticians so far, and that there is no prospect of finding a purely physical solution as the temperatures required to reunite the four forces experimentally are so high....

We have seen that the universe does not have a purely finite, physical origin that can be explained by a Theory of Everything expressible in a formula. I have not said that an infinite singularity is a finite, physical singularity as Materialist reductionists do, having it both ways – having the benefits of the infinite alongside a view of physics as exclusively finite. I have argued that the universe began from the infinity before the Big Bang.

The Secret American Destiny also has things to say about assembling a Theory of Everything:[76]

Scientists' theories have impressively been confirmed by experiments and reductionism, which sees the universe in purely physical and Materialistic terms, does not merely co-exist with the tradition of metaphysical Reality, it now claims to have superseded it. Empirical science has risen high and there are claims that it is on the verge of a Materialistic Theory of Everything, although more sober scientists who have worked on the Standard Model that unites forces and particles deny that this is near. Contemporary science is dismissive of the traditional metaphysical outlook, and has led to the views of the British academic, who would not look beyond Materialist evidence and regarded a whole view of the universe as ambitious.

In our time the two conflicting approaches have expressed themselves in the sciences as holism versus reductionism....

A reunification of all four forces, including gravity, would create a Theory of Everything, but the temperature required, 10^{32}K, cannot be replicated in existing particle accelerators. A Theory of Everything must include quantum physics (which explains much of the particle) and the general theory of relativity (which deals with the universe and gravity). It would put together the very large (relativity) and the very small (quantum mechanics) to explain quantum gravity. It would show an expanding force of light within the first singularity, in which quantum gravity acts as a contracting force – a pair of opposites, light and gravity $(+A + -A)$ present in the first singularity before the Big Bang, within the infinite surrounding it.

The universe is finite and surrounded by infinity, which must therefore pervade the finite and can be experienced now, in the here and now. The infinite is with us now. Within the Void (or quantum vacuum) were the seeds of the order of the universe.

We can glean that +A (a pre-photon of light) + –A (a pre-proton of gravity) = 0 (the singularity before the Big Bang, when the universe was dot-sized).

Thus: +A (the expanding force of light pushing photons) + –A (the contracting force of gravity pulling gravity/hypothetical gravitons) = 0 (the One expanding universe after the Big Bang).

To put it another way, +A (the metaphysical oneness, a non-duality that gave rise to both matter and consciousness, and to humankind) + –A (the entire physical universe) = 0 (the One, unity of the physical and metaphysical universe and humankind).

In assembling a Theory of Everything I would do well to recall Whitehead's focus on metaphysical philosophy's being a system of general ideas that includes *all* known concepts and can interpret *all* experience:[77]

From the point of view of the metaphysical emphasis, philosophy frames and rationally states a system of general ideas which can interpret *all* our experience of the universe, including *all* known concepts. There must be a new framing of a system of general ideas which includes *all* known concepts, including the metaphysical. This must be done *after* a new scientific interpretation of the universe and Nature, so the metaphysical can rise naturally out of the scientific.... Whitehead's words at the beginning of *Process and Reality*,... are appropriate: "Speculative philosophy" (i.e. metaphysics) "is the endeavour to frame a coherent, logical, necessary system of general ideas in terms of which every element of our experience can be interpreted. By this notion of 'interpretation' I mean that everything of which we are conscious, as enjoyed, perceived, willed, or thought, shall have the character of a particular instance of the general scheme."[78] "Every element of our experience" means *every* experience, as Whitehead himself wrote in *Adventures of Ideas*:[79]

"In order to discover some of the major categories under which we can classify the infinitely various components of experience, we must appeal to evidence relating to every variety of occasion. Nothing can be omitted, experience drunk and experience sober, experience sleeping and experience waking, experience drowsy and experience wide-awake, experience self-conscious and experience self-forgetful, experience intellectual and experience physical, experience religious and experience sceptical, experience anxious and experience care-free, experience anticipatory and experience retrospective, experience happy and experience grieving, experience dominated by emotion and experience under self-restraint, experience in the light and experience in the dark, experience normal and experience abnormal."

Whitehead could have included: experience of the Greek boundless and infinite, and experience of hidden Reality.

It is possible to set out a Theory of Everything of four variables that are confined to their respective conditions, that conforms to Whitehead's strictures.

Setting out a Theory of Everything

A Theory of Everything confining itself to four variables and drawing on and including (1) the 20 stages of the origin, development and end of the universe in pp.12–220, (2) all the variations of 0 = +A + −A = 0, the algorithm of Creation, and (3) what the applications and variations of the algorithm of Creation have in common, summarised on pp.227–229, must attempt to arrive at a "system of general ideas" that includes "*all* known concepts" and "every element of our experience", that can encompass the totality of the origin, development and end of the universe and provide a cosmological framework for *all* known concepts and every element of human experience.

Only a Universalist approach including all disciplines and all known concepts can arrive at a Theory of Everything for four variables and 100 conditions. No methodology that is not Universalist can set out a Theory of Everything. It is extremely important that Universalism should be taught in all universities throughout the world as a subject in its own right, and that its methodology should be applied to a wide range of world problems, including world government and a Theory of Everything.

Below are the mathematical symbols denoting the 100 key conditions for the four variables, and what they represent, for a final statement of a Theory of Everything, in alphabetical order:

1. +A, –A = opposites within the algebraic formula/algorithm
2. ACC = acceleration
3. AD = all differences
4. AT = atoms
5. B = Being
6. BB = Big Bang
7. BC = Big Crunch
8. BF = Big Freeze
9. BH = black holes
10. BHE = black holes evaporate
11. BIFC = bio-friendly conditions
12. BIS = belief in survival (after death)
13. C = cells
14. CG = contracting gravity (in quantum gravity and relativity)
15. CIV = civilisations
16. CMBR = cosmic microwave background ripples
17. CW = cosmic web
18. D = DNA
19. DMPBS = dark matter particles build structures
20. DU = dark universe
21. E = Existence
22. EAR = earth
23. ECO = ecology (Nature's self-running, self-organising order)
24. EF = elements form (helium and hydrogen)
25. ENL = expanding natural light (and energy/mass in relativity)
26. ENV = environment
27. EV = evolution
28. EXP = expansion
29. F = 3 forces
30. FM = Form from Movement
31. FN = Fibonacci numbers
32. FO = forms of the One
33. G = gravity
34. GAL = galaxies
35. GAS = gas
36. GCS = galaxies become clusters and superclusters
37. GR = golden ratio
38. HC = higher consciousness (evolving human consciousness)
39. HD = Heat Death
40. HEXT = humans extinct
41. HOM = homeostasis (in body, brain and consciousness)
42. HSS = *Homo sapiens sapiens*
43. I = inflation
44. IAR = intuitive approach to Reality

45. INT = intuition
46. IO = infinite order
47. ISI = inflation smooths irregularities
48. IV = infinite Void
49. L = life
50. LO = Law of Order
51. LR = Law of Randomness
52. LS = Law of Survival
53. LU = Law of Unity
54. M = movement
55. MAD = metaphysical approach in disciplines
56. MAT = matter
57. ML = metaphysical Light
58. MOL = molecules
59. MPDMS = matter particles into dark matter structures
60. MS = mathematical symbols (for example ∞, +, −, ×, ÷, =, ≠, ≈, >, <, →)
61. M–S = irregular spiral movement within infinite movement
62. N(ABS) = Nothingness/the Absolute
63. NAH = national approach to humankind
64. NB = Non-Being
65. ND = non-duality of mind and matter
66. NDA = no differences, Absolute
67. NL = natural light
68. NNL = no natural light
69. NS = natural selection
70. O = the One universe
71. OPP = order principle and purpose, self-organising and self-replicating
72. P = particles
73. +p, −p = first particles from singularity/last particles before singularity
74. PC = physical constants
75. PDL = particles decouple from light
76. PFQ = particles form from quarks (protons and neutrons)
77. PHL = physical love
78. PL = planets
79. PO = pairs of opposites
80. QV = quantum vacuum (with lowest energy and no physical particles)
81. R = rational thinking
82. RAR = rational approach to Reality
83. RAT = ratios
84. S = spiral, irregular movement
85. SAD = secular approach in disciplines
86. SAH = supranational approach to humankind
87. SF = sun forms
88. SI = self-improvement
89. SL = spiritual love
90. SP = singularity particle
91. SPT = space-time
92. SS = solar system
93. STI = stillness
94. STQF = stars and quasars form
95. U = universe

96. USML = universe seen as a system of metaphysical Light

97. USNL = universe seen as a system of natural light

98. WS = World State

99. 0 = zero/the One uniting the algebraic formula/algorithm

100. 5D = fifth dimension

The final statement of a Theory of Everything that combines the metaphysical, infinite and shaping order (in +A) and the scientific, finite, physical process of creating (in −A), developing from Nothingness (the Absolute) to Nothingness (the Absolute) in step-by-step algorithmic instructions/rules in accordance with 0 = +A + −A = 0 (see pp.220–222), and uses all the above mathematical symbols, is as follows:

0 =	(N(ABS) + M + S + (M–S) + NB + ∞ + IV + 5D + IO + SP + (+p + −p) + B + FM + O)
+A	(BB + SPT + CMBR + FO + U + I + ISI + EXP + E + QV + ML + ENL + NDA + OPP + 5D + ND + ACC + BIFC + PC + RAT + GR + FN + LO + LU + L + HSS + HOM + LS + MAD + SL + INT + PO + IAR + USML + SI + BIS + CIV + HC + SAH + WS + ACC + HEXT + HD + DU + NNL + STI + G + BHE + BF + BC)
+ −A	(E + FO + CG + F + PFQ + PDL + DMPBS + EF + GAS + AT + P + MOL + MAT + BH + STQF + NL + GAL + GCS + PL + MPDMS + CW + SS + EAR + AD + SF + C + L + D + EV + ENV + ECO + NS + LR + PHL + R + RAR + USNL + SAD + MS(+, −, ×, ÷, =, ≠, ≈, >, <, →) + NAH + ACC + HEXT + HD + DU + NNL + STI + G + BHE + BF + BC)
= 0	(O + FM + B + (+p + −p) + SP + IO + IV + 5D + ∞ + NB + (M–S) + S + M + N(ABS))

We can see that +A has the metaphysical context and conditions for life on earth, and that −A has the materialistic events that made life possible. Each item in +A and in −A appears to be separated from the One but is in fact derived from the One as it is part of the One. The One is like the *Tao*, an emptiness and a fullness, a Nothingness that is also a plenitude of potential energy and potentialities. Regarding everything before the Big Bang and after the Big Crunch as part of the underlying One is valid so long as it is remembered that everything between the Big Bang and the Big Crunch is also part of the underlying One, which is a continuum from N(ABS) to N(ABS), from Nothingness (the Absolute, a plenitude) to Nothingness (the Absolute, a plenitude).

We can dwell on the symmetry of the equation. Boxes 1 and 4 each have 14 conditions, and boxes 2 and 3 each have 50 conditions. There is also a balanced rise and fall of the universe in a parabola that is reflected in the parabola of each of 25 civilisations (see pp.155, 158) and in the growth and decay of all living things, including the parabolic childhood and old age of humans. The algorithm of Creation, 0 = +A + −A = 0, applies to civilisations, humans and all living things as well as to the universe in which they grow and decay, so my Theory of Everything applies to the whole of Creation.

An alternative way of presenting our one universe and its respective mathematical symbols could be based on the following matrix:

$$A = \begin{pmatrix} a_{11} & a_{12} & \cdots & a_{1j} & \cdots & a_{1c} \\ a_{21} & a_{22} & \cdots & a_{2j} & \cdots & a_{2c} \\ \vdots & & & & & \\ a_{i1} & a_{i2} & \cdots & a_{ij} & \cdots & a_{ic} \\ \vdots & & & & & \\ a_{r1} & a_{r2} & \cdots & a_{rj} & \cdots & a_{rc} \end{pmatrix}$$

$$= (a_{ij}) \qquad i = 1, ..., r; j = 1, ..., c$$

r is the number of rows

c is the number of columns

aij is the (i, j) entry or element of the matrix A.[80]

A matrix showing elements in rows and columns.

A 'matrix' is defined by the *Concise Oxford Dictionary* as "a rectangular array of elements in rows and columns that is treated as a single element". So my Theory of Everything can be presented as a single equation of four matrices for our universe.

In place of A on the left could be 0 = +A + −A = 0, and the four columns could be four matrices side by side with headings 0 =, +A, −A and the last = 0, with the ingredients represented by elements in the matrices (the mathematical symbols) listed in four columns beneath. 0 = +A + −A = 0 can be broken down into four matrices with the contents of the four boxes on p.292 turned sideways and listed downwards as follows:

Our universe			
0 =	+A	+ −A	= 0
N(ABS)	BB	E	O
M	SPT	FO	FM
S	CMBR	CG	B
M–S	FO	F	+p + −p
NB	U	PFQ	SP
∞	I	PDL	IO
IV	ISI	DMPBS	IV
5D	EXP	EF	5D
IO	E	GAS	∞
SP	QV	AT	NB
+p + −p	ML	P	M–S
B	ENL	MOL	S
FM	NDA	MAT	M
O	OPP	BH	N(ABS)
	5D	STQF	
	ND	NL	
	ACC	GAL	
	BIFC	GCS	
	PC	PL	
	RAT	MPDMS	
	GR	CW	
	FN	SS	
	LO	EAR	
	LU	AD	
	L	SF	
	HSS	C	
	HOM	L	
	LS	D	
	MAD	EV	
	SL	ENV	
	INT	ECO	
	PO	NS	
	IAR	LR	
	USML	PHL	
	SI	R	
	BIS	RAR	
	CIV	USNL	
	HC	SAD	
	SAH	MS	
	WS	NAH	
	ACC	ACC	
	HEXT	HEXT	
	HD	HD	
	DU	DU	
	NNL	NNL	
	STI	STI	
	G	G	
	BHE	BHE	
	BF	BF	
	BC	BC	

Four matrices for our universe.

The above table in matrix form is on p.295.

$$
\begin{pmatrix} N(ABS) \\ M \\ S \\ (M-S) \\ NB \\ \infty \\ IV \\ 5D \\ IO \\ SP \\ (+p + -p) \\ B \\ FM \\ O \end{pmatrix}
=
\begin{pmatrix} BB \\ SPT \\ CMBR \\ FO \\ U \\ I \\ ISI \\ EXP \\ E \\ QV \\ ML \\ ENL \\ NDA \\ OPP \\ 5D \\ ND \\ ACC \\ BIFC \\ PC \\ RAT \\ GR \\ FN \\ LO \\ LU \\ L \\ HSS \\ HOM \\ LS \\ MAD \\ SL \\ INT \\ PO \\ IAR \\ USML \\ SI \\ BIS \\ CIV \\ HC \\ SAH \\ WS \\ ACC \\ HEXT \\ HD \\ DU \\ NNL \\ STI \\ G \\ BHE \\ BF \\ BC \end{pmatrix}
+
\begin{pmatrix} E \\ FO \\ CG \\ F \\ PFQ \\ PDL \\ DMPBS \\ EF \\ GAS \\ AT \\ P \\ MOL \\ MAT \\ BH \\ STQF \\ NL \\ GAL \\ GCS \\ PL \\ MPDMS \\ CW \\ SS \\ EAR \\ AD \\ SF \\ C \\ L \\ D \\ EV \\ ENV \\ ECO \\ NS \\ LR \\ PHL \\ R \\ RAR \\ USNL \\ SAD \\ MS \\ NAH \\ ACC \\ HEXT \\ HD \\ DU \\ NNL \\ STI \\ G \\ BHE \\ BF \\ BC \end{pmatrix}
=
\begin{pmatrix} O \\ FM \\ B \\ (+p + -p) \\ SP \\ IO \\ IV \\ 5D \\ \infty \\ NB \\ (M-S) \\ S \\ M \\ N(ABS) \end{pmatrix}
$$

Our known universe in matrix form.

Part of the James Webb space telescope's image the size of a grain of sand held at arm's length, showing thousands of early galaxies shortly after the Big Bang

On 11 July 2022 the NASA/ESA (European Space Agency)/CSA (Canadian Space Agency) James Webb space telescope has delivered the sharpest-ever infrared image of the distant universe. It is the deepest image of the universe yet and is the size of a grain of sand held at arm's length, and reveals thousands of galaxies in a tiny area of the immense universe soon after the Big Bang. It has renewed speculation that there might be other universes before or after ours, despite what Universalists believe. In 2009 the Stanford physicists Andrei Linde and Vitaly Vanchurin calculated from quantum fluctuations that the number of all possible universes is $10^{10^{16}}$, and that the true figure (which is beyond the power of the human brain to grasp) might be $10^{10^{10^7}}$. There is no conclusive evidence that there are any other universes beyond ours, but it is worth emphasising that $0 = +A + - A = 0$ covers these speculative universes before and after ours and parallel universes to ours, as we can now see.

If (contrary to what Universalists believe) there was a universe before ours, from which our singularity, +p, came, and there will be another universe after ours, into which our singularity, +p, will pass, then $0 = +A + -A = 0$ can be expressed in six matrices as follows, columns 1, 2–5 and 6 denoting the universe before ours, our universe and the universe after ours, with three downward dots counting events in the universe before ours (column 1) and after ours (column 6):

The universe before ours, our universe and the universe after ours					
Universe before ours	Our universe				Universe after ours
$0 = +A +$ $-A = 0$	$0 =$	$+A$	$+ -A$	$= 0$	$0 = +A +$ $-A = 0$
. . .	N(ABS) M S	BB SPT CMBR	E FO CG	O FM B	. . .

M–S	FO	F	+p + –p	
NB	U	PFQ	SP	
∞	I	PDL	IO	
IV	ISI	DMPBS	IV	
5D	EXP	EF	5D	
IO	E	GAS	∞	
SP	QV	AT	NB	
+p + –p	ML	P	M–S	
B	ENL	MOL	S	
FM	NDA	MAT	M	
O	OPP	BH	N(ABS)	
	5D	STQF		
	ND	NL		
	ACC	GAL		
	BIFC	GCS		
	PC	PL		
	RAT	MPDMS		
	GR	CW		
	FN	SS		
	LO	EAR		
	LU	AD		
	L	SF		
	HSS	C		
	HOM	L		
	LS	D		
	MAD	EV		
	SL	ENV		
	INT	ECO		
	PO	NS		
	IAR	LR		
	USML	PHL		
	SI	R		
	BIS	RAR		
	CIV	USNL		
	HC	SAD		
	SAH	MS		
	WS	NAH		
	ACC	ACC		
	HEXT	HEXT		
	HD	HD		
	DU	DU		
	NNL	NNL		
	STI	STI		
	G	G		
	BHE	BHE		
	BF	BF		
	BC	BC		

Four matrices for our universe between a universe before and a universe after ours.

The above table for our universe and a universe before and a universe after ours in matrix form (a being anything in the first universe, b being anything in the third universe) is on p.298.

$$
\begin{pmatrix} a_{11} \\ \vdots \\ \vdots \\ a_{i1} \end{pmatrix}
=
\begin{pmatrix} N(ABS) \\ M \\ S \\ (M-S) \\ NB \\ \infty \\ IV \\ 5D \\ IO \\ SP \\ (+p+-p) \\ B \\ FM \\ O \end{pmatrix}
=
\begin{pmatrix} BB \\ SPT \\ CMBR \\ FO \\ U \\ I \\ ISI \\ EXP \\ E \\ QV \\ ML \\ ENL \\ NDA \\ OPP \\ 5D \\ ND \\ ACC \\ BIFC \\ PC \\ RAT \\ GR \\ FN \\ LO \\ LU \\ L \\ HSS \\ HOM \\ LS \\ MAD \\ SL \\ INT \\ PO \\ IAR \\ USML \\ SI \\ BIS \\ CIV \\ HC \\ SAH \\ WS \\ ACC \\ HEXT \\ HD \\ DU \\ NNL \\ STI \\ G \\ BHE \\ BF \\ BC \end{pmatrix}
+
\begin{pmatrix} E \\ FO \\ CG \\ F \\ PFQ \\ PDL \\ DMPBS \\ EF \\ GAS \\ AT \\ P \\ MOL \\ MAT \\ BH \\ STQF \\ NL \\ GAL \\ GCS \\ PL \\ MPDMS \\ CW \\ SS \\ EAR \\ AD \\ SF \\ C \\ L \\ D \\ EV \\ ENV \\ ECO \\ NS \\ LR \\ PHL \\ R \\ RAR \\ USNL \\ SAD \\ MS \\ NAH \\ ACC \\ HEXT \\ HD \\ DU \\ NNL \\ STI \\ G \\ BHE \\ BF \\ BC \end{pmatrix}
=
\begin{pmatrix} O \\ FM \\ B \\ (+p+-p) \\ SP \\ IO \\ IV \\ 5D \\ \infty \\ NB \\ (M-S) \\ S \\ M \\ N(ABS) \end{pmatrix}
=
\begin{pmatrix} b_{11} \\ \vdots \\ \vdots \\ b_{j1} \end{pmatrix}
$$

$$i = 1,...,n$$
$$n \in [1,\infty)$$

$$j = 1,..,m$$
$$m \in [1,\infty)$$

Our universe between a universe before and a universe after ours, in matrix form.

Alternatively, if (contrary to what Universalists believe) there are more than one or an infinite number of universes before and after ours, 0 = +A + −A = 0 can be expressed in four matrices with downward and sideways dots on either side. The multi-universe concept can be conveyed by regarding column 1 as counting events in more than one or an infinite number of universes before ours (sideways dots), and column 2 as counting events in the one universe immediately before ours (downward dots); columns 3-6 as four matrices for our universe; and column 7 as counting events in the one universe immediately after ours (downward dots), and column 8 as counting events in more than one or an infinite number of universes after ours; with the downward dots counting events in the universe immediately before and immediately after ours, and sideways dots counting events in more than one or an infinite number of universes before and after ours, as follows:

More than one or an infinite number of universes before ours		Our universe				More than one or an infinite number of universes after ours	
0 = +A + −A = 0	0 =	+A	+ −A	= 0		0 = +A + −A = 0	
...	.	N(ABS)	BB	E	O	.	...
	.	M	SPT	FO	FM	.	
	.	S	CMBR	CG	B	.	
		M–S	FO	F	+p + −p		
		NB	U	PFQ	SP		
		∞	I	PDL	IO		
		IV	ISI	DMPBS	IV		
		5D	EXP	EF	5D		
		IO	E	GAS	∞		
		SP	QV	AT	NB		
		+p + −p	ML	P	M–S		
		B	ENL	MOL	S		
		FM	NDA	MAT	M		
		O	OPP	BH	N(ABS)		
			5D	STQF			
			ND	NL			
			ACC	GAL			
			BIFC	GCS			
			PC	PL			

			RAT	MPDMS			
			GR	CW			
			FN	SS			
			LO	EAR			
			LU	AD			
			L	SF			
			HSS	C			
			HOM	L			
			LS	D			
			MAD	EV			
			SL	ENV			
			INT	ECO			
			PO	NS			
			IAR	LR			
			USML	PHL			
			SI	R			
			BIS	RAR			
			CIV	USNL			
			HC	SAD			
			SAH	MS			
			WS	NAH			
			ACC	ACC			
			HEXT	HEXT			
			HD	HD			
			DU	DU			
			NNL	NNL			
			STI	STI			
			G	G			
			BHE	BHE			
			BF	BF			
			BC	BC			

Four matrices for our universe with more than one or an infinite number of universes before ours and more than one or an infinite number of universes after ours.

It would be possible as an alternative to have B denoting more than one or an infinite number of past universes and C denoting more than one or an infinite number of future universes, and to state the above table in terms of: $...\equiv 0 = +B + -B = 0 \equiv 0 = +A + -A = 0 \equiv 0 = +C + -C = 0$ $\equiv...$ (the equivalence relation, $\equiv$, meaning that universes are equivalent to our universe but not necessarily equal). This would be a variation of $0 = +A + -A = 0 \equiv 0 = +A + -A = 0 \equiv 0 = +A + -A = 0$ (see p.221), emphasising that more than one or an infinite number of past (B) or future (C) universes are different from ours. However, to emphasise the continuity of the algorithm of Creation in all circumstances I prefer to leave the differences from our universe to be denoted by the equivalence relation, ≡, and to stick to the equation on p.221: 0 = +A + −A = 0 ≡ 0 = +A + −A = 0 ≡ 0 = +A + −A = 0.

Roger Penrose, who in 1992 watched me as I wrote out the mathematics for my Form from Movement Theory in Jesus College, Cambridge's dining-hall, in 2010 proposed a theory called "conformal cyclic cosmology" based on general relativity, in which the universe expands until all matter decays into light, which forms the Big Bang for the next universe.[81] It proposes endless – an infinite number of – cycles of expansion and cooling, each beginning with a Big Bang and ending in a Big Crunch, in a process of cosmic evolution. In terms of the universal algorithm, the algorithm of Creation: ...≡ 0 = +A (expansion) + −A (cooling) = 0 ≡... .

Universalists would say that this view is speculative and unevidential except as a mathematical possibility – that it is in the tradition of Nietzsche's "eternal recurrence" expressed in *Thus Spoke Zarathustra*, for which there is no evidence – and that we should focus on what is evidential, the origin and unity of *our* universe and humankind.

The table on pp.299–300 for our universe and for more than one or an infinite number of universes before and after ours in matrix form, the three dots on the left and on the right counting the universes before and after ours, is on p.302.

There is also a view that there could be a parallel universe or more than one or an infinite number of parallel universes to ours, happening right now. One possible explanation sees a parallel universe or more than one or an infinite number of parallel universes as being within one time, while another possible explanation sees them as being within multiple types of time, which give multiple types of universe. So far our three models of our universe in matrix form have described how things happen over time within one type of time, with the death of one universe leading to the birth of another in one type of time, our time.

The first possible explanation for a parallel universe or more than one or an infinite number of parallel universes could show one fourth dimension being compressed into infinitely-occurring universes over

$$
\dots = \begin{pmatrix} a_{11} \\ \vdots \\ \vdots \\ a_{i1} \end{pmatrix} = \begin{pmatrix} N(ABS) \\ M \\ S \\ (M-S) \\ NB \\ \infty \\ IV \\ 5D \\ IO \\ SP \\ (+p+-p) \\ B \\ FM \\ O \end{pmatrix} = \begin{pmatrix} BB \\ SPT \\ CMBR \\ FO \\ U \\ I \\ ISI \\ EXP \\ E \\ QV \\ ML \\ ENL \\ NDA \\ OPP \\ 5D \\ ND \\ ACC \\ BIFC \\ PC \\ RAT \\ GR \\ FN \\ LO \\ LU \\ L \\ HSS \\ HOM \\ LS \\ MAD \\ SL \\ INT \\ PO \\ IAR \\ USML \\ SI \\ BIS \\ CIV \\ HC \\ SAH \\ WS \\ ACC \\ HEXT \\ HD \\ DU \\ NNL \\ STI \\ G \\ BHE \\ BF \\ BC \end{pmatrix} + \begin{pmatrix} E \\ FO \\ CG \\ F \\ PFQ \\ PDL \\ DMPBS \\ EF \\ GAS \\ AT \\ P \\ MOL \\ MAT \\ BH \\ STQF \\ NL \\ GAL \\ GCS \\ PL \\ MPDMS \\ CW \\ SS \\ EAR \\ AD \\ SF \\ C \\ L \\ D \\ EV \\ ENV \\ ECO \\ NS \\ LR \\ PHL \\ R \\ RAR \\ USNL \\ SAD \\ MS \\ NAH \\ ACC \\ HEXT \\ HD \\ DU \\ NNL \\ STI \\ G \\ BHE \\ BF \\ BC \end{pmatrix} = \begin{pmatrix} O \\ FM \\ B \\ (+p+-p) \\ SP \\ IO \\ IV \\ 5D \\ \infty \\ NB \\ (M-S) \\ S \\ M \\ N(ABS) \end{pmatrix} = \begin{pmatrix} b_{11} \\ \vdots \\ \vdots \\ b_{j1} \end{pmatrix} = \dots
$$

$$i = 1, \dots, n$$
$$n \in [1, \infty)$$

$$j = 1, \dots, m$$
$$m \in [1, \infty)$$

Four matrices for our universe with more than one or an infinite number of universes before ours and more than one or an infinite number of universes after ours, in matrix form.

time and in the same instants of time. If a parallel universe or more than one or an infinite number of parallel universes happen within fixed instants of time so that a moment in our universe corresponds to a similar moment in a separate universe, then they would each be in a bubble originating from the same fourth dimension (time), different universes living within the same bubble of time. It would be like an ecosystem, a pond with fish that all grow differently in the same amount of time.

It would be possible to bring time, a fourth dimension, into the matrices by introducing the three spatial dimensions and a fourth dimension of time. I could decompose the current model for our universe in matrix form, where four matrices have four columns, into three columns representing length, width and height (the three spatial dimensions) and a fourth column representing time which would give birth to more than one or an infinite set of matrices representing other universes at the same time as ours, as in the bubble. This bubble would show different universes as dots, where each of the dots is changing differently over the same time. An example of the dots (universes) within the bubble might be:

$$
\begin{array}{ccccccc}
. & . & . & .. & . & . \\
. & . & . & & . & . \\
. & . & .. & . & ... \\
\end{array}
$$

Each universe is different, and the dots are like the electrons orbiting round a nucleus, so the position of each dot could be given by a probability cloud and is apparently random.

However, parallel universes are at present entirely speculative and there is no evidence for them, and I am not showing a matrix for them. But it is important to note that the columns of such a matrix, if shown, would conform to $0 = +A + -A = 0$, the universal algorithm.

The second possible explanation for a parallel universe or more than one or an infinite number of parallel universes is that there are more than one or an infinite number of fourth dimensions (notions of time), and that there should be more than one or an infinite number of different sequences of universes, and therefore more than one or an infinite number of matrices stacked one on top of another, as in the example below:

$$...\equiv 0 = +A+ -A = 0 \equiv 0 = +A+ -A = 0 \equiv ...$$
$$...\equiv 0 = +A+ -A = 0 \equiv 0 = +A+ -A = 0 \equiv ...$$
$$...\equiv 0 = +A+ -A = 0 \equiv 0 = +A+ -A = 0 \equiv ...$$

It would be necessary to prove that time is the same for each system of universes as there could be more than one or an infinite number of varieties of 'time', and that each fourth dimension compresses into similar enough three-dimensional universes to create a chain of links that results in a 'universe' at the same time as ours. It would be necessary to prove that each sequence of universes is occurring in the same instant of time before vertical dots could be put between two lines.

However, the idea of more than one or an infinite number of fourth dimensions (times) is even more speculative than proposing one parallel universe or parallel universes within one fourth dimension (time), and is purely-speculative, unevidential and complicated, and I am not showing matrices for them. But once again, if shown, its columns would conform to $0 = +A + -A = 0$, the universal algorithm.

So whether there is only one universe (ours) as Universalists believe, or more than one or an infinite number of universes, or a parallel universe or more than one or an infinite number of parallel universes in one or more than one or an infinite number of times, all the permutations of created universes, the algorithm of Creation remains $0 = +A + -A = 0$, the universal algorithm. It is a truly universal algorithm that applies to any number of universes and to all the three permutations of our universe's four matrices, a fundamentally universal algorithm whose sweep covers the whole process of the origin, development and end of the universe and Universalism's immense and universal reach, and was only discoverable by my adopting a cross-disciplinary Universalist approach to the universe.

I cannot stress enough that, as I made clear on pp.221–222, Universalism focuses on the unity of our universe and humankind and regards a succession of universes, whether infinitely occurring or not, as speculative and unproved except in the mathematics of possibilities.

What is in the boxes on p.292 and in the columns on pp.294–300 is what seems to have happened, and science is now catching up with some of the evidence, which I expect will one day endorse this pattern.

In mathematics theorems are proved, and before a theorem is proved it is called a 'conjecture'. In the sciences, only well-tested hypotheses can become part of a 'theory', which still has the unproved status of a hypothesis. Mathematicians would regard the above formula as a framework, a 'conjecture', a scientific hypothesis, until it can be proved. Many of my above 100 conditions can be proved. Equations for each of any unproved conditions now need to be proved, and astrophysicists need to firm up my conjecture, my scientific hypothesis, into a theorem. But for the time being, in its broadest sense it can be called Hagger's Theory of Everything. Universalism, the philosophy of the unity of the universe and humankind in all disciplines, considers everything, and it is only natural that it should lead to a Theory of Everything.

Mathematicians would regard the above formula as an algorithm as it is a set of rules of instructions for the origin, development and end of the universe.

There are 100 mathematical alphabetical symbols involved in this Theory of Everything, but for convenience and summary they can be reduced to four variables like $E = mc^2$, and the Theory of Everything can be abbreviated as: $0 = +A + -A = 0$, the universal algorithm, the algorithm of Creation.

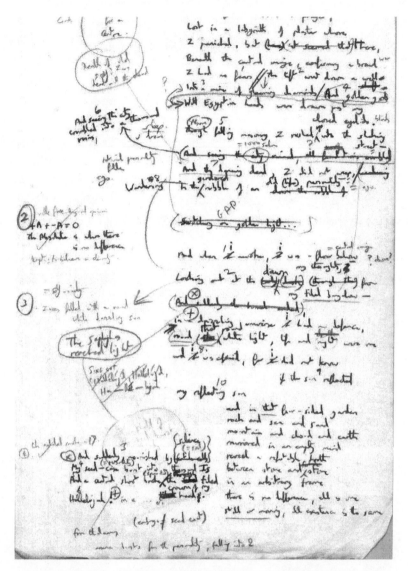

A page (written in 1965) from the original manuscript of 'The Silence' showing under ringed 2 in the left-hand margin "+A + –A = 0./The Absolute is where there is no difference." These two lines appear in the final text as: "(+A) + (–A) = Nothing./The Absolute is where there is no difference." (From Nicholas Hagger's archive in the Albert Sloman Library at the University of Essex.)

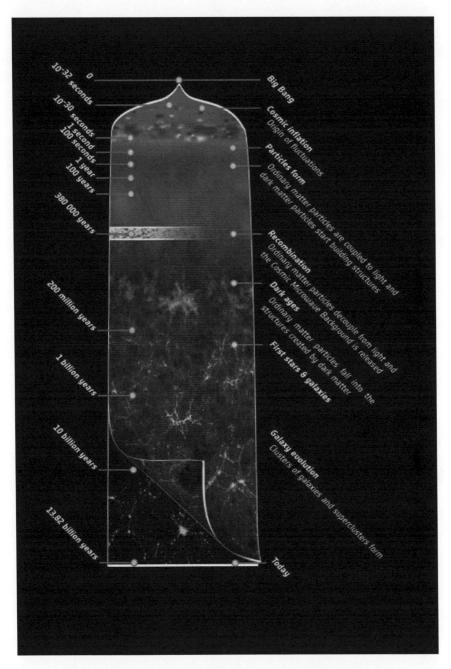

A visual timeline: image of the universe as a narrow
glass, Planck collaboration/European Space Agency.[2]

Timeline

The Origin, Evolution and End of the Universe,
Homo Sapiens and Life on Earth

List of dates of key events relating to *The Algorithm of Creation*

Billion years ago	Event
Before 13.82	Moving infinite Void with ripples
13.82	Big Bang, space and time begin
	Seconds after the Big Bang
	10^{-43} Gravity separates at 10^{32}K, mass-energy becomes the observable universe. Universe begins to expand by a factor of 10^{54} faster than the speed of light
	10^{-36} Inflation begins, lasts until some time between 10^{-33} and 10^{-32}. Expansion in process, inflation smooths out irregularities, quantum fluctuations will develop into structures
	10^{-35} Leptons and quarks form
	10^{-30} Protons and neutrons form from quarks
	10^{-12} Electroweak forces separate from strong at 10^{15}K
	10^{-4} Weak and electromagnetic forces separate from electroweak forces
	10^{28} Strong force separates
	100 seconds after the Big Bang
	Matter particles are coupled to light, dark matter particles build structures
	3 minutes after the Big Bang
	Universe has cooled from 10^{32}K to 10^{9}K (1 billion degrees)
13.773–13.750	*47,000–70,000 years after the Big Bang*
	Matter era begins
13.6	Death of massive star
13.52	*300,000 years after the Big Bang*

	First elements form, helium and hydrogen
	Electrons orbit nuclei and create atoms, cosmic microwave background radiation forms
13.5	Birth of Sagittarius A* (A-star), supermassive black hole at the centre of the Milky Way
13.44	*380,000 years after the Big Bang*
	Matter particles decouple from light; and form cosmic web structures of dark matter
	Electrons orbit nuclei and create atoms, cosmic microwave background radiation is released and glows
13.42	*400 million years after the Big Bang*
	Stars and quasars begin to form, then galaxies
13.42–13.32	*400–500 million years after the Big Bang*
	Galaxies and stars have formed
12.8	*1 billion years after the Big Bang*
	Universe expands, galaxies become clusters and superclusters
5	Expanding universe begins to accelerate
4.6	Formation of the sun
4.6–4.55	Formation of the earth
4.4	Liquid water first flows on earth, first oceans
4.28/3.96/3.9	First rocks
4.28–3.8	Life begins on earth, first cell
3.8	LUCA (the Last Universal Common Ancestor)
3.8–3.5	First prokaryotic cells
3.7/3.5	First fossils of primitive cyanobacteria
2.7	First chemical evidence of eukaryotes, first stable isotopes
2.7–2.3	Huronian Ice Age
2.6	Bacteria living on land
2.5	Great Oxidation Event gives earth an atmosphere
2.5–2	First multicellular eukaryotic cells
1.9	Oxygen atmosphere develops
1.8	Oldest multicellular fossils
1.5	First photosynthesis

Million years ago

850–630	Cryogenian Ice Age
800	First multicellular life
700/575	First animals
543	First shelled animals
533–525	Cambrian explosion
520	First land plants
500–450	First fish, insects and other invertebrates move on land, first colonisation of earth by plants and animals
460–430	Andean-Saharan Ice Age
400	First vascular and seed plants, and insects
365	Tiktaalik moves onto land
360	Four-limbed vertebrates move on land
355	First reptiles
350–260	Karoo Ice Age
320	First amphibians
290	First conifers
250/225	First mammals and dinosaurs
200	Dinosaurs dominate
160	First birds
135	First beaked birds, first flowering plants
65	Large dinosaurs extinct, mammals dominate
60	First primates
40/33.7–present	Our Ice Age
35	First apes
15	Apes diverge from other primates
14	Orangutans diverge from other apes
8–6	Gorillas diverge from chimpanzees, bonobos and ancestors of humans
5.3–5	Ancestors of humans diverge from chimpanzees and bonobos
4.43	First hominids
2.5	Bonobos diverge from chimpanzees, first stone tools
2.3–2	Early *Homo*
1.8	*Homo erectus*

Thousand years ago

400–250	First archaic *Homo sapiens*
250	First *Homo neanderthalensis*
200	Glacier advance peaks, ancestors of modern Cro-Magnon *Homo sapiens* leave Africa
150–130	Anatomically modern humans (*Homo sapiens sapiens*) arise in Africa
115	Glacier advance peaks
100–50	Modern *Homo sapiens sapiens* leave Africa
40/35–10	Cro-Magnon man in France
29/27	*Homo neanderthalensis* extinct
11.5	Holocene interglacial begins
10	First domestic animals
5	First writing and civilisation, and the oldest still-living thing (bristlecone pine)[1]

Events relating to *The Algorithm of Creation*

5 Oct 1965	NH receives $+A + -A = 0$ from Junzaburo Nishiwaki
10 Sep 1971	NH's illumination
11 Apr 1992	NH has breakfast with John Barrow, see p.vi
4 Sep 1992	NH writes out his Form from Movement Theory, reversing $+A + -A = 0$ to $0 = +A + -A$, watched by Penrose at Jesus College, Cambridge
8 Feb 1993	NH in an interview with Tony Mulliken and Melissa Viney of Midas PR at the Charing Cross Hotel, London, on the publication of *The Universe and the Light* states: "I am now seeking to formulate a Theory of Everything."
7 Sep 2021–25 Apr 2022	NH writes *The Algorithm of Creation*
17–19 Sep, 22–26 Oct 2022	NH produces first draft of his Theory of Everything

Million years ahead

7.4	Humans extinct

Billion years ahead

1	Solar luminosity 10 per cent higher than at present, lack of oxygen causes most life to die
2.8–22	Minority view for end of universe. Big Freeze and end of universe according to a minority of theoretical physicists
4	Heating causes earth's surface to melt, all life on earth extinct due to 'Heat Death'
4.5	Andromeda, larger galaxy, collides with the Milky Way
5	Sun runs out of hydrogen fuel in its core, swells to a red giant, and later becomes a white dwarf, then a black dwarf
850	Hypothetical dark energy dominates universe's expansion, universe cold

Trillion years ahead

1	All stars now red dwarfs, which can live 1–20 trillion years
10–100 (10^{13}–10^{14})	Stars form, run out of fuel, every star becomes a black dwarf, a dark universe with no visible stars, no natural light, stillness, after a final burst of light as collapsing causes collisions all stars fade and die, gravitational pull, black holes evaporate, Big Freeze
10^{106} (assuming protons decay)	Majority view for end of universe. Big Crunch, black holes and superclusters of galaxies finish evaporating, the universe collapses back to the singularity (+p) from which it began, space and time cease

Appendices

Appendix 1

Awakening to the Harmony of the Unitive Vision:
New Powers from a Deep Source, Infused Knowledge in
Sleep on Works, the Unity of the Universe, Universalism,
the Algorithm of Creation and a Theory of Everything

Awakening: higher consciousness brings new powers
I have sometimes been asked how I have such a clear view of the
timeline of the origin, development and end of the universe, and it is
now right that I should share an understanding of the new powers I
received from 1964 to 1993 as I progressed towards an instinctive
unitive vision, including infused knowledge fed to me in sleep, which I
dwelt on in the media – in two radio interviews and a TV interview in
the US, a contribution to a US-published book and a US website – while
I was writing *The Algorithm of Creation*.[1] What was fed to me during my
30-year-long Mystic Way (see p.7) and the higher consciousness that
opened to me subsequently came from a source for the reader to bear in
mind along with the Notes, References and Sources on p.363 – to me, a
contributing source that needs to be declared. In particular, from 1965
to 1966 I received the perspective for all of my 60 books, including this
one, and the basis of my Universalism, which took a further 25 years to
come through (see p.xvi).

Japan and the Absolute
I instinctively knew at Oxford that I had to get myself to Japan to discover
the wisdom of the East, but first I had to spend a year lecturing at the
University of Baghdad in Iraq, where I visited Babylon and absorbed
the myth of the death and rebirth of Tammuz under desert palms. I
eventually got to Japan in late 1963, with my wife and young daughter,
and discovered I was Invited Foreign Professor in English Literature at
my main university, Tokyo University of Education, the first Invited
Foreign Professor there since the English poet William Empson – Tokyo
Bunrika Daigaku, Tokyo University of Literature and Science, had been
renamed Tokyo Kyoiku Daigaku, Tokyo University of Education[2] – and
I would be teaching in a particular room where Empson had taught. I

317

lived in a bungalow (at 108 Kohinata Suido-cho, Bunkyo-ku, Tokyo) rented from the university for 3,500 yen (£3.50) a month.

I discovered that my boss, the Representative of the British Council, was the metaphysical philosopher E.W.F. Tomlin, a friend of T.S. Eliot (and later author of *T.S. Eliot: A Friendship*, about the 100 letters he received from Eliot). I had read his *Great Philosophers of the East* and *Great Philosophers of the West* at Oxford, and at an early stage I had ongoing discussions with him on Eastern wisdom and the Absolute, the one Reality "which can exist without being related to anything else" (*Concise Oxford Dictionary*).

My colleague the American poet Tom Fitzsimmons was a sceptical materialist and humanist who had no time for the Absolute. It was a relief to sit with Fitzsimmons and plan a film on the self-unification of a dying man (the idea in my later poem, 'The Labyrinth', 1975). Self-unification would later surface as the theme of my poem, 'The Silence'. Tomlin and Fitzsimmons represented a kind of +A (the Absolute) + –A (humanism) = 0 (self-unification within the One).

Zen Buddhism and a place of silence

At the farewell party for Fitzsimmons I expressed interest in experiencing Zen Buddhism, and the Japanese Professor who looked after me (a pupil of William Empson's in Japan in the 1930s), together with my colleague R.H. Blyth, an authority on Zen who I met once shortly before he died, arranged for me to go with my namesake Professor Haga, who had written a best-selling book on Zen, and one of my graduate students, to a meditation centre in Ichikawa City in mid-July 1964. There we meditated. I wanted to know about *satori* (Zen enlightenment), but that was not mentioned. I had to sit and count my breaths, an early Zen exercise in getting behind words to one's deeper self. We meditated long into the night with the rice-paper *tatami* doors open and flying beetles and moths floating round our heads, and everyone (except me, the foreigner) had to go up and answer a question – and be beaten. On the train on the way home Haga asked me why I thought they had all been beaten. I said, "They made the mistake of speaking. The answer the Master wanted to hear was a silence. As the *Tao te ching* says, 'He who knows does not speak, and he who speaks does not know.'" Haga, the Zen expert, said, "You are right, you have understood. Zen takes you beyond language to a place of silence where you can know *satori* or enlightenment." "A place of silence" resonated and perhaps resurfaced as the title of my long poem, 'The Silence'.

Following my visit to the meditation centre I began what I now regard as my First Mystic Life, which lasted from 20 July 1964 to 18 October 1965. Ten days later I went with my graduate student to Koganji Temple in Tokyo. We meditated in great heat among rows of silent cross-legged Japanese, and then slept. We meditated again just before dawn with the rice-paper doors open to the night and whirring cicadas. Through half-closed eyes I glimpsed the progress of dawn on the polished floor. I went very deep and got below the level of time and difference and becoming to a timeless being in myself, which could have existed forever.

I was working hard. I had been asked to go to the Bank of Japan every Thursday afternoon and correct the Governor's letters and speeches and spend time with the Vice-Governor and an Executive Director (see pp.xxii–xxiv). I was lecturing at three universities now, having been asked to spend a day a week at Tokyo University, where I read Eliot's *Four Quartets* for six months. I climbed Mount Fuji to within a thousand feet of the top, and I visited Hiroshima to give a lecture, and I gazed for a long time at the tangled iron of the dome that had survived the A-bomb. I bought all ten volumes of Toynbee's *A Study of History*, and began a study of civilisations. I was living intensely, and having spent weekends at Nobe, near Kurihama, which was by the sea, staying with my colleague Brian Buchanan (who had known Maud Gonne, for whom Yeats had unrequited love, and who had met Joyce, Shaw and Yeats in his youthful days in Ireland) at his house there, we now rented half a larger house, and I was able to walk alone by the Pacific in a great wind

Nicholas Hagger at Nobe, Japan, holding a whelk shell mentioned in 'The Silence', November 1964.

that flapped my trousers and tore through my hair as the sea flooded in. I recorded in my *Diaries*: "I felt at one with everything."[3] I recorded I had undergone "a development that is obscure, even to myself in its extent".[4]

In December 1964 the novelist and short-story writer Frank Tuohy arrived in Tokyo at a nearby university. I invited him to Nobe. He had

known Wittgenstein and was an anti-metaphysical linguistic analyst and sceptic (like Fitzsimmons), and he criticised Tomlin's Absolute and gave primacy to language, and I was further torn between seeking the Absolute and being humanist.

Ryoanji Stone Garden, Kyoto, Japan.

In January 1965 I went alone to Kyoto and Nara, Japan's cultural centre, and visited the Zen Ryoanji Stone Garden that is on the back cover of my *Selected Poems: A Metaphysical's Way of Fire* – a 15th-century stone garden with raked stones and rocks, swirls that changed with each raking and rocks that never changed, and could be seen as clouds on mountains, sea round rocks, or air round hills, all one essence (stone).

'The Silence', a poem of self-discovery and self-unification

In late January I recorded that I must be prepared to choose myself as a poet, as my poetry was one way of understanding the images that were happening to me. I had begun 'The Silence' in January 1965, and worked on it until June 1966. It was at one level a long poem in the tradition of Wordsworth's 'The Prelude', a poem of self-discovery, and at another level a Modernist poem (in the Eliot tradition) about the inner development of Freeman, a name that was fed to me on 25 April 1965 that echoed Everyman. The first two parts were set in the UK and Iraq. Part Three covered my time in Japan, and reflected my mystic life, my awakening and centre-shift in Japan, and final self-unification. I will quote bits as they become relevant. In 'The Silence' I showed a healthy man opposing a sick and declining society whose religion is dead, and rediscovering spiritual vitality in healthy Japan.

An early passage in 'The Silence' draws on my visits to Nobe, our weekend retreat from hot Tokyo. There was a long sandy and generally deserted beach by day with whelk shells and small sea creatures abandoned by the tide, and there were horseshoe valleys and paddy fields full of croaking frogs, whooping birds and hooting owls – and brightly coloured snakes, which I used to catch with a forked stick, then release. At night there were whirring cicadas and jungle sounds. Passages in 'The Silence' catch the feeling of being close to the universe (thought to be much smaller than we realise now):

> And when I came out of the station I couldn't get a taxi
> And the great sweep of the Milky Way was frozen and crisp
> And my heart welled out and flooded the universe.

On a beach at night he realises that the stars are meaningless without man.

> The sea sighed like a lover,
> Explored the face of the shore with webbed fingers,
> Blind in the dark phase. And in a glittering sweep
> Some of the hundred thousand million suns
> In one of a hundred thousand million galaxies
> Whirled on in a fiery band, shot out rays
> To traverse the silence for a hundred thousand years
> And beam on the man who crouched by a whelk shell.
> On a Japanese beach, my terrestrial illusions shattered,
> I fell headlong into a deep, dark silence
> And floating like an astronaut, I shuddered – for I was
>   nothing.
> Suddenly a great flood of being surged in me,
> Streamed out into the stars, brightened their fires
> And released affirming voices in my blood:
> "Without your creating eye would they exist, those cinders,
> Without your moist cheeks would they have meaning?
> Beside one creator, what are space or number or time?"
> While the sea reached and tenderly smoothed away my
>   footprints.

(lines 893–913)

Advances in cosmology since 1965, during the last 57 years, have updated lines 896–897 from "A hundred thousand million" stars in one of "a hundred thousand million" galaxies to 400 billion stars in 2 trillion galaxies.

On 2 May 1965 I poked among rock-pools at Nobe, watched a starfish and observed a sea-slug spinning a yellow thread of eggs. I had a strong sense that I was on a quest. I wrote in 'The Silence':

> And is this the end of the quest, is this all,
> Is there only this uncontainable complex whole,
> This pattern of Becoming, and nothing more?
> This carpet on this floor?

I shall sink to the floor so, and tear out tufts of hair.

Poet of the Self,
You who gave your youth to questioning despair,
Stare in the mirror, and question each greying hair.

In the early spring,
Observing a sea-slug spinning a yellow thread
I transcended my yearning:
XYZ,
It exists; there is no why; it might be dead.

Sunlight on a wrinkled sea.
The jumping of exploding diamonds and the sparkle of crystal,
I flashed jubilation like a white hot mirror:
Might be dead.

Turning in the seasons of my sunlight
I shall continue to seek,
Transcending the indifference of will-less fatalists:
A luminous sapphire-cell expiring on the beach.

                    (lines 1045–1064)

I found I was increasingly living in extreme situations. For several days I had to walk to work through radioactive rain: the Chinese had exploded an atomic bomb and the rain was 120 times as radioactive as normal rain, according to local American radio. I went snake-hunting with a forked stick and caught a snake that puffed itself up into a menacing red, and I found a centipede five inches long with legs like rose thorns, and a local told me it was deadly. I now found I had become a member of the Imperial Household. I was tutor to Prince Hitachi, the second son of Emperor Hirohito, and was driven to his palace for our first meeting. It was in hexagram 46 that I would be "employed by the King to present [my] offerings" (see p.xxiii), which I would not know until 17 November 1966, and I was now employed by Emperor Hirohito. I had a class with the Prince and his wife Princess Hanako. I was soon going three afternoons a week and (at the chamberlains' request) teaching him the world history of the last 5,000 years, ground I would later cover in my Grand Unified Theory of World History and Religion, *The Fire and the Stones*.

Beyond differences

In July 1965, accompanied by Tuohy, I spent a night at a Zen Buddhist Temple in Kitakamakura: Engakuji. We had to sit and meditate amid a cloud of mosquitoes, and were forbidden to wear socks when we sat in the lotus position, our still bare feet a target for mosquitoes. We were told, "You must achieve *satori* [enlightenment] in spite of the mosquitoes, pass beyond them, be one with them." While we were meditating among the whining mosquitoes a long deadly centipede with fat red legs like the one I had seen at Nobe crawled towards us. Tuohy had seen it and shifted. A man with a stick approached and prodded him in the back and he collapsed in a heap on the floor. Then we sat still and watched the centipede approach one of my knees, and suddenly it veered off to my left.

Next morning before dawn we had to sweep the path near the latrines, which smelt foul, to learn the need for inner purification. We then meditated again. Now I went very deep and as the cicadas scraped I knew among the dawn shadows that all was a unity. I had left the outer world of existence, and had reached an awareness of Being in which all differences were really not different, like the pebbles in the Ryoanji Stone Garden. As I wrote in 'The Silence':

In a Zen temple Freeman is confronted with the spiritual world and with the irrational basis of his existence, the metaphysical ground of his being.	In the meditation hall Each breath is unreal; There on the silence As the dawn shadows fall The empty seekers feel Empty and full existence; As when the ruffled surface calmed – Beyond your reflection You saw the clouded ground And the towering depths around your rational question.

(lines 1157–1166)

I spent a lot of that summer in Nobe among the cicadas and flying beetles, and loved it when a great wind flapped my black shirt and sand trousers. I had dinner with Prince and Princess Hitachi at their palace, and while walking with them in the grounds after going through billowing lace curtains, I saw some nesting-boxes and asked if they were for birds. "No," the Prince replied. "They are my ancestors'." In Shinto belief, the souls of Imperial ancestors become *kami* (divine spirits), and dwell in trees. (These Imperial ancestors were presumably the spiritual beings on Mount Khi to whom I would be making offerings via the Prince when I was "employed by the King", i.e. Emperor Hirohito, according to hexagram 46 – which I would not know about until 17 November 1966.)

In August 1965 I was taken to Kyoto and Nara as a guest of the Bank

Nicholas and Prince and Princess Hitachi in their palace grounds, Tokyo, on 5 August 1965.

of Japan. I visited the thousand statues at Sanjusangendo Temple, Freeman's thousand souls in 'The Silence':

... It seems to him that the power of his Reflection is finally broken.	Leaving the station, I groped down empty streets

Until, near the stadium, I went down a yawning stair
And lost my way into a hall of Buddhas.
I saw them, my thousand selves in tiers
Like a football crowd at prayer.

(lines 1210–1214)

I again visited the Ryoanji Stone Garden. I saw the simple pebbles as revealing that all existence is one underlying unity (being), and that the border or frame round it is quite arbitrary as the stone garden is infinite and boundless: the Absolute.

T.S. Eliot (who had also worked in a bank) died. Tomlin told me he had been invited to dinner by Eliot on 4 September 1963 and was arranging for me to visit Eliot when I was next in London on leave. He asked me to contribute an essay to *T.S. Eliot: A Tribute from Japan*, and during that summer I wrote 'In Defence of the Sequence of Images'. Weirdly, all that summer I was bombarded with images.

Centre-shift and Light, and +A + −A = 0
I was aware I had been undergoing a centre-shift from my ego to my deeper self or universal being. On 15 August 1965 I noted in my *Diaries*, "I have been undergoing some kind of a centre-shift.... I can feel the irrational all round my reason; it is a fact, something I can be aware of half a dozen times a day."[5] On 11 September 1965 I began an intense five weeks when I saw many images behind my closed eyes, a sequence of visions: scrivenings in foreign languages in yellow and blue, a puddle and an orb of fire within it, corn stalks with many ears of corn, a descent into a well.

I had more visions on 13 September: a series of gold heads, diamonds in green and mauve. When I got off to sleep I dreamt of an earthquake and rushed down stairs to a courtyard of fallen masonry. When I awoke I thought I had dreamt of a centre-shift I was undergoing, from my rational, social ego to my deeper spiritual self.

I thought I had opened to my imagination, and that the images were bubbling up from the spring of my Muse.

On 5 October I met the 1920s Japanese poet Junzaburo Nishiwaki (known as Japan's T.S. Eliot) in a small café with sawdust on the floor near one of my universities, Keio University, where he was Professor Emeritus, and over *saké* (rice-wine) we discussed Eliot. He told me that Eliot's approach was "not logic – witty combinations". I asked him to sum up the wisdom of the East as succinctly as he could, and he told me that the Absolute could be expressed as an algebraic formula, a "witty combination". He talked of Confucius's use of *yin* and *yang* and the *Tao*, and, helping himself to a business reply card that was poking out of a copy of *Encounter* I had with me, he said: "The Absolute is where there is no difference." He wrote on the card: "+A + –A = 0." Under +A he wrote "To be" and above the nought he wrote "Great nothing". Across the top of the card he wrote "Algebraic thinking". (For image, see p.4.)[6]

His few words resonated in my soul. They came as a revelation. In a flash I saw the unity in all opposites. I grasped that the wisdom of the East set out a progression, via a reconciliation of opposites, to perceiving the unity of the universe. I immediately saw that a oneness, a unity, reconciles all opposites: day and night, life and death, the finite and the infinite, time and eternity.

My development reached a climax in mid-October. On 8 October I went to a Tutankhamen exhibition, and saw the gold mask which had arrived in Japan from Egypt. It eerily corresponded to one of the Egyptian gold heads I had seen on 13 September. On 10 October I wrote the passage about the Stone Garden in 'The Silence'.

Later that same evening I went for a drink with Tuohy. To my astonishment he invited me to go to China with him the following spring, to write some articles for *AP*.

In the late evening of 11 October I went to the bathroom from our dark bedroom and, turning on the light, was flooded with golden light behind my closed eyes. The pattern was of rings, each linked together into a golden net.

The next day, 12 October, Prince and Princess Hitachi left for a State visit to the UK which I had been helping to plan. Forty of us, mainly old men, lined the walls of two rooms and waited. The Prince entered and bowed to everyone but shook hands with me (the only foreigner), followed by the Princess, who carried a nosegay. Then there was a collective chant of *"Banzai"* ("Victory", a word with wartime connotations).

On Sunday 17 October I was so tired in the afternoon that I slept. I seemed to sink down within myself, and when I awoke at 4.30pm and went to my study and looked out of the bungalow window I seemed to be a floor below my thoughts. I sat down and thought about the centre of my self in relation to the cosmos. My *Diaries* record: "I understood *Tao*, that just as my self-centre unites me, so *Tao* could unite life and death and all cosmic opposites and pluralities, so that all men are brothers."[7]

Light, satori: *round white Light in soul: end of First Mystic Life*

The next morning, Monday 18 October, something extraordinary happened to me. I stayed at home, and my *Diaries* record: "All morning I have been filled with a round white light: I cannot see it, except occasionally when I glimpse it and am dazzled, but I know it is there. It is like a white sun. This is, I suppose, what Christians refer to as the soul – the centre of the self. And the mystical experience is given meaning by the relation between the centre and the sun, so that everything is one."[8] I observed that "it is not the universe that has changed, but my self and my perception of it, so that it now seems more harmonious and my sense of kinship with mankind is stronger".[9]

I did not know it at the time but my experience of the Light as a white sun ended my First Mystic Life. It had taken me to a glimpse of *satori* following my shift from my rational, social ego to my deeper self, my soul. I had glimpsed the Light in my soul. My centre-shift had been bewildering, it had happened to me and I did not understand everything that had happened, as a passage in 'The Silence' (which includes the stone garden) makes clear:

The lift went down a well
Into a mine of gleaming diamonds
And golden gods with Egyptian heads were drawn past
    my closed eyelids.
Through falling masonry I rushed up into the shaking
    street –
And seeing the city crumbled into a thousand ruins,
And the dancing dead, I did not cry out or weep,
Wandering in the girdered rubble of an old personality.

He undergoes a
personality shift
from his Reflection
to the irrational
depths of his
inner being. He
experiences
illumination,

And when i awoke, i was a floor below my thoughts,
Looking out at the dawn as from a tiled bungalow –
And suddenly, nourished by silence
My seed-case burst into a thousand I's
And a central stem broke through the tiled crown of my
    head.
Hallelujah! in a dazzling universe i had no defence,
round that round white light, life and night were one
and i was afraid, for i did not know
                   if the sun reflected
my reflecting sun

and perceives the
unity of the
universe and is face
to face with the
divine world and
with metaphysical
Reality.

and in that four-sided garden
rock and sea and sand
mountain and cloud and earth
mirrored in an empty mind
reveal a refutable truth
between stone and stone
in an arbitrary frame
there is no difference, all is one
still or moving, all existence is the same

An Eastern sage tempts:
"(+A) + (–A) = Nothing,
The Absolute is where there is no difference."
             (lines 1219–1247)

In 1965 I turned +A + –A = 0 into a pentameter: (+A) + (–A) = Nothing. I was not aware of it, but all the strands of my life had been activated while writing my poem on self-discovery 'The Silence', and every theme I have since worked on now reared up like the snakes I caught with my forked stick in Nobe's horseshoe valleys.

A swing back from the metaphysical vision

I recorded that the white sun might explode with energy and cause me to go completely insane.[10] Looking back, I see that I had been given a glimpse, but that the Mystic Way, which I barely knew I was on, protects a new soul from over-exposure until it is ready. I swung back from the metaphysical vision of *The Secret of the Golden Flower* to the materialistic, humanist view of suffering humanity, safe in the view of darkness after exposing my being to the Absolute:

Rationally he resists the prospect of personal immortality, first in terms of death,

the stone garden reflects

what you want to see

a mirror of

subjectivity

and on this hill

bristling with monuments to synthesis

gape, earth, and reveal

a thousand insights into the abyss

and, backslider, do you now deny

between system and finality there is no

difference?

and secondly in terms of his fellow men.

A blossom of white light

illuminates the cosmic night,

my heart welled out –

a volley, and three men bowed low like acrobats

and under this ghastly sky, the condemned of all

nations

are limping and crawling and groaning up the

scorched hill,

writhing, screaming, feel their chopped bleeding,

> be faithful –
> Freeman, be faithful to your fellow men;
> see – some are ignorant of such matters
> but that spurting flesh is yours, you're one of them
>> and whoever they are, would you abandon
>> one?
>> between man and man there is no
>> difference.

And when a headlight lit his stump of an arm

In a graveyard he retreats from the metaphysical, and for the time being rejects immortality as a privilege he has yet to come to terms with, and immediately stakes all on existence, which he now perceives with meaning. He sees things as they are, objectively, without interference from his sense of 'I'. He perceives Being as opposed to Becoming.

In a graveyard of uniform stones, I looked for the
stars,
No matter what escaped old bones, I stood by my
fellow men.
I thought: 'If just one is damned or dealt extinction,
I will be too, and if all men are a part of One
And all shall be eternal, without exception,
Then without effort all are guaranteed,
And I reject a truth that ends all striving.'
And when I had shaken my head, my mind was
calm,
In a long rhythm of shadows, of days and nights
Like the scudding clouds on the unruffled moon-
centred pool,
I mirrored the glory,
I was all existence in the silence of that stone
garden.

> Darkness within
> Throws a chaotic waste on the empty
> screen;
> A disc of white-hot will
> Throws an orderly meaning on the sunlit
> screen;
> And the centre of the universe is in the self

When only what one is is seen.

A red-winged ladybird
        in sunlight,
A meaning of gullwings by the turning tide.
To be
Is to perceive the world
As a stone-like unity.
Therefore, *'percipere est esse'*,
To perceive truly is to be.

I was all 'I', Becoming,
Before the Journey;
Now I share the sky
i perceive truly
        and feel all Being flow like a
              surging tide.

I saw them at the break of day,
All my exorcised ghosts, sullenly skulking away.
And from the baked land joyously burst a
   prepuced palm.

And I was weary, weary – I prayed for rest,
   knowing
$(+A) + (-A) = \text{nothing}$;
Between red and black, the blaze of azaleas
And the dignified weeping of a dying piano,
In a disintegrating summer I became nothing,
I surpassed myself, on my way to becoming
   nothing.

And I was at peace.
        And I had liberated
A great tide of will I did not want to cease.
              (lines 1249–1312)

Two days after I experienced the Light, on 20 October Tuohy and I visited a trader and I bought two strings of baroque pearls for my wife, which found their way into the dedication of 'The Silence':

To Mr. F.T.
This string of baroque pearls,
To be told like a rosary.

I was groping towards a new Baroque principle in art (a reunification of sense and spirit, Classical and Romantic in poetry) and had received the lines:

I heard a cry from the old Professor's darkened room,
"The Age of Analysis is dead!"
Books lined with dust, a buzzing fly....
                                        While, naked on the petalled lawn,
A new Baroque age is born.
                                        (lines 1348–1352)

This is a theme that has been with me for more than 57 years. A selection from my poems, *A Baroque Vision*, will come out 58 years after I received these lines which have charted my poetic direction, for in the 1980s my Baroque turned into Universalism, as *A Baroque Vision* makes clear.

Down in Nobe at the end of October 1965 I wrestled with the phenomenal concealing a spiritual reality, and there being nothing behind the phenomenal. I lay awake in silence, unable to sleep, and imagined myself dying, and then dead, being nothing, and felt a cold terror under the stars.[11] Later I wrote in my *Diaries* of "the vertical vision"[12] and was back to the conflict between my inner sun and social scepticism that had brought about the centre-shift within myself. I resolved the conflict by thinking +A + –A = 0: a Universalist combination of the metaphysical and social perspectives, and their underlying union.

China and the 1966 Cultural Revolution
In March 1966 I made the visit to China with Frank Tuohy. I read 44

books in advance, at his request, and we went in from Hong Kong and visited Canton, Hangchow, Shanghai, Nanking (Nanjing) and Peking (Beijing). The visit was notable as I discovered the coming Cultural Revolution, news of which broke in August 1966. As readers of *The Fall of the West* will know, I was blatantly lied to by the Chinese authorities – when we visited Peking University expecting to see 9,000 students and found only a dozen they at first insisted the students were there. I insisted on speaking to the Vice-President at the university, and two days later established that all the students had been sent out for socialist re-education by the peasants, and that the few students on the campus had been allowed to remain as they had health certificates.

We were invited to the British Legation, and all the staff crowded round the lunch table to hear our account, which was news to them. We had scooped the British intelligence service. We left the Cultural Revolution out of the 14 articles we were commissioned to write, and in the end I received £50 from *Encounter* for our 'China Diary'[13] containing the world scoop of the Cultural Revolution.

On my return from China I edited and revised 'The Silence' and typed it up. The title came to me on 28 April 1966. (I believe I subconsciously chose it after Haga's remark that Zen takes one to a place of silence.) It was about a young man who opposes a sick society and its washed-out religion and finds a new spiritual vitality in the East. I had left the decayed West and had come to the healthy East and had found that my hunch was correct, that inner vitality and growth can be found in a healthy culture. I had to leave the West and go on a journey to discover this vitality within myself in the East, where the stillness I found took me into the silence of eternity and ultimate Reality. I wanted to expose the decay of European civilisation, and I made sure this passage was in the poem:

A vision of
the ruin which is
eventually in store
for European
civilisation

"No crumbling forces are unbreakable," Freeman
    had said.
Grass sprouted from these rotting roofs and walls
But, dictating progress reports, nobody noticed;
And from this Monumental height men are
    crawling like ants

Hither and thither in a pile of rubble and iron,
Hither and thither without a unifying idea
In a heap of crumbled machinery and falling spires
And, exhausted by aimlessness, drag weaker,
　　falter,
Watch from now smouldering compost in the hard,
　　clear dusk,
As with slow, regular flight,
A steel bird soars towards the crimson sun,
And fall, now burning; and, in a crimson
　　conflagration,

expire.
(lines 1414–1426)

A marginal gloss describes this passage as "a vision of the ruin which is eventually in store for European civilisation".

I showed the poem to Frank Tuohy, but he wanted it to be social-rationalist discourse and Wittgensteinian rather than imagistic. He insisted on marginal glosses, as used by Coleridge in 'The Ancient Mariner', with which I complied, wanting to unite the sceptical and metaphysical traditions as another $+A + -A = 0$. I showed the poem to Frederick Tomlin, who as a friend of Eliot immediately understood and warmed to my post-Modernist aims and outlook. My dedication, beginning, "To Mr. F.T." (see p.332), began as Mr Frank Tuohy and turned into Mr Frederick Tomlin, reflecting the tension between the two polarities in my being as I achieved my self-discovery and the self-unification I had planned with Fitzsimmons.

In 'The Silence' I saw my rational, social ego as my "Reflection" (see pp.324, 325, 328), and I included my drive towards to my future self, which I called my "Shadow" as I had often walked by the Edogawa canal near our home in Tokyo, accompanied by my companion, my shadow, in the Eastern heat, and I had seen my shadow as the future self, with qualities I needed, with which I had to unite. I have now become my Shadow, and my 60 works embody all the works that in 1965–1966 I instinctively sensed were ahead, which I would write with my new powers.

Russia, Communism

That experience in China set me on a collision course with Communism. I finished 'The Silence' in June 1966, on my return to Tokyo. I then made two visits to Russia in the coming months. I went back to the UK on leave (fares advanced by the Bank of Japan) and was in Moscow on the way home in June 1966, and again on the way back in August 1966.

On the second day of my first visit to Moscow, 9 June 1966, I visited the Archangel Cathedral, whose walls are covered in murals and icons. I recorded in my *Diaries*,[14] "I got the idea for a 'new' poetic form, the pictures on the wall." It was there, at the height of the Cold War, that I glimpsed the reunification of Eastern and Western Europe, and of the world.

While I was in England we lived in Loughton and between 20 June and 23 July 1966 I wrote 'Archangel', a poem on Communism in China and Russia, a theme I would return to many times during the next 60 years. Thinking of $+A + -A = 0$ I asked in 'Archangel', at the height of the Cold War:

How but by containing both sides, can we heal
This split down the mind of Europe and the world?

(lines 31–32)

And I ended 'Archangel' with a vision of a coming World State:

As I stared at the murals' centre
In this Cathedral-tomb,
The Archangel became a Shadow
With a sword and wings outstretched,
And I saw in the second icon
The future of the West,
From the Atlantic to the Urals:
Into the People's Square,
From the Cathedral gates,
File in the morning rush-hour
An *élite* of self-made Saints
Each still on the last hour's quest.

They reach the central banner
In the forum of statues and graves,
The great mazed *mandala*
Under which the supplicants wait;
Decades of contemplation
Show in their white-haired peace
As, trusting to perfect feelings,
They value each equal they greet;
Until, whispering on silence,
They glide to the Leaders' Hall,
Their hearts, with a World-Lord's wholeness,
At the centre of life, of all,
Their hearts where all past and future meet.

(lines 272–300)

This was a theme I was still writing in *The World Government* (2010), *World State* and *World Constitution* (both 2018) more than 50 years later.

New powers: shown future works

I now found that, following my centre-shift in 1965, I had the seeds of 60 coming works within me, and I was shown locations I would need when I came to write these future works.

While on leave I travelled to Ireland and Scotland, visiting historical places. Communism had set me thinking about revolutions generally, and Cromwell's revolution. On 8 July 1966 I visited the Banqueting House in Whitehall, London to see where Charles I walked under the ceiling painting of James I by Rubens (now on the cover of *A Baroque Vision*) to go upstairs and step through an open floor-to-ceiling sash window and be beheaded.

In this future-shaping year, 1966, I was shown future works I should do. My third masque, *The Coronation of King Charles*, was written in 2019 and published in November 2021, while I was writing *The Algorithm of Creation*. It is set in the Banqueting House and uses details I observed on 8 July 1966, 55 years earlier.

While in England I visited the library in West Dulwich where I had worked briefly before leaving for Japan, and in the course of chatting to me the warden told me a story about his life in Canada. That night I had the idea of writing 100 stories – prose poems that would focus on intense experiences that shape people's lives – and on 15 July 1966 I wrote the first of these, 'Limey', which was based on the warden's story. The first volume of my stories, *A Spade Fresh with Mud*, included some told me by Brian Buchanan in Nobe, on whom I modelled Brewer – and I have now completed seven volumes totalling some 1,400 stories. Again, I was shown future works in the ferment of this time by being made aware of new powers.

My study of all the revolutions since 1453, *The Secret History of the West*, came out of my thinking at this time, and on the way back to Japan (on 29–30 August 1966) I wrote a poem, 'An Epistle to an Admirer of Oliver Cromwell', which later led to my verse play *The Rise of Oliver Cromwell* (2000).

Back in Japan I continued to read world history with the Prince. I was aware of new powers. I was sensing the oneness of the earth and the universe, from which selfhood separates, and I was aware of the dying away from self I had partially undergone. I was convinced that world history was a history of many different civilisations, and my continuous dipping into the ten volumes of Toynbee's *A Study of History* found its way into remarks I made to my postgraduate students, who reported them back to Professor Irie, Empson's pupil who was in charge of me. One day Irie asked me if I would teach a course on the decline of the West to the postgraduates. I said, "What if I think the West is not declining?" He said, "Oh, but we would still like a course of lectures entitled 'The Decline of the West'." I promised to begin this in the second semester, which would begin in February 1967, and I began assembling material on Gibbon, Spengler and Toynbee. The decline of the West was not a subject I could tell Tomlin, the representative of the British Council, I was teaching, and I only found out later that Toynbee visited Japan and Tomlin visited him in his Yokohama Hotel on 9 November 1966.

That course I was asked to teach was Providential, as while I was lecturing on Gibbon, Spengler and Toynbee I saw a fourth way of understanding the rise and fall of civilisations which would take me a further 25 years of research to set out in *The Fire and the Stones* (later updated to two separate volumes as *The Light of Civilization* and *The Rise and Fall of Civilizations*).

I had come to see the unity of the universe as 'the All' (an Eastern concept), my understanding of which had been extended by discussions on 'the Will of Heaven' with the Executive Director at the Bank of Japan, one of my pupils, on 17 November. He explained that Zoroastrianism had left an impact on China in *yin* and *yang*, which are united within the *Tao* whereas Ahriman and Ahura Mazda had no equivalent unity in ancient Persia. He told my 'Will-of-Heaven' fortune as *Shang*, 'earth-wind', as "ascent, advancing upwards blindly" (see pp.xxii–xxiv), which I now saw as an ascent to a union between action and contemplation in my Shadow, my future self, and I began writing 100 stories (two stories a day in November and December) as contemplations of images of action within the All. I now see myself as having united action and contemplation within my +A + –A = 0 perspective.

I was invited to Prince Hitachi's birthday party in his palace on 28 November 1966. He came to meet me and shook hands, and I walked down and stood near the birthday cake, the icing on which said, 'Happy Birthday, Prince Masahito', and at the end of the occasion we all sang the Japanese National Anthem (I being the only foreigner among about 50 Japanese). I had drafted a poem, 'Epistle to his Imperial Highness on his Birthday'. It began:

Imperial Highness, on your birthday,
Spare a thought for the poor of the world.

The poem tells the Prince what his influence could have been about if he used it in a humanitarian way. This was another theme that would haunt me for nearly 60 years and result in *World State* and *World Constitution* (both 2018), and my urging of King Charles to be humanitarian in my second masque, *King Charles the Wise* (2018).

On 8 December 1966 I was asked by the Bank of Japan to rewrite the Governor's speech that would be declaimed at the opening ceremony of the Asian Development Bank on 19 December. I spent five hours rewriting it with input from a male secretary, and in discussion learned that Japan would be pouring money into the Asian Development Bank that would benefit the whole of Asia. I wrote my speech with the poor in mind. I saw Japan's role as working tirelessly for a united Asia that would lead eventually to a world government.

I was becoming an insider. This came home to me on 14 December (after my writing of the speech for the opening ceremony of the Asian Development Bank). The Prince rang me about 9.30pm and, after umming and ahing, expressed the Japanese Government's displeasure at an article written by Fred Emery, the *Times* correspondent in Tokyo, on the corruption of Prime Minister Sato – would I communicate this to Emery (who I knew)? I did, and Emery said, "What a splendid response." It seemed that Prime Minister Sato, who I had met recently at a gathering, had rung the Prince to ask him to ring me to ring Emery. I was now a link in a chain.

Turned down Chair for life

Soon afterwards I was offered a Chair (professorship) for life at my main university in Tokyo, where Empson had taught in the 1930s, by Professor Narita and Professor Irie. I wondered whether this was a reward for being a go-between between Prime Minister Sato and *The Times*. I had gone to the East to escape the decaying West and experience a healthy culture that would revitalise my growth and communicate the wisdom of the East and the plan had worked. I had works to write in the West, and having found what I had sought, I needed to return to the West and fulfil what I was supposed to do. I knew I had to pursue a writing career and set out the oneness of the universe in several disciplines. I knew I had to prioritise becoming cross-disciplinary like a Renaissance man over being a one-discipline Professor, and that I should apply myself to the cross-disciplinary research into the Light that would result in my writing 60 books. In January 1967 I announced that I would be leaving Japan in October.

In December 1966 I had started 'Old Man in a Circle', about the decline of old Europe. It fitted in with my reading of Toynbee, and, despite one visionary and prophetic passage anticipating the coming

EU ("from the mountainside I saw a giant Europe burst into light"), I address the decline of the UK head-on with images of the UK's having entered a Silver Age, a theme that has dominated my work for nearly 60 years:

> The golden sun set and a moon tinted all silver.
> I heard a voice cry, "Watchman, what of the night?"
> "Rise and fall," the watchman said, "rise and fall:
> The morning cometh, and also the night –
> The Habsburgs rose and sank into the night,
> The British rose and shone against the night."

> O Churchill.
> There is silence in Whitehall
> Save for the muffled drums;
> The flag-draped gun-carriage crawls;
> Sailors with reversed guns;
> The launch moans up the river,
> The cranes dip in homage
> Under a fly-past to Rule Britannia
> And the passing of an Age.

> Put another nickel in
> In poor old Winnie's treacle tin
> Screeched "a Jutland veteran" with a black peg leg
> And all down Piccadilly, the indomitable Grand Fleet steamed,
> 71 battleships and battlecruisers, 118 cruisers,
> 147 destroyers and 76 submarines;
> And on the dreadnoughts our guardian angels sang
> "Rule Britannia, Britannia rules the seas."
> O 15 St James's Square, O Edward the Seventh,
> Clyde, Scott and Franklin, Burgoyne and Lord Lawrence,
> Like a gallery of summer ghosts in the winter dark.
> Ah Palmerston! Leaving the HQ of de Gaulle's Free French,
> Sauntering down the Waterloo steps
> And meandering along the Mall to the Admiralty Arch –

skyyyyboltbluewaterstreaktsrtwoooo,
A pair of F-111s skimmed like swallows underneath
Where Vulcan and Victor, coupling blue steel wings,
Droned towards the warning web at Fylingdales,
And, near Holy Loch and closed-down shipping yards,
Disregarded the fish with the independent fins,
The Polaris sub with the interdependent scales.

Aegospotami and Midway. Ah the maritime:
One blink and a whole armada is knocked to bits
Or sold abroad, or stored, as "obsolete",
And shipless Admirals' voices float from aerials
To 4 aircraft carriers, 2 commando ships,
2 cruisers and a few destroyers and frigates –
O Senior Service.
As I left Downing Street during the Seamen's Strike,
Big Ben peered over the trees and pulled a face
And Nelson raised an arm.

(lines 67–112)

I have just completed *The Fall of the West*, which is a kind of coda to the lament in 'Old Man in a Circle'.

The Mystic Way from centre-shift to full illumination

My awakening and centre-shift had resulted in a flooding into my deeper self of aspects of the unity of the universe which gave me direction and growth towards my Shadow, the self I am now which has absorbed the universe, world history, world religions, the end of Communism, a political World State, the decline of European civilisation, the Baroque and Universalism, and set me seeking out the causes of revolutions, writing stories that are contemplative images of action, and pointing me to my masques at the Banqueting House. 'The Silence' had set me thinking about the universe, the stars, its metaphysical and physical aspects, the underlying unification of all differences, which has come to fruition in my present work, *The*

Algorithm of Creation. And I did not know that by early 1967 I was just two and a half years along a Mystic Way I had embarked on without fully realising and was bewildered by, that would last nearly 30 years.

I did not fully understand then that the Mystic Way begins with awakening and continues with purgation and a First Mystic Life and centre-shift so the deeper self can open to the Light in a first illumination, like a puddle reflecting the sun. I did not know I had a long and anguished Dark Night of the Soul ahead of me when my life would seem to implode and I would find myself on my own, and that after further purges there would be a Second Mystic Life and full illumination. I must now fast-forward six years into my Mystic Way to the beginning of my Second Mystic Life on 3 September 1971: a number of intense images and visions, far more intense and profound than I had had until then.

I have described in *My Double Life 1: This Dark Wood*, pp.363–371, the build-up to this intense Mystic Life. I was seeing with a new intensity: a brilliant sun, a weekend by the sea in Worthing, waves like shoals of fish, trees bent in the flowing wind, and the moon with a strange white glow round it. I wrote in my *Diaries*: "Something is happening to my way of *seeing* and I do not understand."[15]

On Friday 10 September 1971, a day that has meant as much to me as Monday 23 November 1654 meant to Pascal, in the lining of whose doublet that date and the word "FIRE" were found sewn after his death in 1662, I had an overwhelming and momentous experience of illumination for an hour and a half, of a white light flowing upwards, a tree of white fire. Visions wobbled up like bubbles from a spring, and I saw a white flower, like a chrysanthemum or dahlia – my first glimpse of the Golden Flower – and many patterns and old paintings of gods and saints.

For an hour and a half I was flooded with light. I wrote in my *Diaries*:[16]

It came in me: a tree, white against black inside me, a bare winter tree of white fire, flowing, rippling as if in water…. And more. A centre of light shining down as if from a great height, rays coming down like rocket blasts or fireworks. Then a white flower like a chrysanthemum (detailed cells). And so much more.... A sun, breaking through cloud (a moon?). Stars. And a fountain of white light. Patterns like my *mandala*. And I was behind my chest – in my *heart* – lying down, breathing slowly and deeply with half-closed eyes, near sleep as at Worthing when I had two visions.... When I sat up I felt... refreshed, turned inside out, and wobbly at the knees.... A painting with 'God sitting on a throne'.... It was dark, my window was open. The clock ticked. Some traffic roared. I fell on my knees in the dark, hands clasped and there was a white point, then a circle of light that went deep deep up into the heavens. I said, 'I surrender', and the light moved and changed till I felt exhausted.... I want to push it away now, forget about it and have a drink in The Bunch of Grapes. I feel relaxed. After all my seeking, I have found my heart, my centre, my soul, I have found my white light. I feel exhausted but blissfully happy. Full of love.... A round blob, sometimes like a jellyfish, sometimes like a celestial curtain blown in a wind – the *aurora borealis* in *Marvels of the Universe*. That is what I have found. Feel too limp and exhausted to write any more. Or to communicate my weary jubilation.

I began to interpret:

Me and my Flowing Light. Nothing else. Everyone has it.... I was like a child that does not know it can walk until it has been told it can..., until it has been taught.... I am quite drunk with it. So wobbly on my knees. Drunk with love. Can't take any more now. Too tired. At great peace.

The experience was accompanied by many images and visions, and further details can be found in *My Double Life 1: This Dark Wood*, pp.372–380. My Second Mystic Life lasted until 28 April 1972.

30-year-long Mystic Way to instinctively seeing the unity of the universe as an interconnected Oneness

I then underwent a further long period of darkness (the Dark Night of the Spirit in which new powers flow in, infused knowledge about the universe and the Light), and then two more Mystic Lives separated by yet a further period of darkness. I emerged from my Mystic Way at the end of my Fourth Mystic Life, on 6 December 1993, to find I was on a Unitive Way. At the end of the Mystic Way is permanent unitive living in which the universe is perceived *instinctively* as a unity, as there has now been a permanent centre-shift to living through the spirit.

Spiritual awakening is a key sequence of events in a long process that in my experience can take the best part of 30 years. The new self then instinctively lives in permanent knowledge of the underlying harmony of the universe and humankind – permanent instinctive awareness of +A + −A = 0 – and this is what I have reflected in my works for a further 30 years.

I had rejected a Chair for life in the Tokyo university where Empson taught in the 1930s, and I left Japan and applied myself to becoming cross-disciplinary like a Renaissance man such as Michelangelo and Leonardo, who were at home in many disciplines. I did not grasp until recently that my Universalism – my philosophy of the fundamental unity of the universe and all humankind, and of all history, philosophy, literature and international politics, which should arguably be a new university subject in its own right – came out of my awakening in 1965, along with all my books. My centre-shift in Japan was a turning-point in my life and, as I grappled to understand what was happening to me and cope with the many images and visions, I knew I had to pursue a writing career and set out in my books the oneness of all the universe I could *instinctively see* in several disciplines.

By the early 1990s I had been through a Dark Night of the Soul and a Dark Night of the Spirit and had emerged into unitive living when I *instinctively saw* the oneness of the universe and of humankind – which form the essence of Universalism. In other words, I was living through my deeper self, my core, which sees the unity of things. The rational, social ego is useful for seeing differences and making distinctions, but the spiritual core, a deeper centre of consciousness known in meditation, reconciles opposites and differences within an underlying unity that includes all pairs of opposites. Day and night, peace and war, life and death, time and eternity, the finite and the infinite are all seen to be part of an underlying unity. This new way of looking is how our deeper self really sees the universe.

The universe is like a broken pot, it is split into specialised areas (all the dozens of disciplines) and is in fragments, whereas Universalism puts the universe back together into one whole, like a curator reassembling the pieces back into a patched-up pot to put on display.

I had come to see that we need to be aware that our analytical, rational, social ego's approach to the world and universe that identifies contradictions and differences, like breaking a pot and reducing it into bits, can be replaced by the deeper vision of our spiritual core, which can be experienced as a unifying "esemplastic power" (Coleridge's word that comes from the Greek 'shape into One') that pieces bits together into a whole like a restored pot. And we also need to rethink our view of philosophy, history and contemporary history, and world politics and see them as a whole and not concentrate on one bit: a current linguistic or analytical movement, one nation's history while ignoring the history of the other 192 nation-states, and divided governments running the world rather than one united democratic world government that can abolish war, enforce disarmament, combat famine, disease and poverty with the money saved, and solve the world's financial, environmental and virological problems.

Universalism sees everything in relation to the oneness of the universe, the One. Mystics have seen the One as the metaphysical Fire or Light. Shelley wrote, "The One remains, the many change and pass." St Augustine wrote in c.400, "I entered (within myself). I saw with the eye of my soul, above or beyond my mind, the Light Unchangeable." Hildegarde of Bingen, who died in 1179, wrote, "The Light which I see... is more brilliant than the sun." As we have seen, Pascal called it the Fire. My works record 112 experiences of the Fire or Light (all dated from diary entries at the time).

My key message in my 60 works is that despite its many contradictions, the universe is a Oneness, that all humankind is fundamentally a Oneness and that all disciplines (including literature, philosophy, history and international politics) are individually a oneness and have to be seen as a whole and are collectively interconnected within a Oneness. My works are within seven disciplines: mysticism (seeing the Fire or Light as the universal mystical experience, seeing the unity of the universe); literature (seeing the fundamental theme of world literature as a quest for the One and as condemnation of follies and vices); philosophy and the sciences (seeing all humankind in relation to the One, the unity of the universe, and so challenging modern philosophy); history (seeing the underlying patterns of world history and of 25 civilisations as beginning with a mystic's vision of the One, seeing all history as one, a whole); religion (seeing the Light as the common essence of all religions, which are therefore one); international politics and statecraft (seeing the benefits of a new democratic World State that can abolish war and enforce disarmament, combat famine, disease and poverty and solve the world's financial, environmental and virological problems); and world cultures (seeing the underlying unity of world culture). All these disciplines are connected by the One they all approach.

I have written about the unity of the universe in books on seven disciplines: on mysticism and comparative religion (*The Fire and the Stones* and *The Light of Civilization*); on philosophy (in *The New Philosophy of Universalism*, and before that in *The Universe and the Light* and *The One and the Many*); in history (in *The Fire and the Stones* and *The Rise and Fall of Civilizations*); in literature (in *A New Philosophy of Literature*, which details the fundamental theme of world literature as a quest for the One and also condemnation of follies and vices as in Horace and Pope, and in my works); in international politics and statecraft (in *The World Government*, *World State* and *World Constitution*); and in world culture (in *The Secret American Destiny*).

These seven disciplines are interconnected like the seven bands of a rainbow: each band is separate yet is part of one rainbow, and behind them all is the Oneness that everyone can experience in their lives, a unity behind all the diversity the rational, social ego sees. My trademark is the stag on my coat of arms, the two seven-branched antlers: one antler representing the seven disciplines, the other antler representing the seven branches of literature in which I have written and worked (my main discipline).

My new philosophy of Universalism surfaced in a time in which philosophy had been dominated by rational, analytical philosophy: logical positivism and linguistic analysis. Universalism draws on the most recent discoveries made by space telescopes while looking back to the Presocratic Greeks, including Anaximander of Miletus, who along with Heraclitus of Ephesus connected philosophy with

Nature. My new philosophy of Universalism returns philosophy to Nature, and philosophical Universalism challenges logical positivism and linguistic analysis, the rational approach to the universe of the Vienna Circle that focused on language and logic more than on the science and beauty of the universe.

New powers from a deep source

My view of the universe in *The Algorithm of Creation* began as the product of a centre-shift from my rational, social ego to my deeper self, which can receive new powers. During the last 30 years my works have been received with infused knowledge as a result of these powers, which have included the gift of healing and of prediction – I foresaw the end of Communism and the coming of a United States of Europe in my contemporary history – and a connection to the beyond which has fed nearly all the titles of my works during my sleep, including *The Algorithm of Creation* (see p.xvii), and regular 'proofreading' of my works. I will be woken with a message to check a page number, perhaps p.148, and I will go to that page the next morning and find there is an error that needs amending.

My deep source is not just an editor and proofreader. Sometimes I am fed quite detailed instructions on how to do my work. *The Promised Land* is a case in point. I was on a ship off Egypt, and I was awoken at 3am with the structure of my (then) new book.[17]

The structure of The Promised Land received in sleep

As Ann was still asleep I breakfasted alone.

I knew I would be writing *The Promised Land* as I stood on Mount Nebo on 1 March [2019], and later as I sat on the coach. I knew I had to structure the book during a previous day at sea (on 9 March), and on 11 March I wrote in my diary: "It's now 8.35, and I have a day at sea ahead – to think out *The Promised Land*." I checked a fact in the ship's library and began a plan.

There was a briefing at 12.30. We were told that a massive storm was crossing Egypt behind us. It covered half Egypt, and it was chasing us, and the harbour at Alexandria had been closed. We would be racing ahead of the storm to Port Said, the other side of the Suez Canal, and would be heading for the safety of Antalya, a port in Southern Turkey, where we would visit Roman ruins.

Ahead would be a second day at sea, keeping ahead of the storm and making for Turkey, and I wrote in my diary, "Worked fitfully on *The Promised Land* plan."

I have often received key ideas in my sleep (see *My Double Life 2: A Rainbow over the Hills*, pp.910–911, 'Sleep inspiration' [and Appendix 2 ('33 Experiences of Sleep Inspiration' on pp.216–218 of *The Promised Land*]) from the idea for *The Fire and the Stones* (1991) – the concept of my study of 25 civilisations as "Light-bearers" (on 24 August 1979) and the beginning of its Preface 'Introduction to the New Universalism' (on 29 April 1989) – to the idea for *A Baroque Vision* (2023) and its theme of the Baroque as the source of Universalism (received in sleep on Christmas morning, 25 December 2019). I attribute this sleep inspiration to the Muses.

On the morning of Friday 13 March 2020 I woke at 3am with the structure of *The Promised Land* in my mind. I later wrote in my diary:[18]

Woke in the middle of the night and scribbled out notes on the structure of *The Promised Land*. Put this into a plan during the journey home [from Perge and Aspendos, see below]. Had lunch later and sat on the back deck for a short while looking at the cranes with magnetic grasps and dozens of containers and then returned to my cabin [419T on the *Serenissima*] and worked on *The Promised Land*. I have finished the plan in draft, incorporating all the bits and pieces I have written at different times and also the restructuring I saw I needed to do between 3 and 3.30am, writing under the small torch light above my bed in my cabin while Ann slept.... It's in five parts. Pt 1, The Promised Land. Pt 2, The Formative Years, 1960–1980, The Mystic Way and the Baroque Vision. Pt 3, The Wilderness Years, 1980–2020, The Unitive Vision and Universalism. Pt 4, Ten Commandments: Ten Universalist Principles. Pt 5, A View of the Promised Land.

It seemed that the Muses might have woken me, their slave, in the middle of the night to tell me what they wanted me to say in response to Israel's closure of its border to our ship.

Other works of mine that came through sleep inspiration include: 12 stanzas of 'Night Visions in Charlestown', the turning-point and link between the Baroque and Universalism (on 6 August 1983); the word Universalism in the title of *The New Philosophy of Universalism* on the morning of 28 February 2007 (see letter of 28 February 2007 in *Selected Letters*); and the structure of *The Secret American Dream*, an early statement of the World State (on 3 December 2009), which I would send to President Obama with the Appendix on bin Laden's 69 attempts to buy nuclear weapons flagged.

I see this ability to see important original ideas in my sleep as being a consequence of my full illumination on 10 September 1971, which opened me to unitive vision and many powers from the beyond, including second sight, the ability to foresee and anticipate future historical events like the Delphic Oracle, such as: the collapse of Communism with the fall of the Berlin Wall; the creation of the European Union as a stage-43 conglomerate; and the break-up of the United Kingdom into a Federation of the British Isles.

I am full of admiration and wonder at the power of inspiration that derives from metaphysical living, which I once described to Colin Wilson as opening one's soul to a spring like the one in Ghadames, and allowing bubbles to wobble up into consciousness from the beyond. I was also thinking of the spring near Horace's villa in Licenza, north-east of Rome, which is thought to be Horace's *"fons Bandusiae"*, his Bandusian spring, and which I squatted next to in 1957 when I was 17 and scooped up water in my cupped hands to sip, and put in my 1973 poem 'Ode: Spring', which refers at the end to "the gushing of this spring between my sleeves".

Such a detailed message from the beyond has been typical of its role in my work.

In short, for the last 30 years I have received all my works from a deep source that is a higher consciousness, and always conveys truth. It is never wrong, and I am its amanuensis. I am scrupulously evidential and follow the scientifically established facts, but my source shapes the clarity of what I recount in a blend, a working partnership, between the metaphysical and scientific approaches, a $+A + -A = 0$.

Without my source, the metaphysical input would be sketchy.

The Algorithm of Creation has come from the same higher source which has a place and a role within the universe, and a Theory of Everything must include it. Just as I asked John Barrow if his mathematical Theory of Everything included love and order (see p.xviii), so it must include the ability to receive information from a higher consciousness that is part of the universe, my deep source. 'High', 'deep' – Heracleitus said in Fragment 60, "The way up and the way down is one and the same"; in other words, +A (up, high) + –A (down, deep) = 0 (the One). In short, the universe is a metaphysical non-duality which gave rise to mind and matter in a +A + –A = 0: +A (mind) + –A (matter) = 0 (a metaphysical non-duality that is the source of both).

I have set out how these new powers have affected my transmission of the algorithm of Creation and my Theory of Everything because this work has come from a deep source and a high consciousness within the universe, and I have needed to make this clear. For a year, in 1965–1966, I received my life's work in transmissions I scarcely understood, and gave up being a one-subject Professor to become a latter-day Renaissance man at home in many disciplines to receive my 60 works from a deep source that can see the universe from the perspective of a high consciousness with exceptional clarity.

New powers and my Form from Movement Theory, and this Appendix

On 1 December 2021 I pondered the sequence of events that led me to write my Form from Movement Theory in the dining-hall at Jesus College, Cambridge on 4 September 1992. Why had I begun with a movement within Nothingness (M)? I went to my library just before bedtime, sensing there was a book I might read, but no book caught my eye and I came away empty-handed.

My deep source was active again in the early morning of 2 December 2021, and as brief and to the point as ever. I woke on 2 December, as my *Diaries* describe, with "Smoot and Mather" in my mind, fed to me in my sleep. Nothing else, no explanation, just "Smoot and Mather". I had not thought of Smoot and Mather for years. I thought they might have co-authored the book I was looking for the previous evening. I looked in my library for a book by Smoot and Mather, but could not find one. In fact, so far as I am aware they never wrote one. I went online and then I realised. Smoot and Mather won the 2006 Nobel Prize for Physics for their work on cosmic microwave background (CMB) radiation from 1989, when NASA launched COBE (Cosmic Background Explorer), to 1992. On 23 April 1992 Smoot announced that they had discovered the blackbody form of CMB, an echo of the Big Bang 400,000 years after the event. The CMB appears black as its cooling surface, which COBE saw 400,000 years after the Big Bang, absorbs rather than rejects radiant energy. Smoot and Mather also claimed to have discovered that CMB is anisotropic, in that it has different temperature properties in different directions. The blue areas in the CMB on the front cover are cooler than the 2.7K norm, and the orange and red areas are warmer. In October 1992 Smoot said he was confident their discovery would be confirmed, but it took a long time to confirm, and they were not awarded the Nobel Prize until 2006.

Then I grasped that on 4 September 1992, when I wrote my Form from Movement Theory, I had long been aware of Penzias and Wilson's accidental discovery of CMB radiation in 1964. It was not until 1992 that the CMB radiation was found to have mass-density fluctuations, ripples or wrinkles, that suggest wells where galaxy clusters could form, and these were thought to be caused by ripples or wrinkles in the energy that became CMB radiation when the Big Bang happened. And I knew of COBE's discovery of CMB, and I had worked out that CMB was significant for two reasons.

First, its faint glow of light that fills the universe and falls on earth from every direction, with uniform intensity and temperature, is the leftover radiation from the Big Bang. It shows – and proves – the expansion of space: shorter gamma wavelengths emitted by the initial explosion of the Big Bang became stretched in the course of inflation into longer-wavelength microwaves. Before the CMB radiation the universe was a hot, dense and opaque plasma containing energy and matter, and as photons could not travel freely no light escaped until the universe cooled 400,000 years after the Big Bang.

The ripples in CMB reflect the architecture of the universe when the light was freed by the Big Bang, and the subsequent cosmic structures – galaxies and clusters of galaxies – as the light passed them on its journey through space and time.

NASA's COBE mission confirmed predictions of the Big Bang and showed hints of cosmic structure not seen before: the ripples or wrinkles. The ripples are generally seen as effects of the Big Bang, but those looking for the still 'ocean of energy' *before* the Big Bang can see the ripples as being movements within the still nothingness, the dark 'ocean of energy' that existed before the Big Bang.

It dawned on me that although in September 1992 I was aware of NASA's COBE measuring and the ripples, the movement within Nothingness (N) was still unproved and would not be confirmed for years. My new powers from my deep source had told me to go ahead and see still Nothingness as a movement (M), following the Greeks Anaximander of Miletus (c.570BC), Heracleitus of Ephesus (c.500BC) and Aristotle (see p.16), and base my view of the origin and creation of the universe on the rippling, wrinkly movement in the 'ocean of energy' that pre-existed the Big Bang (the transformation of energy into matter) and the ripples' appearance in the CMB, by which time inflation had happened and the ripples or wrinkles had helped matter cluster through gravity into 2 trillion galaxies.

No wonder on 4 September 1992 my starting with a movement within Nothingness (N) was being keenly observed in the dining-hall at Jesus College, Cambridge by Roger Penrose (who had asked to borrow my typescript of *The Universe and the Light* so he could read it at our conference), Freeman Dyson and (briefly) John Barrow; my new powers had led me into an innovatory Theory that had the blessing of the very latest scientific probes.

Now I realised I was being told by my deep source to link my Form from Movement Theory (September 1992) to the discovery of Smoot and Mather that was confirmed after October 1992, and that I should conclude this Appendix with this link. I have accordingly done this, and I cannot think of a more appropriate way to end this Appendix on the new powers, deep source and high consciousness that worked through me to achieve this work, its algorithm of Creation and Theory of Everything. Had I not been fed crucial material at key times I would not have written my Form from Movement Theory and there would be no algorithm of Creation or Theory of Everything. My deep source – the source for what I was fed in 1992 – is one of a number of sources for this work, the rest of which are the many scientifically evidential sources.

In the evening of 2 December 2021, while I was working on the early part of this Appendix, I looked at the title of this Appendix, which was then 'New Powers Received from 1964: Awakening to a Higher Consciousness and Source, Instinctive Understanding of the Unity of the Universe, Universalism and the Algorithm of Creation'. I woke next morning, 3 December 2021, with most of the present title (see p.317) in my mind, all the words except for 'on Works' and 'and a Theory of Everything', which I added. My deep source had even insisted that I should include 'Infused Knowledge in Sleep' in the title. On 6 December it even invaded my PA's sleep to tell her not to forget to renumber the notes in section 2 to cope with an inserted note 10A: "Don't forget 10A." I am happy to credit my deep source's contribution to the title of this Appendix in this closing paragraph, and indeed to Universalism's algorithm of Creation and its Theory of Everything.

25 November–3, 6 December 2021

Appendix 2

Chart of 25 Civilizations and Cultures from One to One

Chart of 25 Civilizations and Cultures from One to One

The Fundamental Unity of World Culture

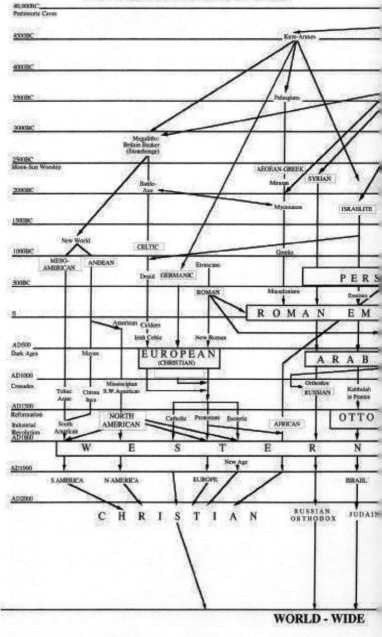

WORLD - WIDE

Appendix 2

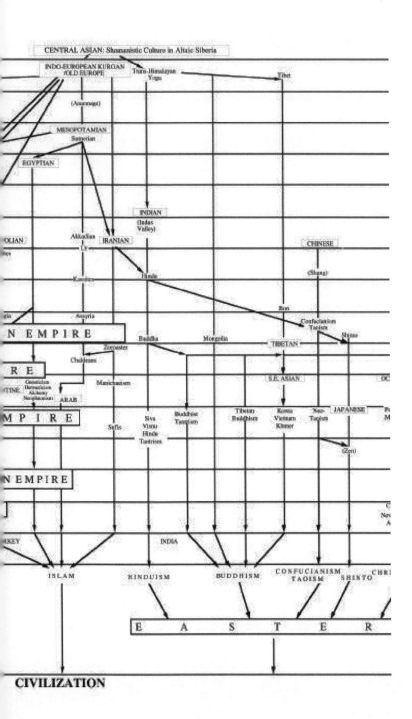

CIVILIZATION

357

Appendix 3

Universalism's View of the Structure of the Universe with its
Co-existing Metaphysical and Physical Perspectives

Universalism's View of the Structure of the Universe with its Co-existing Metaphysical and Physical Perspectives

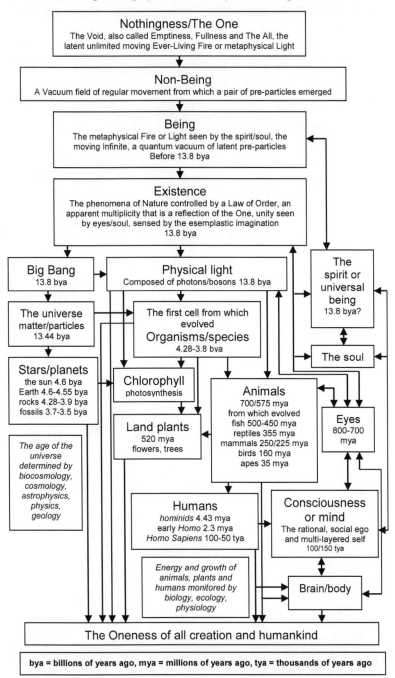

bya = billions of years ago, mya = millions of years ago, tya = thousands of years ago

Notes and References to Sources

within *The Algorithm of Creation*
(including Notes taken from
The New Philosophy of Universalism)

Preface

1. John Barrow, *Theories of Everything*, Preface, paragraph 1.
2. Barrow, *Theories of Everything*, p.200.
3. Barrow, *Theories of Everything*, p.199.
4. Barrow, *Theories of Everything*, p.210.
5. *The I Ching: The Book of Changes*, trans. by James Legge, pp.159–160, 247–248, 307, 324–325; Nicholas Hagger, *Awakening to the Light, Diaries: 1958–1967*, p.378, 17 November 1966.
6. *The I Ching: The Book of Changes, op. cit.*, pp.159–160.
7. *The I Ching: The Book of Changes, op. cit.*, p.251.
8. *The I Ching: The Book of Changes, op. cit.*, p.125.
9. *The I Ching: The Book of Changes, op. cit.*, pp.324–325.
10. *The I Ching: The Book of Changes, op. cit.*, pp.238–239, 307.

1. Algorithms and the Universe

1. https://www.britannica.com/biography/al-Khwarizmi.
2. For a study of Junzaburo Nishiwaki's poetry, see Hosea Hirata, *The Poetry and Poetics of Nishiwaki Junzaburo: Modernism in Translation*, the Princeton Legacy Library, which states that Nishiwaki has been compared to T.S. Eliot, R.M. Rilke and Paul Valéry, see https://press.princeton.edu/books/hardcover/9780691633862/the-poetry-and-poetics-of-nishiwaki-junzaburo.
3. Nicholas Hagger, *The New Philosophy of Universalism*, pp.19–20.
4. For *T'ien* and the *Tao*, see Hagger, *The Light of Civilization*, pp.330–341.

2. The Origin, Development and End of the Universe in 20 Stages

1. Stephen Hawking, *A Brief History of Time*, p.46.
2. Hagger, *Selected Letters*, p.778.

3. Hagger, *The New Philosophy of Universalism*, pp.264–266.

4. See J.J. Williamson, *The Structure of ALL*, lectures delivered at the R.A.F. College, Cranwell, Lincolnshire, UK, which defines metaphysics as "the science of ALL".

5. Planck collaboration, 'Planck 2013 results. XVI. Cosmological parameters', 2013, *Astronomy & Astrophysics*, 571: A16.

6. Neil deGrasse Tyson, Michael A. Strauss and J. Richard Gott, *A Brief Welcome to the Universe*, p.170.

7. Christopher J. Conselice, *et al.*, 'The Evolution of Galaxy Number Density at z < 8 and Its Implications', 2016, *The Astrophysical Journal*, 830(2): 83. Also see Henry Fountain, 'Two Trillion Galaxies, at the Very Least', 17 October 2016, *New York Times*.

8. Hagger, *The New Philosophy of Universalism*, pp.269–273.

9. Clement of Alexandria, *Stromateis*, v,104,1. Quoted in G.S. Kirk, J.E. Raven and M. Schofield, *The Presocratic Philosophers*, pp.197–198.

10. A.A. Long, ed., *The Cambridge Companion to Early Greek Philosophy*, pp.47, 79.

11. Kirk, Raven and Schofield, *op. cit.*, pp.197–198: "Pure or aitherial, fire has a directive capacity.... The pure cosmic fire was probably identified by Heraclitus with *aither*, the brilliant fiery stuff which filled the shining sky and surrounds the world; this *aither* was widely regarded both as divine and as a place of souls. The idea that the soul may be fire or *aither*... must have helped to determine the choice of fire as the controlling form of matter."

12. Aristotle, *Physics*, 8.6; 260a: "But the unmoved mover..., since it remains permanently simple and unvarying and in the same state, will cause motion that is one and simple."

13. W.O.E. Oesterley and Theodore H. Robinson, *An Introduction to the Books of the Old Testament*, pp.24–34; *Encyclopaedia Britannica* (1991), micropaedia, 5.177.

14. Note by Nicholas Hagger: "I announced my Form from Movement Theory over breakfast at a Symposium on Reductionism's Primacy in the Natural Sciences, which was held at Jesus College, Cambridge in September 1992. I had contributed a paper which was circulated among fifteen speakers who included ten Professors of Physics, Biology, Physiology, Neurology, Astronomy and Philosophy in Britain and the

USA. Roger Penrose and John Barrow were among these. About ten of the attendees gathered round as I explained my thinking, and a Norwegian mathematician, Henning Broten, converted what I said to mathematical symbols as I spoke. The mathematics were refined during subsequent discussions, and I was grateful for his expertise."

15. Planck collaboration, 'Planck 2013 results. XVI. Cosmological parameters', 2013, *Astronomy & Astrophysics*, 571: A16.
16. Tyson, Strauss and Gott, *op. cit.*, p.170.
17. Hagger, *The New Philosophy of Universalism*, pp.57–60, 61–63.
18. See Julian Barbour, *The End of Time*, p.12, for the originality of this idea.
19. John Barrow, *The Infinite Book*, pp.122–123.
20. Simon Singh, *Big Bang*, pp.156–161.
21. Singh, *op. cit.*, pp.149–156.
22. Singh, *op. cit.*, pp.214–229, 248–261. The discovery in 1998 that the universe's expansion is accelerating means that some calculations based on "Hubble time" are now unsafe. The oldest stars may not now be older than the universe.
23. Singh, *op. cit.*, pp.341–347, 418–421, 438–440.
24. Singh, *op. cit.*, pp.422–437.
25. Singh, *op. cit.*, pp.453–461, 471–473.
26. Roger Penrose, *The Emperor's New Mind*, p.543. For Dante's infinitesimal "point", see *Paradiso*, canto 28.
27. Penrose, *op. cit.*, pp.435–447.
28. Martin Rees, *Just Six Numbers: The Deep Forces That Shape the Universe*, p.44.
29. Penrose, *op. cit.*, pp.541–544.
30. Penrose, *op. cit.*, pp.436, 470–471; Hawking, *op. cit.*, pp.88–92, 133–136.
31. Barrow, *op. cit.*, p.38.
32. Singh, *op. cit.*, p.489.
33. Hawking, *op. cit.*, pp.137–138.
34. Hawking, *op. cit.*, pp.116, 133–136.
35. Hawking, *op. cit.*, pp.140–141.
36. Penrose, *op. cit.*, p.471.
37. Barrow, *op. cit.*, pp.103–109.
38. John Gribbin, *In the Beginning*, p.xii.
39. Gribbin, *op. cit.*, p.229.

40. Singh, *op. cit.*, p.482.
41. Hagger, *The New Philosophy of Universalism*, pp.80–81.
42. For early quantum theory, see Tom Duncan with Heather Kennett, *Advanced Physics*, pp.391–392.
43. *Encyclopaedia Britannica*, 1991 ed., micropaedia, 9.841, 'Quantum mechanics'.
44. *Encyclopaedia Britannica*, 1991 ed., micropaedia, 9.840, 'Quantum electrodynamics'.
45. Hagger, *The New Philosophy of Universalism*, pp.86–88.
46. David Bohm, *Wholeness and the Implicate Order*, pp.149, 190–193.
47. Renée Weber, *Dialogues with Scientists and Sages: The Search for Unity*, pp.25–27. Also David Bohm in conversation with Nicholas Hagger.
48. Bohm, *op. cit.*, pp.189, 193, 211.
49. Bohm, *op. cit.*, p.149.
50. Bohm, *op. cit.*, pp.ix–x.
51. Bohm, *op. cit.*, p.190.
52. Bohm, *op. cit.*, p.191.
53. Bohm, *op. cit.*, p.191.
54. Bohm, *op. cit.*, p.208.
55. Bohm, *op. cit.*, pp.191–192.
56. Bohm, *op. cit.*, p.193.
57. Rees, *op. cit.*, p.12.
58. Peter Hewitt, *The Coherent Universe*, p.2.
59. Edgard Gunzig is Professor of General Relativity at the Free University, Brussels. This and the next paragraph are based on a lecture he gave in April 1991.
60. Hagger, *The New Philosophy of Universalism*, pp.63–68.
61. To check facts and timings in this section, see Duncan with Kennett, *op. cit.*, pp.493–507.
62. Penrose, *op. cit.*, p.419.
63. Gribbin, *op. cit.*, p.11. For inflation by a factor of 10^{54}, faster than the speed of light, in the next paragraph see http://abyss.uoregon.edu/~js/21st_century_science/lectures/lec25.html.
64. Hagger, *The New Philosophy of Universalism*, pp.81–84.
65. David Castillejo, *The Expanding Force in Newton's Cosmos*, p.108.
66. Castillejo, *op. cit.*, p.108.

67. Castillejo, *op. cit.*, p.109. For ether in the next two paragraphs, see Singh, *op. cit.*, pp.93–98.

68. Hagger, *The New Philosophy of Universalism*, pp.84–85.

69. *The Born-Einstein Letters*, trans. by Irene Born, p.91.

70. Albrecht Fölsing, *Albert Einstein*, pp.587–592.

71. Duncan with Kennett, *op. cit.*, p.507. Freeman Dyson questioned whether any experiment in the real universe can detect one graviton. If not, he held that it is meaningless to talk about gravitons as physical entities.

72. Fölsing, *op. cit.*, pp.704–705.

73. Brian Green, *The Elegant Universe*, p.15.

74. Bohm, *op. cit.*, pp.65–110; and Penrose, *op. cit.*, p.362.

75. Penrose, *op. cit.*, p.362.

76. Hagger, *The New Philosophy of Universalism*, pp.66–67.

77. G. van der Marel, 'The M31 Velocity Vector. III. Future Milky Way M31–M33 Orbital Evolution, Merging, and Fate of the Sun', 2012, *The Astrophysical Journal*, 753(1): 9.

78. R. Cowen, 'Andromeda on collision course with the Milky Way', 31 May 2021, *Nature*.

79. Hagger, *The New Philosophy of Universalism*, pp.112–113.

80. Adams, *Origins of Existence*, p.66. For 10^{23} grains of sand, see http://en.wikipedia.org/wiki/1000000000000_(number).

81. Gribbin, *op. cit.*, p.255. For 90,000 years, see Jack Phillips, *Suppressed Science*, p.32.

82. See https://exoplanets.nasa.gov/exoplanet-catalog/7005/55-cancri-e/#:~:text=55%20Cancri%20e%2C%20also%20known,hangs%20in%20a%20dark%20sky.

83. For 228 planets see http://www.physorg.com/news124121930.html.

84. John Gribbin and Martin Rees, *Cosmic Coincidences*, pp.11–14.

85. John Barrow and Frank Tipler, *The Anthropic Cosmological Principle*, p.385. For the Milky Way containing 200–400 billion stars, see http://en.wikipedia.org/wiki/Milky_Way.

86. James N. Connelly, Martin Bizzarro, Alexander N. Krot, Åke Nordlund, Daniel Wielandt, Marina A. Ivanova, 'The Absolute Chronology and Thermal Processing of Solids in the Solar Protoplanetary Disk', 2 November 2012, *Science*, 338(6107): pp.651–655.

87. Hagger, *The New Philosophy of Universalism*, pp.67–68.

88. For 10^{23} stars see Adams, *op. cit.*, p.66.

89. For 10^{23} grains of sand see Fred Adams and Greg Laughlin, *The Five Ages of the Universe*, p.xxiii. And http://en.wikipedia.org/wiki/1000000000000_ (number).

90. Hagger, *The New Philosophy of Universalism*, p.127.

91. For 4.6 billion years ago, see Adams, *op. cit.*, p.183; C.J. Clegg with D.G. Mackean, *Advanced Biology: Principles and Applications*, p.76; *Encyclopaedia Britannica* (1991), macropaedia, 19.773. For 4.55 billion years ago see, for example, http://www.talkorigins.org/faqs/faq-age-of-earth.html. Paul Davies, *The Goldilocks Enigma*, p.23, puts the age of the earth at 4.56 billion years.

92. Clegg with Mackean, *op. cit.*, p.664.

93. See http://en.wikipedia.org/wiki/Origin_of_life.

94. See https://en.wikipedia.org/wiki/Big_Bang.

95. Hagger, *The New Philosophy of Universalism*, pp.68–72.

96. The acceleration was discovered from studies of distant supernovae and results announced by two international teams of researchers. For acceleration, see Martin Rees, *Our Cosmic Habitat*, pp.104–108.

97. For acceleration beginning 8 billion years after inflation, 5 billion years ago, see Barrow, *op. cit.*, p.147. Barrow claims that dark energy took over 8 billion years after the first 379,000 years, that deceleration soon changed to acceleration and that no more galaxies could form.

98. Davies, *op. cit.*, pp.47–49.

99. Barrow, *op. cit.*, p.xv.

100. Barrow, *op. cit.*, p.96.

101. Barrow, *op. cit.*, p.22.

102. Barrow, *op. cit.*, p.22.

103. Barrow, *op. cit.*, p.146; Davies, *op. cit.*, p.26.

104. For WMAP putting the age of the universe at 13.7 (since updated to 13.82) billion years ago, see http://map.gsfc.nasa.gov/news/ and http://en.wikipedia.org/wiki/Age_of_the_universe.

105. Barrow, *op. cit.*, p.149.

106. John Gribbin, *The Universe: A Biography*, p.224.

107. Adams, *op. cit.*, p.66.

108. Adams and Laughlin, *op. cit.*, p.xxiii.

109. Adams, *op. cit.*, p.218.

110. Adams, *op. cit.*, p.204.

111. Adams, *op. cit.*, p.63.

112. See https://physics.stackexchange.com/questions/659661/is-space-really-expanding.

113. Rees, *Our Cosmic Habitat*, pp.102–104.

114. Hagger, *The New Philosophy of Universalism*, pp.107–108.

115. Gribbin and Rees, *op. cit.*, pp.23–26.

116. Hagger, *The New Philosophy of Universalism*, pp.97–98.

117. I have broadly followed, with some refinements, the scheme in Davies, *op. cit.*, pp.295–302.

118. Hagger, *The New Philosophy of Universalism*, pp.98–99.

119. Hagger, *The New Philosophy of Universalism*, pp.99, 101–102.

120. Fred Hoyle, 'The universe: past and present reflections', *Annual Review of Astronomy and Astrophysics*, vol. 20, 1982, p.16.

121. Hagger, *The New Philosophy of Universalism*, pp.102–122.

122. Hagger, *The New Philosophy of Universalism*, pp.122–123.

123. Adams, *op. cit.*, pp.84–85.

124. Adams, *op. cit.*, p.55.

125. See http://www.leaderu.com/offices/billcraig/docs/teleo.html.

126. Barrow and Tipler, *op. cit.*, pp.401–403.

127. Hagger, *The New Philosophy of Universalism*, pp.372–387.

128. Hagger, *The New Philosophy of Universalism*, pp.388–393.

129. Hagger, *The New Philosophy of Universalism*, pp.394–395.

130. Hagger, *The New Philosophy of Universalism*, pp.127–133.

131. For 4.6 billion years ago, see Adams, *op. cit.*, p.183; Clegg with Mackean, *op. cit.*, p.76; *Encyclopaedia Britannica* (1991), macropaedia, 19.773. For 4.55 billion years ago, see, for example, http://www.talkorigins.org/faqs/faq-age-of-earth.html. Davies, *op. cit.*, p.23, puts the age of the earth at 4.56 billion years.

132. Clegg with Mackean, *op. cit.* https://physics.stackexchange.com/questions/659661/is-space-really-expanding, p.664.

133. See http://en.wikipedia.org/wiki/Origin_of_life.

134. Adams, *op. cit.*, p.182. Also as 121.

135. See http://en.wikipedia.org/wiki/Abiogenesis.

136. The "Primordial Soup Theory" suggested that life began in a pond or ocean as a result of the combination of chemicals. It was so named after

Oparin referred to a "Primeval Soup" of organic molecules. See http://en.wikipedia.org/wiki/Origin_of_life.

137. See http://en.wikipedia.org/wiki/Origin_of_life.

138. See http://en.wikipedia.org/wiki/Origin_of_life.

139. See http://en.wikipedia.org/wiki/Origin_of_life.

140. Clegg with Mackean, *op. cit.*, p.664.

141. See http://en.wikipedia.org/wiki/Origin_of_life.

142. For 10^{130} combinations of amino acids see *Darwinism, Design and Public Education*, ed. by John Angus Campbell and Stephen C. Mayer, p.241. Also see http://leiwenwu.tripod.com/primordials.htm.

143. See http://en.wikipedia.org/wiki/Origin_of_life.

144. See http://en.wikipedia.org/wiki/Origin_of_life.

145. Fred Hoyle, Chandra Wickramasinghe and John Watkins, *Viruses from Space*, and Fred Hoyle and Chandra Wickramasinghe, *Evolution from Space, passim.*

146. See http://en.wikipedia.org/wiki/Origin_of_life. See obituary of Leslie Orgel in *The Times*, 6 December 2007: "The theory that life is RNA-based life was shared by the late Francis Crick."

147. See http://en.wikipedia.org/wiki/Origin_of_life. For a report on the "Lost City" at the bottom of the Atlantic – so-called because of its towering pinnacles and chimneys – see http://www.telegraph.co.uk/earth/main.jhtml?xml=/earth/2008/01/31/scilost131.xml.

148. See http://en.wikipedia.org/wiki/Origin_of_life.

149. See http://en.wikipedia.org/wiki/Origin_of_life.

150. Mary Morrison in conversation with Nicholas Hagger.

151. Mary Morrison in conversation with Nicholas Hagger.

152. Mary Morrison in conversation with Nicholas Hagger. For 10^{14}, see http://en.wikipedia.org/wiki/1000000000000_(number).

153. Mary Morrison in conversation with Nicholas Hagger. For 10^{14}, see http://en.wikipedia.org/wiki/1000000000000_(number).

154. UN Population Division, see https://www.infoplease.com/world/population/world-population-milestones.

155. See http://en.wikipedia.org/wiki/World_population.

156. See http://leiwenwu.tripod.com/primordials.htm. Also, https://cneos.jpl.nasa.gov/about/life_on_earth.html#:~:text=Life%20on%20Earth%20began%20at,period%20of%20late%20heavy%20bombardment.

157. Gribbin and Rees, *op. cit.*, pp.33–34: "We are, quite literally, made from the ashes of long-dead stars." Also see http://books.nap.edu/openbook. php?record_id=12161&page=9.

158. Hagger, *The New Philosophy of Universalism*, p.136.

159. See http://books.google.co.uk/books?id=FTtrd9cl8tAC&pg=PA151&lpg= PA151&dq=3.8+billion+years+ago+fossils&source=web&ots=phr1A6h6I a&sig=-h52e-rCOO3Xko9T2cEeFb56d8E&hl=en..

160. *Encyclopaedia Britannica* (1991), macropaedia, 19.773.

161. See https://www.britannica.com/science/Precambrian/Precambrian-life.

162. Hagger, *The New Philosophy of Universalism*, pp.138–139.

163. Hagger, *The New Philosophy of Universalism*, pp.116–117.

164. Rees, *Just Six Numbers*, pp.18–19.

165. Barrow and Tipler, *op. cit.*, pp.544–545.

166. Hagger, *The New Philosophy of Universalism*, p.139.

167. Hagger, *The New Philosophy of Universalism*, pp.139–140.

168. George Greenstein, *The Symbiotic Universe*, pp.256, 63–65.

169. Hagger, *The New Philosophy of Universalism*, pp.158–159.

170. Hagger, *The New Philosophy of Universalism*, pp.143–144, 148.

171. Clegg with Mackean, *op. cit.*, p.647.

172. Clegg with Mackean, *op. cit.*, p.657.

173. Mary Morrison in conversation with Nicholas Hagger.

174. Hagger, *The New Philosophy of Universalism*, pp.149–152.

175. Clegg with Mackean, *op. cit.*, p.669 for apes and for Africa joining with Eurasia 16 or 17 million years ago.

176. See http://en.wikipedia.org/wiki/Human_evolution#Comparative_table _of_Homo_species.

177. Clegg with Mackean, *op. cit.*, p.669. Douglas Palmer, *Neanderthal*, pp.65–66, 160.

178. Palmer, *op. cit.*, pp.157–161, 162. For 400,000/250,000 years ago, see *Encyclopaedia Britannica*, 1991 ed., micropaedia, 6.28.

179. See http://en.wikipedia.org/wiki/Human_evolution#Comparative_table _of_Homo_species.

180. Palmer, *op. cit.*, pp.180–181.

181. Palmer, *op. cit.*, pp.165, 182–183. For *Homo antecessor*, see *The Daily Telegraph*, 27 March 2008, p.36. For 400,000/250,000 years ago, see *Encyclopaedia Britannica*, 1991 ed., micropaedia, 6.28.

182. Palmer, *op. cit.*, pp.157–159.
183. See http://en.wikipedia.org/wiki/Human_evolution#Comparative_table
_of_Homo_species.
Also see
http://johnhawks.net/weblog/reviews/genomics/divergence/dawn_
chumans_patterson_2006.html.
184. See http://en.wikipedia.org/wiki/Human_evolution#Comparative_table
_of_Homo_species.
Also see Frans de Waal, *Our Inner Ape*, p.14, http://johnhawks.net/
weblog/reviews/genomics/divergence/dawn_chumans_patterson_2006.
html.
185. See http://en.wikipedia.org/wiki/Human_evolution#Comparative_table
_of_Homo_species. Also see de Waal, *op. cit.*, p.14.
186. See de Waal, *op. cit.*, pp.13–15.
187. Hagger, *The New Philosophy of Universalism*, pp.152–155.
188. Palmer, *op. cit.*, pp.160–162.
189. Palmer, *op. cit.*, pp.157–159.
190. Palmer, *op. cit.*, p.162.
191. Palmer, *op. cit.*, p.162.
192. Palmer, *op. cit.*, pp.156–159.
193. Palmer, *op. cit.*, p.157.
194. Palmer, *op. cit.*, p.158. Evidence for interbreeding is within the multi-regional theory.
195. Palmer, *op. cit.*, pp.157, 165, 180–181.
196. Palmer, *op. cit.*, p.157.
197. See http://en.wikipedia.org/wiki/human_evolution for 70,000/50,000 years ago. See Bryan Sykes, *The Seven Daughters of Eve*, p.278 for 100,000 years ago.
198. See https://australian.museum/learn/science/human-evolution/when-and-where-did-our-species-originate/.
199. Hagger, *The New Philosophy of Universalism*, pp.155–157, 160–161.
200. See http://www.newscientist.com/article/dn3744-chimps-r-human-gene-study-implies.h.
201. See http://www.newscientist.com/article/dn3744-chimps-r-human-gene-study-implies.h.
202. Sykes, *op. cit.*, pp.50–51.

203. *Encyclopaedia Britannica* (1991), micropaedia, 4.140–141.

204. See http://en.wikipedia.org/wiki/Toba_catastrophe_theory.

205. Sykes, *op. cit.*, ch.1 and p.277.

206. Sykes, *op. cit.*, p.278.

207. Sykes, *op. cit.*, pp.50–51.

208. Clegg with Mackean, *op. cit.*, p.651.

209. Hagger, *The New Philosophy of Universalism*, pp.161–165.

210. Clegg with Mackean, *op. cit.*, pp.648–650.

211. Clegg with Mackean, *op. cit.*, pp.651–655.

212. Denyse O'Leary, *By Design or by Chance?*, p.86.

213. O'Leary, *op. cit.*, p.86; *Darwinism, Design and Public Education, op. cit.*, pp.324–326.

214. O'Leary, *op. cit.*, p.274, notes 8 and 9.

215. O'Leary, *op. cit.*, p.86.

216. Gribbin and Rees, *op. cit.*, p.289.

217. O'Leary, *op. cit.*, p.93.

218. O'Leary, *op. cit.*, p.97.

219. O'Leary, *op. cit.*, pp.98–99.

220. [The original website is now hard to access: http://members.iinet.net.au/~sejones/histry2a.html#hstrydrwndshnstystrtgyws.] See https://www.ntu.ac.uk/about-us/news/news-articles/2014/06/did-darwin-lie-about-discovery-of-natural-selection.

221. Hagger, *The New Philosophy of Universalism*, pp.170–176.

222. Compare Willis W. Harman, *A Re-examination of the Metaphysical Foundations of Modern Science*, p.87, where "integrative" is used as the opposite of "reductionist".

223. Samuel Taylor Coleridge, *Biographia Literaria*, ch.13, p.356, published 1854, original from the University of Michigan, digitized 23 November 2005. http://books.google.co.uk/books?id=5xg5G-4ai3oC&pg=PA356&dq=esemplastic+power+of+the+imagination.

224. For example Geoffrey Read, who attacks the holistic principle in Hewitt, *op. cit.*, pp.143–144.

225. Lee Spectner, *Not by Chance!*

226. See http://harvardscience.harvard.edu/foundations/articles/j-craig-venter-named-visiting-scholar.

227. Adams, *op. cit.*, p.189.

228. Hagger, *The New Philosophy of Universalism*, pp.181–184.
229. Clegg with Mackean, *op. cit.*, p.656.
230. Idea developed in Nicholas Hagger's discussions with Mary Morrison.
231. Nicholas Hagger's research in the Galapagos Islands.
232. Clegg with Mackean, *op. cit.*, p.516.
233. Clegg with Mackean, *op. cit.*, pp.516–517.
234. Hagger, *The New Philosophy of Universalism*, pp.192–193.
235. Clegg with Mackean, *op. cit.*, pp.44–45.
236. Clegg with Mackean, *op. cit.*, pp.46–47. Also see http://en.wikipedia.org/wiki/Food_chain. Also http://waterontheweb.org/under/lakeecology/11_foodweb.html. For 10^{18} see http://en.wikipedia.org/wiki/1000000000000_(number).
237. Hagger, *The New Philosophy of Universalism*, pp.197–198.
238. Clegg with Mackean, *op. cit.*, pp.45, 65.
239. Clegg with Mackean, *op. cit.*, p.65.
240. Hagger, *The New Philosophy of Universalism*, pp.199–200, 205–206, 209.
241. Clegg with Mackean, *op. cit.*, p.67.
242. Clegg with Mackean, *op. cit.*, p.66.
243. Clegg with Mackean, *op. cit.*, p.67.
244. Hagger, *The New Philosophy of Universalism*, pp.211, 216–217.
245. See Dieter Podlech, *Herbs and Healing Plants of Britain and Europe*.
246. See https://www.ncbi.nlm.nih.gov/pmc/articles/PMC3847409/. Also see https://www.intechopen.com/chapters/64420.
247. Hagger, *The New Philosophy of Universalism*, pp.347–355.
248. Clegg with Mackean, *op. cit.*, p.89; Jeff Rubin, *Antarctica*, p.230.
249. See https://discoveringantarctica.org.uk/oceans-atmosphere-landscape/atmosphere-weather-and-climate/the-ozone-hole/.
250. See http://inventors.about.com/library/inventors/blfreon.htm.
251. See which states that ozone forms 0.00006 per cent of the atmosphere.
252. Rubin, *op. cit.*, p.98.
253. Clegg with Mackean, *op. cit.*, p.247.
254. Clegg with Mackean, *op. cit.*, p.88. Also see http://www.ucar.edu/learn/1_6_1.htm.
255. Clegg with Mackean, *op. cit.*, p.89.
256. Patrick Abbott's research. It should however be pointed out that in 2002 the ozone hole over the Antarctic was much smaller than in 2000 and 2001, and had split into two holes. The opening was the smallest

since 1988. See https://earthobservatory.nasa.gov/images/2832/the-antarctic-ozone-hole-in-2002. Also see http://www.sciencedaily.com/releases/2006/06/060630095235.htm

257. Patrick Abbott's research.

258. Patrick Abbott's research.

259. Henry N. Pollack, *Uncertain Science... Uncertain World*.

260. See https://www.ft.com/content/c70b7cfc-9ce4-11db-8ec6-0000779e2340.

261. Pollack, *op. cit.*, pp.219–220.

262. Patrick Abbott, quoting James P. Kennett, author of *Marine Geology*.

263. Pollack, *op. cit.* Also see http://flood.firetree.net/?ll=33.8339,129.7265&z=12&m=7

264. Pollack, *op. cit.*, p.224.

265. Pollack, *op. cit.*, pp.225–226.

266. Pollack, *op. cit.*, pp.219–226.

267. See https://www.un.org/development/desa/pd/sites/www.un.org.development.desa.pd/files/files/documents/2020/Jan/un_1999_6billion.pdf, figures of the UN Population Division.

268. Charles W. Fowler and Larry Hobbs, 'Limits to Natural Variation: Implications for Systemic Management', 2002. Also, 'Is Humanity Sustainable?', 2003, see https://www.ncbi.nlm.nih.gov/pmc/articles/PMC1691539/pdf/14728780.pdf.

269. Patrick Abbott's research.

270. See Hagger, *The Syndicate*, pp.267–274.

271. See http://en.wikipedia.org/wiki/Little_Ice_Age.

272. Patrick Abbott's research.

273. See http://news.bbc.co.uk/1/hi/world/americas/6999078.stm.

274. Nicholas Hagger's research in Antarctica. For the Wilkins Ice Shelf, see *The Daily Telegraph*, 26 March 2008, p.34.

275. Hagger, *The New Philosophy of Universalism*, p.218.

276. Adams, *op. cit.*, p.189.

277. Adams, *op. cit.*, p.189.

278. See Adams, *op. cit.*, p.202, for 10^{29}. For (7×10^{27}), for 10^{15} (twice) and 10^{12} see http://en.wikipedia.org/wiki/1000000000000_(number).

279. Hagger, *The New Philosophy of Universalism*, pp.218–221.

280. Robert Brooker, Eric Widmaier, Linda Graham and Peter Stiling, *Biology*, pp.4–5.

281. For homeostasis, see Clegg with Mackean, *op. cit.*, pp.502–515; and Brooker, Widmaier, Graham and Stiling, *op. cit.*, pp.856–859.

282. Clegg with Mackean, *op. cit.*, pp.503–504; Brooker, Widmaier, Graham and Stiling, *op. cit.*, pp.858–859.

283. Brooker, Widmaier, Graham and Stiling, *op. cit.*, p.858.

284. Jason Cook's research shared with Nicholas Hagger.

285. Hagger, *The New Philosophy of Universalism*, pp.231–234, 237.

286. Adams, *op. cit.*, p.202.

287. Clegg with Mackean, *op. cit.*, p.465.

288. A general, fairly reductionist non-medical book on the brain is Susan Greenfield, *The Human Brain: A Guided Tour*.

289. See http://en.wikipedia.org/wiki/Brain: "Evidence strongly suggests that developed brains derive consciousness from the complex interactions between the numerous systems within the brain. Cognitive processing in mammals occurs in the cerebral cortex but relies on midbrain and limbic functions as well." Also see http://en.wikipedia.org/wiki/Human_brain.

290. See http://en.wikipedia.org/wiki/Brain: "Evidence strongly suggests that developed brains derive consciousness from the complex interactions between the numerous systems within the brain. Cognitive processing in mammals occurs in the cerebral cortex but relies on midbrain and limbic functions as well." Also see http://en.wikipedia.org/wiki/Human_brain.

291. Clegg with Mackean, *op. cit.*, p.467.

292. See http://en.wikipedia.org/wiki/Famine_response.

293. Hagger, *The New Philosophy of Universalism*, pp.241–242, 243–244, 246–246.

294. See https://www.britannica.com/technology/holography.

295. See https://www.zdnet.com/article/could-quantum-networking-rescue-the-communications-industry-status-report/.

296. See https://arxiv.org/abs/2106.12295.

297. See https://www.researchgate.net/publication/3242347_Slow-light_optical_buffers_Capabilities_and_fundamental_limitations

298. See https://physicsworld.com/a/photon-shape-could-be-used-to-encode-quantum-information/.

299. See http://www.trnmag.com/Stories/2002/071002/Photons_heft_more_data_7-10-02.html.

300. See http://www.americanscientist.org/template/AssetDetail/assetid/162

18?fulltext=true&print=yes.

301. See http://www.psych.ndsu.nodak.edu/mccourt/Psy460/Anatomy%20 and%20physiology%20of%20the%20retina/Anatomy%20and%20 physiology%20of%20the%20retina.html.

302. See https://www.ncbi.nlm.nih.gov/books/NBK10885/.

303. See http://arxiv.org/ftp/quant-ph/papers/0208/0208053.pdf.

304. See https://www.ncbi.nlm.nih.gov/books/NBK10885/

305. See https://www.ncbi.nlm.nih.gov/books/NBK10885/

306. See https://www.ncbi.nlm.nih.gov/books/NBK10885/

307. See https://www.ncbi.nlm.nih.gov/books/NBK10885/

308. See http://en.wikipedia.org/wiki/Photoreceptor_cell.

309. See https://www.sciencedirect.com/topics/nursing-and-health-professi ons/doppler-flowmeter.

310. Hagger, *The New Philosophy of Universalism*, pp.246–248.

311. See https://wuttkeinstitute.com/eeg-details/. Also W. Grey Walter, *The Living Brain*. Also see diagram of electromagnetic spectrum on p.400 of Hagger, *The New Philosophy of Universalism*.

312. Owen J. Flanagan, *Consciousness Reconsidered*, p.234; quoted in Malcolm Hollick, *The Science of Oneness*, p.266.

313. Roger Penrose, *Shadows of the Mind*, p.366.

314. Hagger, *The New Philosophy of Universalism*, pp.249–253.

315. William Blake, letter to Thomas Butts, 22 November 1802; in *Poetry and Prose of William Blake*, ed. by Geoffrey Keynes, pp.859–862.

316. Samuel Taylor Coleridge, *Collected Letters II*, p.709.

317. Hagger, *The New Philosophy of Universalism*, p.253.

318. Hagger, *The Secret American Destiny*, pp.3–4, 24–26, 28.

319. Hagger, *The Fire and the Stones*, pp.378–414; *The Rise and Fall of Civilizations*, pp.11–64.

320. Hagger, *The Fire and the Stones*, pp.350–353; *The Light of Civilization*, pp.502–506.

321. Hagger, *The Rise and Fall of Civilizations*, pp.11–56.

322. Hagger, *The Light of Civilization*, p.506.

323. Hagger, *The New Philosophy of Universalism*, pp.334–337.

324. See Hagger, *The Syndicate*, pp.30–32.

325. Arnold Toynbee, *A Study of History*, one-volume edition, p.97. Quoted in Hagger, *The Light of Civilization*, pp.525–527.

326. Hagger, *The Rise and Fall of Civilizations, passim.*

327. Hagger, *The Fire and the Stones; The Light of Civilization; The Rise and Fall of Civilizations.*

328. See Hagger, *The Secret History of the West, passim;* and *The Syndicate.*

329. Hagger, *The New Philosophy of Universalism,* pp.338–339.

330. Hagger, *The Secret American Destiny,* p.7.

331. Hagger, *The Secret American Destiny,* pp.8–9, 11–13.

332. Evelyn Underhill, *Mysticism,* p.169.

333. See Hagger, *My Double Life 1: This Dark Wood,* pp.372–380.

334. Hagger, *The Light of Civilization,* pp.29–355, which contain all sources reflected in this section.

335. Hagger, *The Universe and the Light,* p.8; *The Light of Civilization,* pp.498–499, which cite all sources reflected in this section.

336. Hagger, *The Light of Civilization,* pp.7–10.

337. St Augustine, *Confessions,* 7.10.

338. Hildegarde von Bingen quoted in John Ferguson, *An Illustrated Encyclopaedia of Mysticism and the Mystery Religions,* p.77.

339. Hagger, *My Double Life 1: This Dark Wood,* pp.53, 60–61, 92–93 and 110–111.

340. William Wordsworth, 'Lines composed a few miles above Tintern Abbey', lines 100–102.

341. Hagger, *The Secret American Destiny,* pp.14, 15, 18–19.

342. Hagger, *The New Philosophy of Universalism,* pp.344–346.

343. Hagger, *The Secret American Destiny,* pp.21–22, 22–23, 24.

344. Hagger, *The New Philosophy of Universalism,* pp.11–22, 23–54, which cite all sources reflected in this section.

345. Aristotle, *Metaphysics IV,* 1–2/1003a.

346. Hagger, *The Secret American Destiny,* pp.24–25, 28.

347. Hagger, *The Fire and the Stones,* pp.350–353; *The Light of Civilization,* pp.502–506.

348. Hagger, *The Rise and Fall of Civilizations,* pp.11–56.

349. Hagger, *The Light of Civilization,* p.506.

350. Hagger, *The Secret American Destiny,* pp.29–30, 31.

351. Hagger, *The Universe and the Light,* pp.139–161.

352. Hagger, *The Universe and the Light,* pp.140–161.

353. Hagger, *The Universe and the Light,* p.140.

354. Hagger, *The Light of Civilization*, pp.523–525.

355. Hagger, *The Fire and the Stones*, pp.317–324; *The Light of Civilization*, pp.216–218.

356. Hagger, *The Fire and the Stones*, pp.743–753; *The Rise and Fall of Civilizations*, pp.412–428.

357. Hagger, *The New Philosophy of Universalism*, pp.357–358.

358. Hagger, *The Fire and the Stones*, and *The Light of Civilization*.

359. Hagger, *The Secret American Destiny*, pp.32–36.

360. Hagger, *The World Government*, pp.122–158.

361. Hagger, *The Secret American Destiny*, pp.36–38.

362. Hagger, *The Light of Civilization*, pp.534–535.

363. Hagger, *The New Philosophy of Universalism*, pp.339–340.

364. Hagger, *World State*, p.153.

365. Hagger, *World State*, pp.165–182.

366. See list, http://en.wikipedia.org/wiki/List_of_countries_by_population #cite_note-unpop-3 .

367. See http://answers.google.com/answers/threadview?id=55124 for 12 3 conflicts and http://www.war-memorial.net/wars_all.asp for 162 conflicts.

368. Hagger, *World State*, p.173.

369. See Hagger, *The Syndicate*, pp.257–260.

370. The 69,401 figure was in *The Guardian*, http://www.guardian.co.uk/ news/datablog/2009/sep/06/nuclearweapons-world-us-north-korea-russia-iran. (This website is no longer available, and the Bulletin of the Atomic Scientists' figures now apply.) For the Federation of American Scientists' figures, see https://fas.org/issues/nuclear-weapons/status-world-nuclear-forces/.

371. Hagger, *The Syndicate*, pp.110–111, citing *Spotlight*, 22 April 1996, pp.4–5: "Sources have told *The Spotlight* Chernomyrdin and David Rockefeller are secret partners in Russian energy combines."

372. Hagger, *World State*, pp.182–185.

373. Hagger, *World Constitution*, p.3.

374. Hagger, *The New Philosophy of Universalism*, p.251.

375. Hagger, *The New Philosophy of Universalism*, pp.281, 284–287.

376. Willis Harman in *Global Mind Change*, pp.34–35, distinguishes the first three. I have added the fourth.

377. For example, in 1 *Thessalonians* 5.23. See also http://books.google.co.uk/ books?id=Hk4iTvy4Gx8C&pg=PA353&lpg=PA353&dq=st+paul+spirit+p neuma&source=web&ots=skidBEIBG9&sig=L3XQDhf5hkUyZ5MGzSfD D7zxVtc&hl=en.

378. See Mircea Eliade, *Shamanism: Archaic Techniques of Ecstasy*, pp.91–93.

379. Z'ev ben Shimon Halevi, *Tree of Life*, p.18. Abraham is reputed to have received the Kabbalah from Melchizedek, King of Salem or Jerusalem.

380. Hagger, *The New Philosophy of Universalism*, pp.303, 308–309, 311–312, 323–324.

381. Shelley, 'Adonais', LII.

382. *Encyclopaedia Britannica*, micropaedia, II.176. For photons as bosons see Barbour, *op. cit.*, p.191.

383. Hagger, *The New Philosophy of Universalism*, pp.86, 359–360, 366–367.

384. Bohm, *op. cit.*, pp.189, 193, 211.

385. James N. Connelly, Martin Bizzarro, Alexander N. Krot, Åke Nordlund, Daniel Wielandt, Marina A. Ivanova, 'The Absolute Chronology and Thermal Processing of Solids in the Solar Protoplanetary Disk', 2 November 2012, *Science*, 338(6107): pp.651–655.

386. WMAP – 'Shape of the Universe', map.gsfc.nasa.gov; Lauren Biron, 'Our universe is flat', 7 April 2015, symmetrymagazine.org, Fermilab/SLAC.

387. Steven Frautschi, 'Entropy in an Expanding Universe', 13 August 1982, *Science*, 217(4560): pp.593–599; Fred C. Adams and Gregory Laughlin, 'A dying universe: the long-term fate and evolution of astrophysical objects', 1997, *Reviews of Modern Physics*, 69(2): pp.337–372.

388. James Glanz, 'Breakthrough of the year 1998. Astronomy: Cosmic Motion Revealed', 1998, *Science*, 282(5398): pp.2156–2157.

389. Paul Davies, *The Last Three Minutes: Conjectures about the Ultimate Fate of the Universe*.

390. WMAP – 'Shape of the Universe', map.gsfc.nasa.gov.

391. Richard Gott in Tyson, Strauss and Gott, *op. cit.*, p.192.

392. Richard Gott in Tyson, Strauss and Gott, *op. cit.*, p.199.

393. Richard Gott in Tyson, Strauss and Gott, *op. cit.*, p.193.

394. Adams and Laughlin, 'A dying universe: the long-term fate and evolution of astrophysical objects', 1997, *Reviews of Modern Physics*, 69(2): pp.337–372.

395. Frautschi, 'Entropy in an Expanding Universe', 13 August 1982, *Science*, 217(4560): pp.593–599.

396. Anders Andreassen, William Frost and Matthew D. Schwartz, 'Scale-invariant instantons and the complete lifetime of the standard model', 12 March 2018, *Physical Review*, D97(5), 056006.

397. Richard Gott in Tyson, Strauss and Gott, *op. cit.*, p.194.

398. Richard Gott in Tyson, Strauss and Gott, *op. cit.*, pp.209–210.

399. Gregory Laughlin, Peter Bodenheimer, Fred C. Adams, 'The End of the Main Sequence', 1997, *The Astrophysical Journal*, 482(1): pp.420–432.

3. The Algorithm of Creation and a Theory of Everything

1. Hagger, *The New Philosophy of Universalism*, pp.89, 90–96.

2. *The Daily Telegraph*, 26 February 2008, p.27.

3. *The Sunday Telegraph*, 6 April 2008, p.31. For Higgs boson field being like *aither*, see Davies, *The Goldilocks Enigma*, pp.178–179; *The Times*, 8 April 2008, p.18.

4. Barrow, *The Infinite Book*, pp.186–188.

5. Hawking, *op. cit.*, p.74.

6. Hawking, *op. cit.*, p.136.

7. Barrow, *op. cit.*, p.xiv.

8. See Hewitt, *op. cit.*, pp.55 and 106, for Geoffrey Read's Leibnizian view that Newton was wrong about the primary nature of time and that Einstein's view of time was closer to Leibniz's than to Newton's.

9. Hagger, *The New Philosophy of Universalism*, p.85.

10. Brian Greene, *The Elegant Universe*, p.15.

11. Hagger, *The New Philosophy of Universalism*, pp.262–263.

12. Hagger, *The New Philosophy of Universalism*, pp.266–269.

13. For Uncreated Light see Robert Grosseteste, *On Light (De Luce)*, trans. by Clare Riedl.

14. For Universal Light see Plato, *Republic*, book VII, 540a: "The time has now arrived at which they must raise the eye of the soul to the universal light which lightens all things, and behold the absolute good" (trans. by Benjamin Jowett). See http://classics.mit.edu/Plato/republic.8.vii.html.

15. St Augustine, *op. cit.*, 7.10, pp.146–147.

16. Mechthild of Magdeburg, *The Flowing Light of the Godhead*. See Hagger, *The Light of Civilization*, pp.142–143 for context.

17. Dante, *Paradiso*, canto 33, line 124; p.346.

18. Dante, *op. cit.*, canto 30, line 40; p.319.

19. Dante, *op. cit.*, p.323.

20. Adams, *op. cit.*, p.37.

21. Adams, *op. cit.*, p.41.

22. Duncan with Kennett, *op. cit.*, p.405.

23. Duncan with Kennett, *op. cit.*, p.484. For a millionth, see Davies, *The Goldilocks Enigma*, p.135. For neutrinos travelling at the speed of light, see Adams, *Origins of Existence*, p.44.

24. Adams, *op. cit.*, p.66.

25. Hagger, *The New Philosophy of Universalism*, pp.273–277.

26. Nicholas Hagger's research in China.

27. Duncan with Kennett, *op. cit.*, p.497; Singh, *op. cit.*, pp.430–431. See http://www.nersc.gov/news/nerscnews/NERSCNews_2005_02.pdf

28. Lars Olof Björn, *Light and Life*, p.17.

29. Björn, *op. cit.*, pp.46–47.

30. Björn, *op. cit.*, pp.81, 86, 101.

31. Björn, *op. cit.*, p.101.

32. Nicholas Hagger's research in the Galapagos Islands.

33. Björn, *op. cit.*, pp.81–90, 101.

34. Björn, *op. cit.*, pp.81–83, 101–127.

35. Björn, *op. cit.*, ch.4.

36. Björn, *op. cit.*, ch.5.

37. Björn, *op. cit.*, ch.6.

38. Björn, *op. cit.*, ch.7.

39. Hagger, *The New Philosophy of Universalism*, pp.277–280.

40. See https://smartenergyeducation.com/law-of-entropy/. Also *Encyclopaedia Britannica*, micropaedia, 111.911.

41. Aristotle, *Metaphysics*, p.xxviii.

42. Hagger, *The New Philosophy of Universalism*, pp.280–284.

43. Willis Harman in *Global Mind Change*, pp.34–35, distinguishes the first three. I have added the fourth.

44. Plato held that all knowledge can be derived from a single set of principles, perhaps even from a single principle. In *Republic* VI–VII, Plato describes the philosopher's knowledge as "synoptic", taking in the whole of reality, and resting on The Good, or knowledge of The Good. Aristotle in *Metaphysics* IV, 1/1003a, writes: "There is a science which

studies Being as Being.... This science is not the same as any of the so-called particular sciences, for none of the others contemplates Being generally as Being; they divide off some portion of it and study the attribute of this portion, as do for example the mathematical sciences." Also see http://books.google.co.uk/books?id=8tYcp0vYd5EC&pg=RA1-PA140&lpg=RA1-PA140&dq=universal+science+plato&source=web&ots=l1ASNzIF8d&sig=VvK5PCviA_86bDy5jUrRNkWZXeo&hl=en.

45. Eleanor Swift, *A Layman's Guide to Neometaphysics*, which is based on the three parts of J.J. Williamson's Cranwell Lectures, particularly on Part Two ('Metaphysical Application') and Part Three ('Metaphysical Analysis').

46. Hagger, *The New Philosophy of Universalism*, pp.288–290.

47. Hagger, *The New Philosophy of Universalism*, pp.292–295, 299–301.

48. Hagger, *The New Philosophy of Universalism*, pp.303–306.

49. Based on Nicholas Hagger's thinking. Many traditions have similar applications, for example New Thought.

50. Hagger, *The New Philosophy of Universalism*, pp.307–312.

51. For the *chakras* or *cakras*, see *Encyclopaedia Britannica*, micropaedia, II.445. Also K.M. Sen, *Hinduism*, pp.61, 69, 70. For a theosophical view of the *chakras*, see C.W. Leadbeater, *The Chakras*.

52. Shelley, 'Adonais', LII.

53. *Encyclopaedia Britannica*, micropaedia, II.176. For photons as bosons see Barbour, *op. cit.*, p.191.

54. Renée Weber, *op. cit.*, p.48 and also p.44 (which reports Bohm referring to G.N. Lewis). For matter as frozen light, see p.45. Also David Bohm in conversation on light with Nicholas Hagger. For matter as frozen light, see Weber, *op. cit.*, p.45. For matter being formed by collision of photons and therefore by light, see http://abyss.uoregon.edu/~js/21st_century_science/lectures/lec25.html.

55. Rees, *Just Six Numbers*, p.52.

56. Weber, *op. cit.*, p.45.

57. Björn, *op. cit.*, p.16.

58   Hagger, *The New Philosophy of Universalism*, pp.319–320.

59. Hagger, *The New Philosophy of Universalism*, pp.321–322.

60. Hagger, *The New Philosophy of Universalism*, pp.324–326.

61. Hagger, *The New Philosophy of Universalism*, pp.326–327.

62. Hagger, *The New Philosophy of Universalism*, p.327.

63. Hagger, *The New Philosophy of Universalism*, pp.327–329.

64. Atomic theory was proposed in the 5th century BC by the Greek philosophers Leucippus and Democritus and was revived by Lucretius in Roman poetry in the 1st century BC. Modern atomic theory began with John Dalton in the 19th century, and his scientific theory was confirmed in the 20th century. See *Encyclopaedia Britannica*, II.346–351.

65. Hagger, *The New Philosophy of Universalism*, pp.330–331.

66. Alfred North Whitehead, *An Introduction to Mathematics*, pp.41–42.

67. Hagger, *The New Philosophy of Universalism*, pp.84–85.

68. Fölsing, *op. cit.*, pp.587–592.

69. Duncan with Kennett, *op. cit.*, p.507. Freeman Dyson questioned whether any experiment in the real universe can detect one graviton. If not, he held that it is meaningless to talk about gravitons as physical entities.

70. Hagger, *The New Philosophy of Universalism*, pp.92–94, 110, 99, 255.

71. Barrow, *op. cit.*, pp.186–188.

72. Hawking, *op. cit.*, p.74.

73. Hawking, *op. cit.*, p.136.

74. Barrow, *op. cit.*, p.xiv.

75. Gribbin and Rees, *op. cit.*, pp.177–178.

76. Hagger, *The Secret American Destiny*, pp.24, 53.

77. Hagger, *The New Philosophy of Universalism*, p.50.

78. Whitehead, *op. cit.*, p.567.

79. Whitehead, *op. cit.*, p.845.

80. Matrix as a model for a universe. Image taken from https://biometrics. ilri.org/Publication/Full%20Text/Linear_Mixed_Models/AppendixC. htm, Neter *et al.* (1990), Searle (1971, 1982, 1987) and Searle *et al.* (1992).

81. Roger Penrose, *Cycles of Time: An Extraordinary New View of the Universe*.

Timeline

1. Nicholas Hagger saw a cypress tree in Abarkuh, Iran (pictured in Hagger, *The Last Tourist in Iran*, p.87), that was reputed to be 4,500–5,000 years old, as old as, or older than, the Great Pyramid.

2. See https://www.google.com/search?q=Planck+collaboration/esa+scienc e+photo+library+image+of+the+universe&hl=en-GB&sxsrf=APq-WBtoz

cAVZtzR3iRt8ev7XJhTx3hJMw:1647260590803&source=lnms&tbm=isch
&sa=X&ved=2ahUKEwjc4_fyy8X2AhVlmVwKHQAuCIIQ_AUoAXoEC
AIQAw&biw=1280&bih=587&dpr=1.5#imgrc=yeIPmbPapNrQeM.

Appendices

1.   Tom   Sumner,   https://rss.com/podcasts/tomsumnerprogram/312036/;
     Paul   Dolman,   https://www.paulsamueldolman.com/parent/podcasts/
     pauly-cast/nicholas-hagger-919/; and Paula Vail, https://youtu.be/tFTsV_
     gf_Eg.
2.   Tokyo Bunrika Daigaku (Tokyo University of Literature and Science)
     was renamed Tokyo Kyoiku Daigaku (Tokyo University of Education),
     and Nicholas Hagger taught in a particular room where Empson had
     taught.
3.   Hagger, *Awakening to the Light*, p.105.
4.   Hagger, *Awakening to the Light*, p.111.
5.   Hagger, *Awakening to the Light*, p.184.
6.   Hagger, *Awakening to the Light*, pp.197–198.
7.   Hagger, *Awakening to the Light*, pp.200–201.
8.   Hagger, *Awakening to the Light*, p.201.
9.   Hagger, *Awakening to the Light*, p.201.
10. Hagger, *Awakening to the Light*, p.201.
11. Hagger, *Awakening to the Light*, p.234.
12. Hagger, *Awakening to the Light*, pp.236–238.
13. See https://www.nicholashagger.co.uk/78-79-80-81-82-83-84-encounter-
     december-1966-from-a-china-dia.
14. Hagger, *Awakening to the Light*, p.301.
15. Hagger, *My Double Life 1: This Dark Wood*, p.372.
16. Hagger, *My Double Life 1: This Dark Wood*, pp.372–380.
17. Hagger, *Diaries*, 11 March 2019.
18. Hagger, *Diaries*, 13 March 2019.

Bibliography

Adams, Fred, *Origins of Existence*, The Free Press, 2002.

Adams, Fred and Laughlin, Greg, *The Five Ages of the Universe*, Touchstone, 2000.

Aristotle, *Metaphysics*, books I–IX, trans. by Hugh Tredennick, Harvard University Press, 1933/2003.

Augustine, St, *Confessions*, trans. by R.S. Pine-Coffin, Penguin, London, 1961.

Barbour, Julian, *The End of Time*, Phoenix, 2000.

Barrow, John, *The Infinite Book*, Jonathan Cape, 2005.

Barrow, John, *Theories of Everything*, Clarendon Press, Oxford, 1991.

Barrow, John and Tipler, Frank, *The Anthropic Cosmological Principle*, Oxford University Press, 1986.

Björn, Lars Olof, *Light and Life*, Hodder and Stoughton, 1976.

Blake, William, *Poetry and Prose of*, ed. by Geoffrey Keynes, The Nonesuch Library, London, 1956.

Bohm, David, *Wholeness and the Implicate Order*, Routledge & Kegan Paul, 1980.

Brooker, Robert J., Widmaier, Eric P., Graham, Linda E., Stiling, Peter D., *Biology*, McGraw-Hill, 2008.

Capra, Fritjof, *The Tao of Physics*, Wildwood House, 1979; Wildwood House, 1975; Fontana/Collins, 1976.

Castillejo, David, *The Expanding Force in Newton's Cosmos*, Ediciones de Arte y Bibliofilia, Madrid, 1981.

Clegg, C.J. with Mackean, D.G., *Advanced Biology: Principles and Applications*, 2nd ed., Hodder Murray, 2000.

Coleridge, Samuel Taylor, *Biographia Literaria*, published 1854, original in the University of Michigan, digitized 23 November 2005.

Coleridge, Samuel Taylor, *Collected Letters II*, ed. Earl Leslie Griggs, Oxford University Press, 2000.

Dante, *The Divine Comedy*, Book 3: *Paradise*, trans. by Dorothy L. Sayers and Barbara Reynolds, Penguin Classics, 1962.

Dante Alighieri, *The Divine Comedy*, trans. by Allen Mandelbaum, Everyman's Library, 1995.

Darwinism, Design and Public Education, ed. by John Angus Campbell and Stephen C. Mayer, Michigan State University Press, 2003.

Davies, Paul, *The Goldilocks Enigma*, Allen Lane, 2006.

Davies, Paul, *The Last Three Minutes: Conjectures about the Ultimate Fate of the Universe*, Basic Books, 1997.

De Waal, Frans, *Our Inner Ape*, Granta Books, 2006.

Duncan, Tom with Kennett, Heather, *Advanced Physics*, fifth edition, Hodder Murray, 2000.

Eliade, Mircea, *Shamanism: Archaic Techniques of Ecstasy*, Princeton University Press, USA, 1964/Routledge & Kegan Paul, 1974.

Ferguson, John, *An Illustrated Encyclopaedia of Mysticism and the Mystery Religions*, Thames and Hudson, 1976.

Flanagan, Owen J., *Consciousness Reconsidered*, MIT Press, Cambridge, 1992.

Fölsing, Albrecht, *Albert Einstein*, trans. by Ewald Osers, Viking, 1997.

Greene, Brian, *The Elegant Universe*, Vintage, 2000.

Greenstein, George, *The Symbiotic Universe*, William Morrow, New York, 1988.

Gribbin, John, *In the Beginning*, Viking, 1993.

Gribbin, John, *The Universe: A Biography*, Allen Lane, 2006.

Gribbin, John and Rees, Martin, *Cosmic Coincidences*, Black Swan, 1990.

Grosseteste, Robert, *On Light*, trans. by Clare Riedl, Marquette University Press, 1942.

Guth, Alan, *The Inflationary Universe*, Vintage, 1998

Hagger, Nicholas, *A New Philosophy of Literature*, O Books, 2012.

Hagger, Nicholas, *Awakening to the Light, Diaries: 1958–1967*, Element, 1994.

Hagger, Nicholas, *My Double Life 1: This Dark Wood*, O Books, 2015.

Hagger, Nicholas, *My Double Life 2: A Rainbow over the Hills*, O Books, 2015.

Hagger, Nicholas, *Selected Letters*, O Books, 2021.

Hagger, Nicholas, *The Fire and the Stones*, Element, 1991.

Hagger, Nicholas, *The Last Tourist in Iran*, O Books, 2008.

Hagger, Nicholas, *The Light of Civilization*, O Books, 2006.

Hagger, Nicholas, *The New Philosophy of Universalism*, O Books, 2009.

Hagger, Nicholas, *The Rise and Fall of Civilizations*, O Books, 2008.

Hagger, Nicholas, *The Secret American Destiny*, Watkins Publishing, 2016.

Hagger, Nicholas, *The Secret History of the West*, O Books, 2005.

Hagger, Nicholas, *The Syndicate*, O Books, 2004.

Hagger, Nicholas, *The Universe and the Light*, Element, 1993.

Hagger, Nicholas, *World Constitution*, O Books, 2018.

Hagger, Nicholas, *World State*, O Books, 2018.

Halevi, Z'ev ben Shimon, *Tree of Life*, Rider, 1972.

Harman, Willis W., *A Re-examination of the Metaphysical Foundations of Modern Science*, The Institute of Noetic Sciences, California, 1991.

Harman, Willis W., *Global Mind Change*, Knowledge Systems/The Institute of Noetic Sciences, USA, 1988.

Hawking, Stephen, *A Brief History of Time*, Bantam Press/Transworld, 1988.

Hewitt, Peter, *The Coherent Universe*, Linden House, 2003.

Hirata, Hosea, *The Poetry and Poetics of Nishiwaki Junzaburo: Modernism in Translation*, the Princeton Legacy Library, Princeton University Press, 2016.

Hollick, Malcolm, *The Science of Oneness*, O Books, 2006.

Hoyle, Fred and Wickramasinghe, Chandra, *Evolution from Space*, Grenada, 1983.

Hoyle, Fred, Wickramasinghe, Chandra and Watkins, John, *Viruses from Space*, University College Cardiff Press, 1986.

Kennett, James P., *Marine Biology*, Pearson, 1981.

Kirk, G.S., Raven, J.E. and Schofield, M., *The Presocratic Philosophers*, second edition, Cambridge University Press, 1957/1995.

Leadbeater, C.W., *The Chakras*, The Theosophical Publishing House, Adyar, 1927.

Long, A.A., ed., *The Cambridge Companion to Early Greek Philosophy*, Cambridge University Press, 1999.

Magdeburg, Mechthild of, *The Flowing Light of the Godhead*, Paulist Press International, USA, 1998.

Oesterley, W.O.E. and Robinson, Theodore H., *An Introduction to the Books of the Old Testament*, Macmillan, 1934–1958.

O'Leary, Denyse, *By Design or by Chance?*, Augsburg Books, Minneapolis, 2004.

Palmer, Douglas, *Neanderthal*, Channel 4 Books, 2000.

Penrose, Roger, *Cycles of Time: An Extraordinary New View of the Universe*, Bodley Head, 2010.

Penrose, Roger, *Shadows of the Mind*, Oxford University Press, 1994.

Penrose, Roger, *The Emperor's New Mind*, Vintage, 1989.

Phillips, Jack, *Suppressed Science*, John J. Phillips Jr, 2006.

Plato, *The Collected Dialogues*, ed. by Edith Hamilton and Huntington Cairns, Princeton University Press, 1961/1982.

Podlech, Dieter, *Herbs and Healing Plants of Britain and Europe*, Collins, 1996.

Pollack, Henry N., *Uncertain Science... Uncertain World*, Cambridge University Press, new edition, 2005.

Rees, Martin, *Just Six Numbers: The Deep Forces That Shape the Universe*, Basic Books, 2000.

Rees, Martin, *Our Cosmic Habitat*, Phoenix, 2001.

Rubin, Jeff, *Antarctica*, Lonely Planet, 2005.

Sen, K.M., *Hinduism*, Penguin Books, 1961.

Shelley, *A Selection* by Isabel Quigly, The Penguin Poets, 1956.

Singh, Simon, *Big Bang*, Fourth Estate, 2004.

Spectner, Lee, *Not by Chance!*, Judaica Press Inc., New York, 1997.

Swift, Eleanor, *A Layman's Guide to Neometaphysics*, The Society of Metaphysicians, Hastings, 1993.

Sykes, Bryan, *The Seven Daughters of Eve*, Bantam Press, 2001.

The Born-Einstein Letters, trans. by Irene Born, Macmillan, London, 1971.

The I Ching: The Book of Changes, trans. by James Legge, Dover Publications, 1963.

Toynbee, Arnold, *A Study of History*, revised one-volume edition, OUP/ Thames and Hudson, London, 1972.

Tyson, Neil deGrasse, Strauss, Michael A. and Gott, J. Richard, *A Brief Welcome to the Universe*, Princeton University Press, 2021.

Underhill, Evelyn, *Mysticism*, Methuen, 1911, 1960.

W. Grey Walter, *The Living Brain*, W.H. Norton and Company, 1963.

Weber, Renée, *Dialogues with Scientists and Sages: The Search for Unity*, Routledge & Kegan Paul, 1986.

Whitehead, Alfred North, *An Introduction to Mathematics*, Oxford University Press, 1958.

Williamson, J.J., *The Structure of ALL*, The Society of Metaphysicians, Hastings, 1986.

Index

268, 277

Enlightenment, the 159, 160, 177, 178, 253

environment xii, 94, 98, 99, 115, 116, 119, 120, 121, 123, 125, 127, 131, 135, 136, 138, 139, 142, 147, 154, 187, 199, 203, 205, 226, 243, 255, 290, 345, 346

and human population 131

glacial-interglacial cycles of our Ice Age 135

harsh 98, 119, 127, 154, 226, 255

'Environmental Universalism' (Hagger) 127

epistemology 251, 252, 253, 254, 259, 260, 273, 278

"esemplastic power of the imagination" (Coleridge) xv, 7, 113, 266, 345

eternity 5, 6, 163, 209, 211, 218, 220, 223, 254, 268, 326, 333, 344

evolution xii, xviii, 13, 27, 48, 87, 94, 96, 97, 98, 99, 102, 103, 108, 110, 111, 112, 113, 114, 115, 116, 117, 118, 119, 142, 148, 160, 209, 214, 216, 226, 238, 239, 245, 246, 249, 250, 251, 254, 257, 269, 276, 280, 286, 290, 301, 309, 364, 367, 370, 380, 389

and environment: Expansionism and Darwinism xii, 94, 113, 226

existence 12, 13, 14, 17, 19, 20, 23, 25, 32, 38, 207, 208, 209, 210, 211, 213, 215, 224, 225, 227, 228, 229, 237, 239, 241, 242, 246, 249, 253, 254, 255, 256, 257, 258, 263, 264,

268, 273, 274, 281, 290

expansion 15, 27, 28, 32, 38, 39, 41, 54, 56, 58, 60, 61, 62, 63, 66, 105, 113, 132, 156, 172, 174, 216, 217, 218, 219, 220, 225, 227, 228, 230, 237, 240, 290, 301, 309, 313, 351, 365

Expansionism xii, 94, 115, 116, 226, 238

'Expansionism: "Mechanism" of Order' (Hagger) 113

'mechanism of order' 113

Fibonacci numbers 86, 290

fifth dimension 35, 214, 215, 216, 227, 228, 229, 292

survival after death 214, 216, 227, 228, 229

finite, the 5, 6, 12, 13, 14, 30, 32, 57, 62, 63, 173, 207, 208, 209, 211, 212, 223, 235, 238, 240, 242, 251, 252, 254, 255, 256, 257, 258, 259, 264, 266, 268, 269, 272, 275, 276, 278, 288, 326, 344

Fire or Light 21, 161, 171, 173, 174, 176, 256, 257, 259, 260, 261, 345

"ever-living Fire" 16, 17, 20, 21, 158, 159, 160, 161, 169, 174, 176, 237, 239, 250, 256, 257, 259, 260, 261, 269, 270, 273, 274, 279

Form from Movement Theory (Hagger) vii, xii, xiv, 8, 14, 15, 16, 20, 21, 22, 23, 301, 312, 350, 351, 352, 353, 364

forms of the One 290

forces, four vi, xx, 64, 65, 230, 231,

universe 206, 227, 228

survival after death of xii, 206,
207, 216, 227, 228, 229

humans 92, 99, 100, 102, 103, 104,
105, 106, 107, 109, 110, 111, 112,
113, 114, 118, 123, 125, 126, 127,
132, 136, 137, 140, 142, 147, 178,
208, 215, 219, 229, 244, 245, 246,
254, 257, 259, 268, 270, 271, 275,
276, 279, 290, 311, 312, 361

descent of *Homo*/modern humans
106

emergence of modern from
Africa 107

extinct 96, 100, 102, 106, 112, 219,
290, 312, 313

family tree or bush: descent
of modern humans from 4.5
million years ago 104, 107

Homo sapiens 96, 99, 100, 101,
102, 103, 104, 105, 106, 108, 109,
118, 132, 133, 238, 260, 309, 312,
361

Homo sapiens sapiens 99, 100, 102,
106, 107, 108, 290, 312

symmetry of rise and fall of
universe, civilisations and all
living things 293

Ice Ages 95, 96, 97, 98, 108, 128, 133,
134, 135, 209, 226, 238, 310, 311, 375

Andean-Saharan 96, 97, 311

Cryogenian 96

glacial-interglacial cycles

Huronian 95, 310

Karoo 96, 98, 311

present Ice Age began 40 million
years ago 97

infinite, the xii, xxv, 5, 6, 12, 13, 14,
15, 16, 18, 19, 22, 24, 25, 29, 30, 32,
35, 37, 38, 54, 56, 57, 58, 62, 63, 64,
65, 152, 167, 173, 206, 208, 209, 210,
212, 213, 214, 215, 216, 221, 224,
225, 234, 235, 238, 241, 243, 247,
249, 250, 251, 251, 263, 254, 255,
256, 257, 258, 259, 260, 264, 265,
266, 268, 269, 271, 272, 273, 274,
275, 276, 277, 278, 279, 280, 281,
284, 285, 286, 287, 288, 289, 326, 344

density 29, 30, 34, 233, 284

order xii, 32, 34, 35, 37, 210, 225,
264, 265, 291

Void 14, 31, 57, 63, 221, 227, 229,
237, 241, 277, 288, 291, 309

infinity 12, 21, 24, 25, 29, 30, 31, 35,
44, 58, 60, 61, 63, 234, 235, 254, 285,
287, 288

mathematical 24, 30, 31

metaphysical 30

inflation 15, 20, 38, 39, 51, 54, 56,
59, 60, 68, 92, 93, 225, 235, 237, 240,
256, 286, 290, 291, 351, 352

smooths irregularities 291, 309

'Instinct: Inherited Reflex or
Transmitted Order?' (Hagger) 119

instinctive perception of unity xi,
xxiv, 7

intellect 133, 159, 173, 214, 215, 239,
260, 261, 263, 267, 268, 271, 272,
273, 274, 277, 278, 279, 289

'Interaction of Brain and Levels of

O-BOOKS

O is a symbol of the world, of oneness and unity; this eye represents knowledge and insight. We publish titles on general spirituality and living a spiritual life. We aim to inform and help you on your own journey in this life.

If you have enjoyed this book, why not tell other readers by posting a review on your preferred book site?